MANAGEMENT STRATEGY FOR INFORMATION TECHNOLOGY

About the Author

Utpal K. Banerjee took his honours degree in Pure Physics (1955) from the University of Calcutta and his Ph.D. from the University of Manchester, UK (1972) where he was awarded the Commonwealth Scholarship.

After serving for 17 years in the Ministries of Defence, Industry and Finance, Dr. Banerjee worked with Tata Steel as Chief of Management Services; the Administrative Staff College of India as Director of Computer Division; Electronics India as its Senior Executive Director; and finally, All India Management Association as its Director General. He has also considerable experience of teaching at the IITs, IIMs Jawaharlal Nehru University, International Management Institute, FORE School of Management, and Management Development Institute. Apart from many research papers, Dr. Banerjee has 15 books to his credit.

MANAGEMENT STRATEGY FOR INFORMATION TECHNOLOGY

(IT and Management Convergence)

Based on the Internationally Class-Tested Course,
Conducted in China and India

With Seven Case Studies

Dr. Utpal K. Banerjee
Consultant and Adviser on Information Technology

CONCEPT PUBLISHING COMPANY, NEW DELHI-110059

Cataloging in Publication Data—DK
Courtesy: D.K. Agencies (P) Ltd. <docinfo@dkagencies.com>

Banerjee, Utpal K. (Utpal Kumar), 1935-
Management strategies for information technology (IT and management convergence) : based on the internationally class-tested course, conducted in China and India : with seven case studies / Utpal K. Banerjee.
p. cm.
Includes index.
ISBN 13: 9788180695414
ISBN 10: 8180695417

1. Information technology—Management—Study and teaching (Higher). 2. Information technology—India—Management. 3. Electronic data interchange—Case studies. I. Title.

DDC 004.068 22

ISBN-13: 978-81-8069-541-4 (HB)
ISBN-10: 81-8069-541-7 (HB)

First Published 2008

Published and Printed by
Ashok Kumar Mittal
Concept Publishing Company
A/15-16, Commercial Block, Mohan Garden
New Delhi-110059 (INDIA)

Phones : 25351460, 25351794
Fax : 091-11-25357103
Email : publishing@conceptpub.com

Dedicated
To

Professors, Scholars and Academics of China's premiere universities
Who made possible as well as pleasurable
My extended visits and interactive Post-Graduate Courses and Examinations:

TSINGHHUA UNIVERSITY, Beijing
Prof. Liu Qiang, Assistant Dean, School of Software

BAYHANG UNIVERSITY, Beijing
Prog. Yao Shuzhen, Deputy Dean, Computer Science and Engineering Dept.
Prof. Jingnong Weng, Technical Director, UNIWARE Co. Ltd

SOUTH-EAST UNIVERSITY, Nanjing
Prof. Luo Janzhou, Head, Dept. of Computer Science and Engineering
Prof. Michael Y. Qu, Dept. of Computer Science and Engineering
Prof. Yuequan Wu, College of Software Engineering

JIAO TONG UNIVERSITY, Shanghai
Prof. Dr. Shen Weiping, Vice President
Prof. Dr. Li Miglu, Vice Chair, Dept. of Computer Science and Engineering
Prof. Yuxi Fu, Dept. of Computer Science
Prof. Chen Yue, Division of International Cooperation & Exchange

DALIAN UNIVERSITY OF TECHNOLOGY, Dalian
Prof. Shen Hongshu, Vice President
Prof. Bai Zhongxiao, Director, Division of International Cooperation and Exchange
Prof. Xue Qiang, Deputy Head, School of Computer Software

YUNNAN UNIVERSITY, Kunming
Prof. WangShi Pu, Chair, School of Software
Prof. Liao Hongzhi, Vice Dean, School of Software
Prof. Li Tong, Vice Dean, School of Software

And

My amazingly affectionate students
Who
Shared with me their numerous queries, curiosities and personal predilections,
Held my hands to navigate through the maze of bicycle traffic within the huge campuses,
Urged me to give them Christian names in addition to their Chinese ones,
Had tearful eyes when I left,

And who must remain nameless...

NOTE TO READERS

BACKGROUND

IT revolution has reached such a pinnacle today that it is no longer enough to be IT-numerate, but also to be highly aware of the management issues and concerns: at every stage of IT usage. This has to be a mutually enriching process of interaction between the end-users and managers on one hand and the IT professionals at the other end, — so as to increase comprehension of IT benefits, enhance IT efficiency and implement long-term strategy for IT.

OBJECTIVE

The book provides for a platform for the interfacing of management and IT community to evolve a skill-set: for using IT on a planned basis in diverse areas and obtain value for money. The emphasis is throughout on interaction and dialogue on a cross-disciplinary basis.

TARGET-AUDIENCE

1. MBA students in the final year
2. Young managers, shaping up for higher responsibilities
3. MCA students in the final year
4. Young IT practitioners, trying to come to terms with management expectations.

NATIONAL AND INTERNATIONAL CLASSROOM-TESTING

- International management Institute: 30-hour course for MBA Students (Final Year, seconded from Industry): with real-life projects in both years: 1999 and 2000
- Jawaharlal Nehru University: 30-hour course for MCA (Final Year): 2001.

- Beijing University of Aeronautics and Astronautics, Beijing; Dalian University of Technology, Dalian; Jiao Tong University, Shanghai; and South-East University, Nanjing; China: Guest Lectures Series to UG and PG students from Computer Sc. and Management (Final Year): May-June 2002
- Bayhang University, Beijing — two 30-hour courses for full-time and part-time PG students (Final Year); Ching Hua University, Beijing – one 12-hour course for PG students (Final Year); and South-East University, Nanjing – two 30-hour courses for full-time and part-time PG students (Final Year): all courses in their Schools of Software Engineering: September-October 2002.
- Yunnan University, Kunming – two 30-hour courses for PG students (Final Year): in the School of Software Technology: September 2003

All course-participants in every institute were evaluated based on objective tests and given credits accordingly.

ACHIEVEMENT AT THE END OF STUDY

The readers are expected to:

- Judge their self-knowledge by going through structured objective tests;
- Use IT in working life: productively and purposively: in diverse management applications;
- Achieve better rapport among the IT and Management disciplines and reduce communication gap to a substantial extent.

Utpal K. Banerjee

CONTENTS

Part Six
Seven Case Studies

LIST OF TABLES

LIST OF FIGURES

PART ONE

MANAGEMENT CHOICE

1

MANAGEMENT CHOICE OF HARDWARE TECHNOLOGY

INTRODUCTION

We introduce in this chapter the Indian scenario of available hardware, with the cautionary note that there are changes taking place all the time: in keeping with technology and the market demand. Specific management concerns are then taken up that need be addressed—before dealing with eight experience-based selection and evaluation criteria for hardware, when decisions are to be made by management to make the best buy for the organisation: in consultation with the computer professionals.

INDIAN IT MARKET

IT scene—it is a-changing! The booming market scenario of the end-nineties had undergone a recessionary change and the first few years of the twenty-first century saw a halting recovery. There has, however, been now a definite promise of a healthy growth as the millennium has progressed. Table 1.1 provides the IT spending and covers the following:

- *Total Hardware*: comprising all hardware devices and components;
- *Packaged Software*: covering all software packages and products marketed;
- *IT Services*: comprising IT-enabled services like medical transcription, call centres; and
- *Others*: covering IT-training and education consumables.

Table 1.1: Indian IT Market

	IT Spending ($ M)				*IT Growth Rate*		
	2000-01	2001-02	2002-03	2003-04	2001/ 2000	2002/ 2001	2003/ 2002
Total HW	$3,051.00	$2,919.00	$3,355.00	$4,301.00	– 4%	15%	28%
Package SW	$409.00	$467.00	$546.00	$674.00	14%	17%	23%
IT Services	$932.00	$1,067.00	$1,301.00	$1,555.00	14%	22%	20%
Others (Trg., Consummables)	$409.00	$467.00	$576.00	$629.00	14%	17%	15%
Total IT Market	$4,801.00	$4,920.00	$5,752.00	$7,150.00	2%	17%	24%
Annual Growth	33%	2%	17%	24%			

HARDWARE BRANDS

Among the foreign brands from multinational corporations (MNC), the dominant players include—

1. IBM-NetVista 'M' Series
2. IBM-NetVista 'A' Series
3. DELL Optiplex
4. HP-Compaq Presario

Tables 1. 2, 1.3, 1.4, 1.5 and 1.6 illustrate some of their key-features.

Table 1.2: IBM-NetVista Desktops

	Features
NetVista A Series Rs. 40,600.00 – Great Technology and affordable prices, IBM Desktops	– Flat Panel Monitor
NetVista M Series Rs. 57,600.00 – Stability and manageability for the enterprise	– SO DIMM, DIMM, Cache – Keyboards, Mouse, Spaceball 3D
NetVista X Series Rs. 130,300.00 – Integrated flate panel PC, Intel PIV processors, an integrated 15 inch or 17 inch TFT flat panel display	

At the lower end, for instance, IBM NetVista 'A' and 'M' series desktops are priced at Rs. 40,000 and Rs. 57,000 at the minimum configuration—with a fair amount of stability and manageability for the enterprise. They all have flat panel monitor, DIMM, S0 DIMM, cache, keyboard, mouse and 3-dimensional space ball. At the higher end, IBM NetVista X series is priced at Rs. 130,000 at the lowest configurations—

Table 1.3: IBM-NetVista M Series

Models	*Price*	*Processor*	*Memory/ Graphic*	*Hard Drive*	*Ethernet*	*OS*
Economy	Rs. 66,100.00	1.8 GHz	128 MB/Intel Extreme Graphics	40GB	10/100 Ethernet	Windows XP Home Edition
Value	Rs. 69,800.00	2 GHz	128 MB/Intel Extreme Graphics	40 GB	10/100 Ethernet	Windows XP Home Edition
Performance	Rs. 79,400.00	2 GHz	128 MB/32 MB intel Extreme Graphics	40 GB	10/100 Ethernet	Windows XP Professional

Table 1.4: IBM-NetVista A Series

A Series						
Models	IBM Web Price	Processor	Graphic/Ha	Ethernet	OS	Warranty
NetVista A22p 6049C5A	Rs. 46,240.00	1.66 GHz/ 128 MB	32 MB Sis	10/100 Ethernet	Windows XP Home Edition	3 yrs parts/1 yr. labour
NetVista A22p 6049C2A	Rs. 52,750.00	1.66 GHz/ 128 MB	32 MB Sis 350/40 GB	10/100 Ethernet	Windows 2000	3 yrs parts/1yr labour
NetVista A30 682443A	Rs. 56,400.00	1.7 GHz/ 128 MB	32 MB Sis 305/40 GB	10/100 Ethernet	XP Home Edition	3 yrs parts/1 yr labour
NetVista A22p 6049D1A	Rs/ 59,400.00	1.7 GHz/ 128 MB	32 MB Sis 305/40 GB	10/100 Ethernet	Windows 98	3 yr parts/1yr labour

Table 1.5: DELL Optiplex

Models	*Processor*	*Memory*	*Hard Drive*	*Warranty*
OptiPlex GX50	Celeron/1.4 GHz	512 MB PC 100 shared SDRAM	40 GB	3 yr next business on site service
OptiPlex GX240	PIV/2.4 GHz	1 GB PC 133 SDRAM ECC	160 GB	3 yr next business on site service
OptiPlex GX260	PIV/2.53 GHz	1 GB 266 DDR SDRAM Non ECC	160 GB	3 yr next business on site service

with integrated flat panel PC, Intel P-4 processor and an integrated 15-inch or 17-inch flat panel display.

Table 1.6: HP Compaq Presario

Model	*Processor*	*Memory*	*Hard Drive*	*Modem*
Presario 6125AI	PIV/1.7 GHz	256 KB/128 MB PC 2100 DDR SDRAM	40 GB	56 Kbps

Among the Indian brands, which can be called semi-multinational corporations (semi-MNCs), the prominent ones include HCL. Wipro, Zenith, Samsung and LG My PC. Their prices start at Rs. 300,000 for Intel Pentium 4-based systems, against the MNC brands beginning at Rs. 40,000, as seen above.

MANAGEMENT CONCERNS

(i) Brand War

There is a fair amount of competition prevalent among brands. The cutting edge is usually the price, easy availability and marginal differences in technology. A good example is the competition offered by AMD in constantly undercutting price and coming up early with new technology like 'Athelon' chip in the market, although their production volume is low—in contrast to Intel's higher-priced, high-volume marketing strategy for Pentium III, IV and V and then associated chips of Celeron.

(ii) Grey Market

In the East Asian scenario, there are several countries like India, Taiwan, Singapore and Hong Kong who have a thriving industry of assembled products. The latter are virtual copies of the brands, but offered in the market with the following advantages and disadvantages:

- Prices are low, since the establishment overloads of the brands are not incurred;
- Quality is comparable, although not quite guaranteed, as the brands;
- No warranty is offered unlike the brands and
- Maintenance can be dicey, as there are many fly-by-night operators.

(iii) Bankruptcy

Some Indian brands, which indulged from time to time in high-voltage marketing blitzkrieg: to tempt customers with low price and high technology, could not sustain their operations. Consequently, they went into bankruptcy, making customers lose money and shaking their faith in all brands. Also, some grey-market assembling outfits (mostly found in Mumbai and Delhi) also could not gain customer confidence and went into liquidation in the long run.

(iv) PCs Versus Comparable Products

While IBM PCs (along with their branded as well as unbranded "clones") virtually monopolise the market, there are others who have offered them a stiff run for money. These are Apple Computer's iMax and Microsoft's advocacy of Net Computers (NCs) that use the Internet-based disk storage and major "intelligence".

(v) OS Dependency

Operating System (OS), since its appearance in 1964, has played a dominant role to run computer hardware efficiently and productively. This fact led to the development of many operating systems, which were offered by hardware vendors that remained proprietorial to individual machines. This was the case till UNIX came up from the Bell laboratories—written in C language—which could operate on almost all platforms (except IBM's own XENIX). UNIX is today virtually an industry standard, but has its own problems due to difficulty in handling its shell and kernel programming. Now, thanks to Linus Tarbolds and his band of devoted followers, LINUX has come up as an OS with all the UNIX-like features and yet easy to operate. Moreover, LINUX is offered free of cost and is poised to acquire application software support from diverse user-groups, which has still remained a strong point with UNIX. The latest development is in dispensing with OS altogether, such as, Oracle Corporation working for their very popular RDBMS being embedded in the hardware!

(vi) Application support on hardware

Most application software today depends on the back-end RDBMS for support. A large variety of RDBMS packages like SQL Base, Oracle, Ingres, DB-2, bTrieve and others have made their appearances in the market, generating varying demands; the first two so far have survived well. DBMS needs strong administration (DBA): to take backups, do maintenance and updating, and provide front-end support. Further, RDBMS standards, like the fourth-generation structured-query languages (SQLs), are important to provide user confidence in the hardware.

(vii) User Communication

Users need graphical capability available in the front-end, comprising the graphic user interfaces (GUIs). The well-known GUIs couple quite well with their back-end DBMS counterparts, such as, SQL Base with SQL Window, Oracle with Developer 2002, Gupta Base with Power Builder, and so on. A "tight coupling" between the back-end DBMS and the front-end GUI is again a welcome feature to render satisfactory service to the user.

(ix) Post-Y2K Scene

The year 2000 had come and gone, and the large-scale disruption and malfunctioning feared in the computers scene, as the 'Y2K' problem, did not take place. But a new concern has come up, namely, that of linkage between the third-generation legacy systems (of very large programmes mostly written in COBOL language and quite often "patched" by several programmemers) and the fourth-generation databases. This, as we shall see later, has become the subject matter of data mining and data warehouses.

(viii) Super-Computing System

Scientific computing needs the very high end of computing power and efforts have been on globally to have super computers for this purpose. In the absence of the supply of the supercomputing technology like Cray X/MP 24 or the IBM, India has made consistent endeavour to develop such technology. Bhava Atomic Research Centre (BARC) has now produced ANUPAM, a customised supercomputer: with 300 MIPS for integer processing and 2 Giga Flop for scientific processing involving decimals. With OS as LINUX and using FORTRAN, the range of applications includes: molecular dynamics; crystal structure analysis; gamma ray simulation; electromagnetic 3-D plasma simulation; finite element analysis; remote-sensing satellite data processing (1 sq km * 1 sq km); and weather forecasting.

SELECTION AND EVALUATION OF HARDWARE

A legitimate concern of management today is not to be overawed (or even "brainwashed") by the glittering presentations of the hardware vendors and end up by buying machines not entirely suitable to the uses organisation. It is the responsibility of computer professionals to provide assistance to managers in making appropriate selection of hardware and for this purpose, a comprehensive evaluation strategy is very desirable to adopt.

Based on the author's three-decade-long experience, associated with selection and evaluation of computer hardware for as varied organisations as the public sector giant National Thermal Power Corporation (NTPC); Rollatainers (a leading packaging and container company); Bhadrachalam Paper Board (an ITC concern); the state government of Andhra Pradesh, and the Sangeet Natak Academy (the central apex body for performing arts, with its counterparts in the states), to name a few—management choices were consolidated in the following eight evaluating criteria:

(i) Performance

The most important criterion remains the performance: according to the chosen configuration: of the hardware. While the performance parameters are specified by the vendors and are usually above board, the needs for a comparison in performance with other similar systems is often necessary. Two factors are helpful to the management in this regard:

First, there are benchmarks available: to compare with the best standards: from impartial sources like *Computer World* magazine from the USA, *Computer Weekly* magazine from the UK, and India's own *Dataquest* and *PCquest*. Their comparative figures can be fully depended upon. These computer benchmarks are small programmes: aimed at measuring the performance of computer systems and their components. They execute 'workload' tasks and report the time taken for such execution. Low-level benchmarks measure performance of hardware components, such as, CPU clock, cycle-times of different levels of cache, hard-disk access time, latency, etc. In contrast, high-level benchmarks evaluate combined effect of hardware, drivers, OS and compilers.

Second, *Gibson's Ratio* for the system can also be obtained from these impeccable sources. This ratio aims at comparing computer's efficiency for processing scientific application workload versus commercial application workload. Scientific workload uses data with floating point operands, whereas commercial workload uses main data-types as integers and strings. Example of the former is decision support system, used for business analysis: to understand business trends and forecasting business directions: by raising long queries that search through large portions of the database. Examples of the latter are airlines reservations and banks, where a large number of clients continually access and update small portions of the database through short transactions. *Gibson's ratio* looks at the computer's performance by: (a) running a deck of programmes which are CPU-intensive, largely using CPU after suppressing the I/O devices; (b) running again programmes which are I/O-intensive (utilising mostly I/O devices), after suppressing CPU; and (c) taking a ratio of their pro-rata run-times. If this ratio is high on the CPU side, the hardware is deemed suitable for academic, scientific and research institutions where calculation-needs remain high, with negligible data-volume. On the contrary, if the ratio turns out to be high on the I/O side, the hardware is appropriate for commercial data processing in business and industry, where data-volume is large but calculations are elementary. *Gibson's ratio* is a useful *a priori* way to determine the overall suitability of computer systems to specific organisations and is a useful management tool.

(ii) User Feedback

Most vendors would be willing to furnish, on request, a list of their clients—as a matter of self-pride. During evaluation, such a request should be made: with complete address and telephone-numbers. Thereafter, discrete enquiries should be made about the experience of other clients with the vendor regarding: system performance, timely availability of support-services, problem solving during the teething-trouble phase, and so on. While such queries need not be formal (or in writing), most clients would not mind sharing information in an informal manner, perhaps on telephone. It would be surprising to guess what a wealth of information can be gleaned about the potential of the system under evaluation as well as the attitude-pattern of the vendor, by launching such informal enquiries across their client-base! Some remarks heard by the author are: on hardware, "the system is very good, but often the disk-head floats," or, "the printer gives occasional problems"; and, on attitude, "Oh, they are never available, once the sale is over," or, "their support is excellent!"

(iii) Upgradability of Computing Power

While the evaluation process is no doubt with reference to the current workload of the organisation, much of this load keeps growing. Applications have a way of turning up and it is exciting for management to evince their "Computer-literacy" by suggesting new areas of data-processing or problem-solving or both, as time passes. It is, therefore, a good principle to enquire whether the proposed system can be expanded in near future in terms of: RAM (for the CPU), multiple CPUs, disk-storage, multiple printers, more terminals, and so on. This is because it is the usual practice for vendors to shy away from pointing out their system's limitations, while extolling its virtues! On the contrary, most systems have inherent limits in upgradability, which is essential for the potential user to know. Lack or more disk-connectivity is a common phenomenon, as is the problem of adding on new printers.

An interesting case arose in evaluation for the Sangeet Natak Academy, when the Academy needed 16-terminal systems in the different Indian states for the start-up jobs. The possibility of future expansion was then posed to the vendor who confidently responded that the system could be upgraded to have 32 terminals. With the existing CPU capability of the vendor, this seemed highly doubtful, but the vendor glibly stated that the Academy may not be using all 32 terminals at the same time and, in any case, up to 24 terminals, concurrency of use should be no problem for the system. Now, this was not quite the same answer furnished earlier about a true 32-terminal machine! What came out clearly from this experience was that the likely degradation in 'response time' of the terminals could be a critical question to ask, when the increase in terminal-numbers was being mooted. If response time becomes too long, it would be clearly unacceptable by the client. If, on the other hand, there is a simultaneous expansion in the CPU at the back-end, the desirable response time can be maintained even after expansion—but this possibility should also be taken into account.

(iv) After-sales Support Services

This is perhaps one single factor based on which most vendors' reputations are made or marred. The 'hard sale' by the vendors in essentially what is today a 'buyers' market' leaves any vendor little scope to sustain a costly, countrywide after-sale service network. But that is no consolation for the large client whose system/s must run smoothly all the time. In today's market-condition, 3 years warranty has almost become a norm, where the vendor provides all the parts, but the labour is charged for, after one year. Four clearly demarkable cases may arise:

- ***When Essential (or core) Spares for Hardware are Involved***: it is in the interest of the client that these are kept with the vendor nearby and made available within 12 hours (or so) turn-around time;
- ***When Non-essential (or non-core) for Hardware Spares are Involved***—these may perhaps be held by the vendor's warehouse and made available within 24 hours (or so);
- ***When Application Software Support is Involved***—This is not so critical as hardware vendors—while 'bundling' application software with their machine—do have a pool of programmers knowledgeable about such software. This knowledge is easily gained and such people are available when the client needs them;
- ***When Systems Software Support is Involved***—This is, however, a different ball-game altogether. Most systems software are developed in the USA and few programmers have any detailed insight in the event of OS showing a 'bug', or, going into a 'hang' stage. Clients have a tough time, as such knowledgeable people are difficult to come by and are quite expensive anyway.

While there is no clear-cut answer in all such cases, a mutually acceptable solution between the vendor and the client should exist—to be worked out before an expensive purchase-deal is stuck.

(v) Documentation

Documentation, accompanying hardware, has not been a strong point for many hardware vendors and, quite often, it has remained at a shoddy level. The international quality of documentation, on the contrary, is consistently high, as may be seen from the IBM manuals. There can be 3 broad categories of documentation: **systems manual** (giving technical details of the system and needed for subsequent maintenance by engineers); **operation manual** (giving I/O instructions to the operators for running the system); and **user manual** (being optional, needed by the user-managers for an overview of the system's architecture and performance characteristics). Out of the three, the first two are absolutely essential.

(vi) Maintenance

Once the warranty period is over, there can be maintenance of any of the four kinds: **vendor maintenance** (where the vendor is asked to maintain, beyond the warranty period); **Maintenance by specialised agency** (CMC being earlier one such specialised agency for maintenance in India); **third-party maintenance** (by contractual agreement with outside agencies who store the spare parts procured from Taiwan, Hong Kong or Singapore and have trained their manpower on maintenance of machines); or **in-house maintenance** (where the user organisation undertakes maintenance by recruiting maintenance personnel). **The annual maintenance contract** (AMC) for outside parties is usually pegged at 10 per cent of the capital value, but often reduced to 8 per cent under negotiation, in a buyers' market!

User organisations often resort to a mix of the above four alternatives to suit their requirements, varying from **round-the-clock service** to **prime time service** (i.e., during the prime shift of the day when the job pressure on the machines is at its peak) to **on-call service** (when outside maintenance engineers are called only when needed). Oil and Natural Gas Corporation (ONGC) at its corporate head-quarters, Dehra Dun, for instance, has followed a judicious mix of such alternatives as in-house engineers round-the-clock and CMC engineers on-call: to derive best value for money.

(vii) Training

Since the turn-over problem of skilled persons exist all over the world, it is a salutary idea to get as many people as possible trained by the vendor at the beginning. Negotiation for training could be on the following counts: free or at-cost training; training in-house or at the vendor-site (the latter likely to be cheaper, as the vendor can get people from many client-organisations together); long-duration or short-duration training; and appreciation-level (for senior managers) and operating level (for junior supervisors). To repeat, any cost incurred at the initial stage on training proves many times its worth in value, later on.

(viii) Vendor's Guarantee

Most vendors are willing to provide guarantee for trouble-free performance of their machines: in percentage terms. While 100 per cent guarantee may be somewhat unrealistic, 95 per cent to 98 per cent guarantees are quite usual. The notion behind such a guarantee is that the CPU should give 'up-time' (remaining available) to that extent during a month. If, however, the actual performance of the CPU falls below the guaranteed stipulation, the vendor remains bound to provide equivalent time from his or her own system, free of cost—to make up for the shortfall.

As can be easily understood in commercial practice, the price quoted would steeply rise if the vendor were pushed to give 100 per cent or near-100 per cent guarantee, as he has to cater for the likely shortfall in the event of a systems failure. In fact, the higher the guarantee, the more is the chance that failure will occur. Hence is the price coverage by the vendor for such a high-risk venture. This should be kept in view by the user organisation, while having a higher and near-perfect expectation of performance by the machine.

CONCLUSION

In today's scenario, apart from dedicated scientific applications, the selection and evaluation strategy for the business world has to be honed on merits—to suit the needs of the user-organisation. It is a good idea to decide on the organisation's total computer-needs, and then do the market-survey among the available vendors and note their price-expectation. Thereafter, the short-listed vendors should arrange a presentation individually and the bulk of the above points can be covered through amicable interrogation during, or after, the presentation.

2

MANAGEMENT CHOICE OF SOFTWARE TECHNOLOGY

INTRODUCTION

We deal here with application software: packages and products. The distinction between the last two is vital, though often omitted. Positive and negative spiralling is discussed in the context of product life-cycle, along with considerations of *de facto* and *de juro* standards of software. Application software has many ramifications: menu-driven versus command-driven; and in-house development versus outsourcing. Emerging application software categories are important to take note of, in the Indian context, before the selection and evaluation criteria are delineated.

SOFTWARE PACKAGES AND PRODUCTS

Historically, software started by being an adjunct to hardware. There was a time before 1960's when even the OS did not exist and rudimentary monitoring instructions were given to the second-generation computers. Once the third-generation computers came along with the OS, there has been no turning back. While most of the software was in the form of application programme *customised* (i.e., made to order, to suit specific customer's needs), there were other application packages, which were of general applicability across multiple machines. These *packages* needed marginal, if at all, customisation to suit user-specific needs—like the IBM's KRAUSS package for machine-shop scheduling in the olden days. As *general application packages* began to be tested with several user-managers and their views obtained on the shortfalls and modification-needs of such packages, those comments were closely worked upon for further development. With recursive testing and feedbacks, the packages could be improved as to have almost universal applicability—in order to become globally available *products*. Examples of such products come famously from Microsoft's stables, as indicated below:

1. Operating Systems

MS Windows 95, Windows 98, Windows 2000, Windows XP—office edition and home edition, Windows Professional, MS NT Workstation 4.0, NT Server 4.0

2. Productivity Tools

MS Schedule+ 7.5, Office 95 7.0b, Office Professional 95 7.0b, Project 4.1a

3. Internet Tool

MS Internet Explorer

4. Development Tools

MS Front Page 95 (& later versions), Visual Studio

5. Server Products

MS Exchange, Proxy Server, SNA Server, SQL Server, Systems Management Server 1.2, Exchange Server 5, Sub-Server 3.

The other notable examples of standard products are from the engineering side: AutoDesk Corporation's Auto-CAD (for engineering design and analysis) and Auto-Sketch (for draftsman's blueprinting work).

Due to the expensive nature of launching and sustaining the above testing and feedback process in a reiterative manner, India has not been conspicuously successful to launch many products. Although help has been organised by the government and private sector to organise extensive testing, de-bugging and, in general, 'hand-holding' for such incipient packages in order to market them well in India and abroad, examples of really successful products have been few and far between. In one such endeavour by the public-sector ET&T Corporation in its heydays—when the author was seminally involved—much hope was raised and a few products did emerge after incorporating detailed comments and feedbacks: This experience will be shared later in this chapter.

SPIRALLING EFFECT

Some Indian software firms did try out on their own to do the needful and to market acceptable products. Vedika Corporation (based in Singapore) came out with a financial package FACT that attracted fresh talents to work with them and drew many customers. As a result, FACT developed into an excellent, internationally acceptable product, the proof of which is its incorporation on Microsoft's Windows platform. Tata Consultancy Service (TCS) product EX is an even better example where the initial accounting package EX, although named a product, had to be withdrawn from the market soon after its launching, as it seemed to have a few logical errors and usage limitations. After extensive testing, EX II version was released in the market, but it still did not come up to the user expectations. After a further trial and extensive development, the new version, still named EX, was eventually launched and became an excellent product—marketed in India and the West. On the whole, really good products attract investors and talents, leading to fund-flow from potential partners and customers—triggering even more efforts towards the next success. This can be termed *positive spiral* of product life-cycle.

On the other hand, a deck of genuinely good products, modelled on Microsoft's analogy, were called Softword, Softspread and Softbase, which were launched by Softek to capture the Indian market. Due to various management reasons, the firm was unable to provide nation-wide maintenance and requests for de-bugging from bonafide clients went somewhat unheeded. In consequence, customers lost confidence and many skilled people left the software house—turning it into a sick company. A well-meaning initial effort was thus caught into what can be termed as *negative spiral* of product life-cycle.

PRODUCT STANDARDS

Software products, with bonafide qualities, do not need any stamp of recognition and the market acceptance is their best hallmark of quality. Most Microsoft products like MS Office 2000 and MS Visual Studio are examples in point. Another illustration can be drawn from IBM's 'Qwerty' keyboard, which every computer has adopted, and that itself is its eloquent recognition. This is known as *de facto standard.*

There are, however, systematic efforts often needed to stamp recognition on a continuous basis to software efforts spouring up in the market from time to time. American National Standards Institution (ANSI) is one such body which has been active since the time of early programming and has given approval for what were ANSI COBOL, ANSI FORTRAN, etc.—apart from ANSI Instructional code use in all machine languages. International Standards Organisation (ISO) provides similar recognition like ISO-9002 to software documentation standards, with its Indian counterpart being Indian Bureau of Standards (IBS)-14002. This is known as *de juro standard.*

Some graphical capabilities have been floated as integral part of proprietorial hardware, which deny their availability for other machines. Apple Corporation of the USA, for instance, incorporated graphics in their Mac I and Mac II personal systems, which could not be run on other machines. On the other hand, Xerox Paolo Alto Research Centre (XEROX-PARC) developed *Windows* (comparable to their other seminal development of *mouse*) which were quietly adopted by others. MS-Office and Auto CAD are other examples of facilities, which run on all computers.

When UNIX (developed by Bell Labs under the AT&T) came to replace a plethora of proprietorial OSs running on a variety of machines, IBM first countered it with its own XENIX: as a clone of UNIX. Later, after UNIX had established its universal worth, IBM veered towards adopting it as a standard platform and founded Open Software Foundation (OSF), together with DEC and C Corporation, to promote it. Today, most academics use UNIX—Berkley Enhancement (developed by the University of California) and the bulk of commercial houses use the Santa Cruz Operations (SCO) version of UNIX: with an abundance of application packages created around it, thanks to its standardisation.

All OSs, in fact, try to keep up with the advances in hardware, communications, infrastructure facilities, and new competition and research. One instance was Sony's efforts to combine computer OS with consumer electronics, so that a TV can replicate as a PC: to send and receive messages, among other things. A second example is Netscape whose Internet browser Navigator was expected to become embedded in its own OS, instead of running on MS-Windows. A third illustration can be drawn from Sun Microsystems which was poised to develop OS out of a large number of its Java language runtime library of programming routines. The common point in all the above developments is that there is a large-scale endeavour going on all the time to bring computers close to the user and remove, if possible, its mystical OS that began in 1964 to support application software.

APPLICATION SOFTWARE

User-managers are primarily concerned with application software and its worth in their organisation. Before the selection and evaluation process begins, the basic information has to be noted as follows for each subject-category:

(i) Technical Information

Minimum hardware requirement; OS; menu-driven or command-driven; the nature of help provided to the user; operating options (additions, deletions and messages); scope of backtrack; frequency of *bugs*; help through windows; likely support from the developer/vendor for future updation; copy protection; special points.

(ii) Market Information

Availability of similar software; competitors' software, with quality and price; corporate image of the developer/vendor; list of other customers (if needed).

(iii) Training Information

Availability of manuals; availability of tutors; availability of demonstration package; expected training time.

(iv) General Remarks

Plus and minus points are to be summarised.

While the above basic information will be handy for selection and evaluation, two special conflict-areas should be noted as follows:

(a) Menu-driven versus Command-driven

User-managers and software professionals have entirely different perceptions here:

Menu-driven

- OK by managers
- Not OK by software professionals
- User-friendly
- Inflexible

Limited in scope, as created by the original developer needs through comprehensive testing to assess range of coverage

Command-driven

- Not OK by managers
- OK by software professionals
- Not user-friendly
- Fully flexible
- Unlimited in scope as it can be developed further
- Can be customised, to enhance coverage.

(b) Developing in-house versus Outsourcing

There is always a tussle about the choice between developing software in-house and *outsourcing* it to a specialised software agency that can develop the same for a consideration. The factors affecting the management choice are:

- *Specialisation Factor*: If specialist professionals are available in-house or can be recruited, the package can be developed in-house. This particularly applies if the level of specialisation is very high (as, say, for power control system by NTPC), since outsiders are likely to make a hash job.
- *Trained Internal Manpower Factor*: If such manpower is available, it is wise to utilise them and spend in-house man days and man months to do the development.
- *External Dependency Factor*: The considerations are the same here as for specialisation factor. Outsourcing leads to outside dependency: for *source* code (the code in which the programmes are written and not easily parted with by the developer) and for future trouble-shooting.
- *Maintenance Factor*: Once in use, application software needs maintenance, with changed perceptions of management or exigencies of the application environment. It remains essential to do maintenance in-house. But the trouble with in-house maintenance professionals is that they are seldom satisfied with such a routine job and would seek creative outlets elsewhere.
- *Cost Factor*: Surprisingly, cost is not found to be the sole important factor for deciding on in-house versus outsourcing development of application software. While it remains in the

reckoning, the other factors as above usually get priority. NTPC spent, for instance, a fortune to develop its critical application software in-house.

APPLICATION SOFTWARE—SUBJECT-CATEGORY AND PRICE-RANGE

Indian market (to which this section is confined) is full of many laudable packages and products. A wide market-survey in the 90's yielded the followings illustrative categories and their approximate price-ranges:

A. Medical and Hospital Package

There seemed to be 3 categories: large hospitals with 200-300 beds or more; nursing homes and polyclinics, with 50-100 beds or so; and doctor's chambers with occasional provision of a few private beds. In the first category, all the necessary modules of outpatient management, in-patient admission management, ward management, paramedical staff management, medicine and stores management (with shelf-life held as critical), doctor's appointments system, facilities management (for X-Rays, etc,), operation theatre management and financial managements were present. Some advanced packages even maintained disease-wise records—indicating diagnosis, therapy, cure (or fatality), and recurrence. The second category was a water-down version of the above, with only chosen modules. Doctor's chambers packages mainly emphasised about patient history and medication, apart from some financial provisions.

In the survey, price-ranges for the above were found to be wide and varied. Centre for Railway Information System (CRIS) had a comprehensive and attractive user-friendly package HMS, priced around Rs. 1.0 lakhs. HOMPAC was a product of Jaypee, Delhi: very user-friendly and fully modular, but quite expensive around Rs. 3.0 lakhs. HMS packages from ACS, Hyderabad, and HMS package from RCC, Chandigarh, were good, but not comprehensive, hence prices were not ascertained. HOMPAC had a nursing home version, too. For individual doctors, DOCPAC by R&D, Market Men from Chennai, although not comprehensive, was priced at Rs. 10,000.

B. Financial Package

There were broadly two varieties. The larger one for business houses covered all heads of financial accounts like journals, ledgers, trial balance, profit and loss account and balance sheet. The smaller category was meant to run by the individual shop and department stores for maintenance of accounts, through simple bookkeeping.

In the first category, TCS, Mumbai, offered the excellent package EX (discussed earlier) for around Rs. 8,000. Pentonics, Chennai, offered Tally: also very good and interfaced with inventory, although priced higher around Rs. 12,000. The third one was 'Touch Key', very user-friendly, from Quiller Corporation, priced around Rs. 11,000. FACT from Vedika Corporation (discussed earlier) was also in the same price-range. In the smaller category, 'Hisab' package from Goenkas, Chennai and 'Munimji' from Saxenas, Mumbai, were found to be good and elegant, priced between Rs. 500 and Rs. 900. The prices have since tended to come down.

C. Educational Package

Out of three available categories, the first concerns administration of universities, colleges and schools with such nodules as fee accounting, library, admission, registration, payroll and time-tabling. The second category is related to examination result processing, and listing of results and mark sheets. One, relatively rare, variation is actually conducting an examination from a randomised question-bank and analytical evaluation of individual results. The third category is Computer-Based Training (CBT)—also known as computer-assisted learning (CAL) and computer-aided instructions (CAI)—where computers are used for directly imparting training and education.

In the first category, MISS package from Computers, Bangalore, offered high promises, with price around Rs. 8,000 and found acceptance by the University Grants Commission (UGC), besides the education authorities of Mauritius. In the second category, while the examination result-processing and mark-sheet packages were mostly found to be *customised* for suiting the needs of individual secondary and higher secondary boards (like UPSEB, Bihar SEB, WBSEB, etc.), the only examination-conducting package from a randomized question-bank was found to be 'Vinyas' package, developed by NCST, Mumbai, but not apparently marketed very well. Among CBT packages, CUBE from Computers was found to be very textbook-dependent and on subject-basis: from the fifth to the twelfth standards. There have been several CBT packages in the market since. Among small CBT packages, U-LEARN (for learning UNIX) and C-LEARN (for learning C language) were priced around Rs. 600 each, but again very textbook-bound. CAD-Teacher from Vision Labs, Hyderabad, was suitable for teaching in engineering colleges and polytechnics for engineering drawings, but not finally priced.

D. Law Package

For allocation of daily business for individual benches and administration of register's office, Supreme Court, High Courts and district courts all qualify. National Informatics Centre (NIC), Delhi, has done exemplary work for the first two. The other level is the chamber of lawyers, where follow-up of cases, references to case laws and amendments to the Indian Constitution and different acts are required.

In the second category, Judipack' package was marketed by R&D Market Men, Chennai, priced around Rs. 10,000, but found to be mostly scratch-pad kind and not comprehensive. 'Lawman' package from Plus Computers was small and sold for around Rs. 8,000. 'Judix' package from DT&T sold databases for Rs. 5,000 each to lawyers—to be accessed through a key-search utility costing about Rs. 15,000.

E. Engineering Package

Engineering packages for design, stress-strain analysis, structural drawings, pipe-laying and instrumentation diagrams and drafting packages are of consistently high demand among engineering industries, process industries, architectural firms, civil construction companies and mechanical equipment firms, besides engineering colleges and professional training institutions.

Vision lab, Hyderabad, launched Vision Designer package at a cost of Rs. 20,000, suitable to cover 2-D and 3-D competition of Auto-CAD. At a lower end, Mac-CAD was introduced by Mac System, Chennai, at Rs. 8,000, which was comprehensive for 2-D alone, and quite user-friendly. Incidentally, IIT Delhi's M.Tech. and research work produced quite efficient packages to design mechanical cans, dies and castings, but they were hardly inclined to market them to such lucrative trades as perfume-bottlers and others.

F. Office Automation Package

Large and medium business offices and offices of senior executives need office automation for high office productivity, follow-up on assigned tasks (and projects) and day-to-day appointments, diaries, directories and reminder services.

CISTEMS, Delhi, offered CISMAIL for Rs. 700 with limited facilities. 'I-Office' from Integra System, Bangalore, was, on the other hand, quite comprehensive and on UNIX platform, with a price tag of Rs. 12,500. In the in-between range 'Executive Desk' package from Softplus, Mumbai, had all the desired features, attractively designed and user-friendly, at a price of Rs. 1,000.

G. Desk-top Publishing Package

DTP packages for book design, page-layout and graphic insertion are needed by large and medium book-publishing houses, newspaper offices and magazine-publishers, apart from business offices and industry houses, all of whom require publication of bulky, yet attractive, documents.

Vision Publisher from Vision Labs, Hyderabad, was offered as a command-driven package (and not menu-driven), hence was not priced. Hindi DTP package 'Lipika', embedded in to the Page Maker package from the West, was doing quite well.

H. Travel and Tourism Package

This is hugely needed by travel agents, in addition to industries and business houses who require travel plans for executives on duty and leave. CRIS, Delhi, attempted at Rs. 120, the SMART package, which was tested and found to be interactive, but it foundered on the problem of Railways and airlines frequently changing their travel-schedules. At that low price, such changes could hardly be updated and sent out to customers. But the package had immense potential and remains to be exploited.

SELECTION AND EVALUATION OF APPLICATION SOFTWARE

Management of user-organisations is faced with the choice among application software packages and products—as illustrated above and many new ones added by the day by vendors (or developers themselves, if they are small) who knock at their doors. Once the subject-category and the price-range is decided upon, the selection of software is broadly determined based on a few management points as well as technical points. As mentioned earlier, the author's involvement in evaluating a few scores of Indian software packages and products—done through a team of qualified software engineers—yielded valuable insight in this direction, which is shared here.

(i) Management Points

Demonstration: The first choice of management is to look at the package first hand. Many vendors bring demonstration ('demo') diskettes for this purpose, containing the main features of their package, if not the whole package. Occasionally, user-organisations are tempted to retain these diskettes: for major use, if not full use. This, of course, is unethical and vendors today counter the tendency by *either* making these diskettes *volatile*, i.e., not effective beyond a specific number of runs, *or* making them *time-bound*, i.e., tied to a calendar-date after which they expire.

Tutors: These are favourite commodities for management, in order to learn how to execute the package. Engineer-managers, for example, find it difficult to operate Auto CAD for engineering design at one go, and deem it convenient to use Auto CAD tutors available in the market in the learning phase. Many packages today offer *on-line* tutors, such as, MS WINDOWS series at the higher end by way of 'help', and it is exceedingly smooth to follow their step-by-step guidance.

Documentation: This again is a major concern of management. Quality manual provide plenty pf graphics, worked-out examples of various steps and "how to" kind of instructions. Again, Autodesk Corporations manuals, for both Auto-CAD and Auto-SKETCH packages are good standards to emulate by vendors.

Training: Training for application software has four aspects, namely, cost (different for in-house and vendor-determined location); duration (usually short-term); level (appreciation for management levels and execution at operating levels); and site (company-site or vendor-site). Training for complex packages, even at a cost, should not be overlooked in any case.

Maintenance and After-sales Support: In cases of large application software purchase, their maintenance and after-sales support may become important, to be negotiated with outsourced agencies.

(ii) Technical Points

Performance: It is important to understand not merely the running features and their productivity for the application software, but also its basic nature, namely, how much user-friendly, menu-driven and code-free the software is, for all of which user-managers have a marked preference. At this stage, the issue regarding availability of *source-code* from the vendor—now or in future—should be sorted out in clear-out monetary terms, to avoid inconvenience and embarrassment later on. If some *customisation* is needed, the same should also be sorted out.

Quality of Menu: Since menu remains the main driving force for application software, its hierarchy from *Main* menu to sub-menu, sub-sub-menu and even sub-sub-sub menu should be checked up. One should also check up features like *pull-down* menu and *pop-up* menu. *Drag and drop* is another feature for joint steps of some complex menus.

Quality of Help: Clear and comprehensive help is very desirable to be available. First, the *main help* should be a simple narrative for basic comprehension. Second, for various topics and areas, *context-sensitive help* should be there for guidance. Finally, there should be *choice-sensitive help* where a number of choices arise in the given context.

Error Messages: Since all users are susceptible to erroneous entries, judicious provision of error messages should be there. An error can be of two types: for *syntax* (where the entry-language has been faulty) and *logic* (where there is a mistake in the logical input). Both must be clearly and instantly *flagged* at the bottom and allowed to be cleared as the next step.

Default Features: Most software packages allow for common sense, by default, as the first choice of use. In printing instructions, for instance, the first choices are *all pages* (in the absence of a choice like "print pages 5 to 7") and *Single copy* (in the absence of a specific choice like "print 2 copies"). Thus, they became the default features. Since default feature save considerable data-entry labour, they should be carefully worked out and catered for.

Validity Check: Beyond simple syntax errors and logic errors, there can be sheer invalid entries, such as, age of office-goers shown as above 70. These are most subtle points and need a lot of attention to incorporate. There should be error messages for them as well.

Looping: Sometimes, the cursor gets struck at the choice-options and does not come out of the loop, even if no choice is executed. This is a sign of bad programming and the only way to escape is to go back to the OS and invoke OS command to re-start the execution. Some software packages show this proclivity and management should guard against it.

Window Boxes: In execution of a package, successive overlapping window boxes (definitely up to 4) should be feasible, as also an easy flexibility to erase them and go back to the original window box.

Updates: All application packages need updation. In the beginning of their marketing operation, for instance, Softek used to furnish free updates to their customers, which was a good business practice. Naturally new releases can be obtained (or negotiated) free, but entirely new versions, like Oracle 7 to Oracle 8i, are priced commodities. The same applies for enhancements.

Security Features: Last but not the least, good security is an integral part of a package. Many security measures have been tried out, like software write-lock, hardware lock, however, is among the best safeguard, but is cumbersome to insert and remove every time. Another innovating feature is to inset a key diskette first (which has some deliberate bad sectors, created by programming) and then begin the software operations. This is substantially foolproof, and breaking the bad-sector logic can be exceedingly difficult. Such 'laser keys' have been developed by a Hyderabad firm and needs encouragement.

CONCLUSION

The above account does not go into systems software, which is a specialised domain for software engineers and not of direct concern to user-managers. It also does not touch upon the software utilities and development of tool packages, which fall into the grey area between system software and application software, and are needed by software vendors and developers, and big industry and business houses that get into in-house software development. In the selection and evaluation exercise, it is again to be done with short-listed software vendors (or developers)—whose presentations should be followed up with all the management and technical quarries for eliciting satisfactory response.

3

MANAGEMENT CHOICE OF COMMUNICATIONS TECHNOLOGY

INTRODUCTION

Here we look at various data communications and data messaging methods: from the end-user point of view, alongside their line and wireless facilities. In today's digital domain, many interesting contenders have arisen, among integrated systems data networks, digital subscriber's lines and satellite transmission. In the latter area, industry has found a very powerful, yet cost-effective, tool in the Very Small Aperture Terminals (VSATs), especially for efficient remote connectivity. Essentials of the omnipresent e-mail services are also seen from the standpoints of standards compliance, reliability and uptime, Fax connectivity, PC user agent, application support, ease of access and good service back-up, among other criteria.

DATA COMMUNICATIONS

Communications of business data among short distances—confined to a limited business domain—have long been supplanted by long-distance needs of expanding business services, where modern-day telecommunications have become relevant. The two broad segments of telecommunications are aimed at: *cost saving* (for business support systems and for operations support systems) and *revenue generating* (for value-added services and operations, and for intelligent networks). The application focus can cover: data network services; database access services; directory services; messaging services; transaction processing; voice activation; multi-media services; and virtual private networks. Data communications can span from data network services at one end to virtual private networks at the other end—with relevant services planned to address the end-to-end network. There must be broad industry acceptance for the standards and platforms used, where critical elements are cohesive visibility in managing across value-added services and the traditional networks and services.

DATA MESSAGES

In a commercial network, where the nodes can be hired on a commercial basis, business messages can be sent from one node to a remote one, in the form of *data messages.* The following facilities are used for this purpose:

(i) Communication in Conversation Mode

Once the connectivity between two nodes is established, messages keyed in the local computer is displayed on the monitor of the remote computer, and *vice-versa.* If sending and receiving such messages—and further replies and rejoinders—are set up at a fixed time for both nodes, *chat line* can be established and a dialogue may ensure across the PCs on a conversational mode. This is becoming a common feature among business-users, and even among friends and colleagues using home computers.

(ii) Remote Login

The local computer can be made as terminal to a remote computer. Thus, a PC located in a remote region can be used as a terminal of a powerful host computer. In the case of multi-user environment (on a UNIX platform), even a terminal of a local computer can act as terminal to the host computer.

(iii) File Transfer

Medium to large files can be transferred in two different modes across computers. In the simple mode, *file transfer protocols* (*FTPs*) (software programmes which set up, execute and control file transfers) can be used to download bulky files (like class-notes, reference material and entertainment programmes such as, pre-recorded plays and films) from the host computer of a university to the local computer of its students, using remote log-in facility, for the students benefit and enjoyment—as are being done by large universities in China. In the more complex mode, the job (data and programmes) can be transferred to the remote computer, in the first place. Then, remote login facility can be used to let the local computer act as terminal to the host computer and, thus, the job could be executed by the local computer on the host computer. Thereafter, the file transfer facility can be used to bring the results of the job back to the local computer.

(iv) Electronic Mail (e-mail)

This has become a very common occurrence for electric transmission of messages, using highly reliable and fast networks. The manner of addressing the message is very similar to the addressing of a postal message—by using the address-box on the screen. Messages previously received can be *saved* in and *forwarded* from the existing files, or *replied* back-to-back to the sender of the original message. Confirmation of receipt can also be obtained electronically. Messages can be *one-to-one* in a personal dialogue mode, or, *one-to-many*, with copies endorsed to many receivers. Business circulars are often composed in the *one-to-many* mode, say, originated by the office-manager. The message travels across networks and finally reaches the destination computer. By connecting the various networks through a *Gateway* switch, international e-mail facilities, out of whom *Hotmail, Rediffmail, Yahoo!*, *gmail, Indiatimes* and quite a few others are available free of cost (with their expenses net from advertisement revenues). Many such e-mails can be both sent and received from any part of the globe by the account-holder, absolutely free.

WIRELESS AND LINE TRANSMISSION

Wireless transmission can be used quite effectively in the form of *microwaves*, up to a distance of some 50 km. Beyond that point, the signal begins to fade quite rapidly and it becomes necessary to use a booster or repeater station beyond 50 km. The other limitation is that microwaves operate based on *line-of-sight* principle, i.e., the sending station should be actually visible by the receiving station (or the booster), without being impeded by any obstacle like a hillock or a high-rise building. Even where the booster is used, it adds substantially to the cost, tending to almost double the original cost. In general, the 50 km data-transmission cost at 64 kbps rate is about Rs. 10 million.

Thus, basic advantage of microwave transmission is that, besides the cost of licensing, the remaining operation and maintenance costs are quite reasonable. If, however, the user organisation wishes to change over from 64 kbps to 128 kbps wireless-connectivity (for, say, starting video-conferences between the corporate office and the branches), most of the equipment would need changing—running into heavy expenditure. Indian

Railways and the state police have been old users of microwave connectivity, apart from private organisations like Bektel Corporation.

Landline (i.e. telephone) connectivity have been typically provided though *dial-up lines* or *leased lines*. For dial up lines, the data transmission rates have been 9.6 kbps, showing signs of increase lately. Added to this low rate is the other factor that unless the source point and the destination point are centrally located, there is the 'last mile' problem, i.e., the connection between the telephone exchange and the user at each end point picks up noise. Further, if their users wish to switch to a higher transmission rate offered by MTNL or DOT, the escalation is to 2 mbps, which may be too much capacity for the user. On the other hand, dial-up lines offer perhaps the lowest cost among all forms of connectivity.

As regards leased lines, leaving aside the trivial data transmission rate of 9.6 kbps, we can move on direct to a 64-kbps leased line. At the prevailing rate for MTNL and DOT at Rs. 1150 per km per year, the cost of using a 64 kbps over 1500 km, say, between Mumbai and Delhi, works out to Rs. 1.7 million per year. While the rate is prohibitively high for using a 2 mbps leased line, the cost of using 64 kbps lines remains reasonable and compares quite favourably to any other form of data communication. The major problem for using leased-line facility is, however, that of reliability, since the amount of downtime has been observed to be rather high, compounded by the fact of usually poor service received from MTNL. The costs are, however, coming progressively down

INTEGRETED SYSTEMS DATA NETWORK (ISDN)

This facility operates under the government of India, Department of Telecommunication (DOT) and Mahanagar Telephone Nigam (MTNL) on landlines. ISDN combines data, text, voice, image and video. It is available among all the metropolitan cities and several other large townships like Hyderabad, Lucknow, Ranchi, Guwahati, Bangalore, Nagpur and Kanpur, among others. ISDN operates on telephone networks of various kinds. The underlying technology for ISDN services is the asynchronous transfer mode (ATM), where all the data are transmitted in small fixed packets, called *cells*, each cell is 53 bytes long: 48 *data* bytes and 5 *header* bytes. The payload size is limited to exactly 48 bits, for carrying out *cell relay* (as opposed to the earlier frame relay). Conceptually, it is a digital bit-pipe between the user and the carrier, where the type of data is irrelevant to the bite. Also, the data-flow is in both directions. In many countries in the Southeast Asia, like China, ISDN has become universally available.

Broadband ISDN services, in particular, have no technical limitation for use in wide area network (WAN). For high-quality 2 mbps ISDN connectivity, the charges are Rs. 2000 per km per year. Thus, the cost of setting up a ISDN network between two cities 1500 km apart—again considering Delhi and Mumbai—the cost would work up to Rs. 3 million. But, of late, the proposed cost of 64 kbps and 2 mbps leased lines are getting sharply reduced and the cost of leased lines is becoming comparable to the 2 mbps ISDN lines. The service provider so far continues to be MTNL and DOT.

DIGITAL SUBSCRIBER'S LINE

Digital subscriber's lines (DSL) and its cheaper version *asynchronous digital subscriber's lines* (ASDL) are sweeping the East Asian countries, like South Korea and mainland China, as a viable alternative to ISDN. China, for instance, has 8 optical fibre-routes spreading from north to south, and another 8 from east to west. These fibres are connected to houses through telephone lines, solving the 'last mile' problem. At the Chinese currency cost of RMB 200 (about Rs. 1200) per month, ASDL provides the Internet connectivity to houses at the uploading speed of 128 bauds and much faster downloading speed of 512 kbps. ASDL provides not merely the text, data, image, sound, and video at an adequate quality level, but also access to Microsoft's sophisticated software packages: Service Park 1 and Service Park 2.

SATELLITE TRANSMISSION

If a communication satellite can be placed in the equatorial plane at a height of 36,000 km above the earth and is made to travel in circular orbits around the earth with a periodicity of 24 hours (synchronising with the rotational speed of the earth)—the satellite remains apparently stationary with respect to any point on the earth. A geostationary satellite thus avoids the problem of being out of communication with its earth station and its use eliminates the complex shifting of antennae: to keep track of the satellite shifting from its position in relation to the ground station. In other words, the height of the geostationary satellite above the earth is fixed and the transmission of the incoming signals is simply responded (by transponders) as outgoing signals—without any amplification or any transmission loss at the satellite. The carrier frequency for the outgoing signals is, however, marginally altered to avoid its interference with the incoming signals.

There are two types of earth stations to link the geostationary satellite to the antennae on earth:

(i) Master Earth Station

It has very large disc-like antennae set up in an open space, main earth station controller and physical cables connecting them. As the name implies, it is the nodal centre to send signals to, and to receive signals from, the satellite and has a rather high cost commensurate to its configuration—dealing with several single-hop and multiple-hop data transmission facilities. National Informatics Centre (NIC), for instance, has a master earth station of large magnitude—located in the CGO complex of Delhi that caters to the all-India network of regional, state, union territory and district computer-terminals.

(ii) Micro Earth Stations

They have small antennae, earth station controllers and physical cables connecting them. The antenna is generally kept on the rooftop and has to be focused on the satellite. The antenna, 1.2 m by 1.8 m, can be easily installed at any location. The rooftop antenna is called a Very Small Aperture Terminal (VSAT). The VSAT controller is kept indoors and connected to the antenna through a twin coaxial cable. There are two ports in each VSAT: for connecting computers, or for connecting other communication devices like FAX machines. A unique 12-digit number identifies each VSAT controller. A number of VSAT nodes are networked together, using an antenna uplinked to the geostationary satellite, as shown in Figure 3.1. The antenna attached to the geostationary satellite serves as repeater.

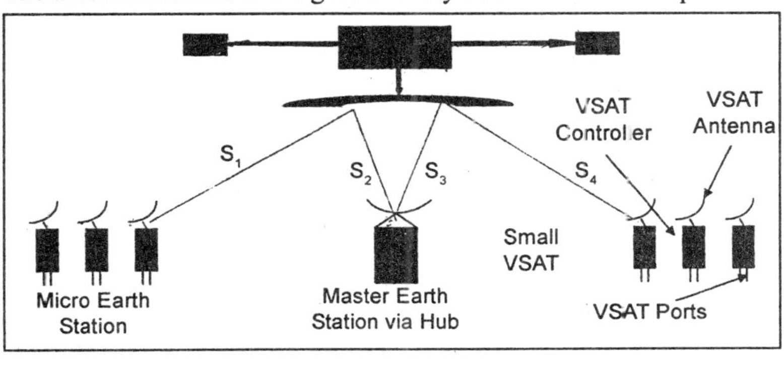

S_1, S_2: Feeble Signals of encoded message
S_3, S_4: Amplified signals by master earth station
[2-Hop Path for signal to travel 144,000 km.]

Fig. 3.1: VSAT Network

Referring to Figure 3.1, let us consider the transmission of messages between computers A and B (On top, left and right, unmarked). The message is encoded in the VSAT controller at A and transmitted from the VSAT antenna in the form of 128-byte packets (with the address of the source and destination). The signal S_1 from VSAT antenna A is feeble in nature, since the antenna is small. This is received at the satellite and retransmitted as signal S_2: with a different carrier frequency by the transponder of the satellite. Signal S_2 is also weak, because of no amplification at the satellite. The master earth station receives this feeble signal and amplifies it (since the antenna is big) as S_3 and transmits it back to the satellite as a powerful signal. The quantum of data transmitted, the source and destination are recorded in the master earth station, but not the actual data. S_3 is now re-transmitted by the satellite as strong signal S_4—to be received by the VSAT antenna B. Powerful signals can thus be received by small antennae and, in this case of a *2-hop path* from antenna A to antenna B, the signal travels (36,000 × 4) or 144,000 km.

VSAT facilities come in two forms—*time-division multiplexing/time-division multiple access* (TDM/TDMA) and *single channel per carrier* (SCPC). SCPC, in turn, comprises *permanently assigned multiple access* (PAMA, not normally used) and *demand-assigned multiple access* (DAMA), This is illustrated in Figure 3.2.

As seen there, there is a permanent *bandwidth* in TDM/TDMA, allocated to the user and consequently the usage charges tend to be high. Although the same *bandwidth* is not required all the time, the user tries to book (and pay for) his maximum *bandwidth* requirement, as a precautionary measure. On the other hand, in DAMA (which came historically later) the *bandwidth* is available on demand. Consequently, the usage charges tend to be lower. The installation costs, however are exactly the opposite: around Rs. 600,000 per site for IDM/TDMA (including the necessary routers and bridges) and around Rs. 1,300,000 per site for DAMA. Comparatively speaking, the operational charges of TDM/TDMA work out to be 5 to 20 times higher than that of DAMA. Further, the *bandwidth* available for TDM/TDMA is of the order of 64 kbps, which may not be enough for activities like video-conferencing. Also, the integration of voice and data channels is not very good in the case of TDM/TDMA. For DAMA, in contrast, the available bandwidth can go up to 8 mbps. All told, there is normally a good case for selecting SCPC (with DAMA) for a VSAT facility due to its much-reduced operational cost—although over a number of years the total cost of TDM/TDMA and DAMA tend to become equal.

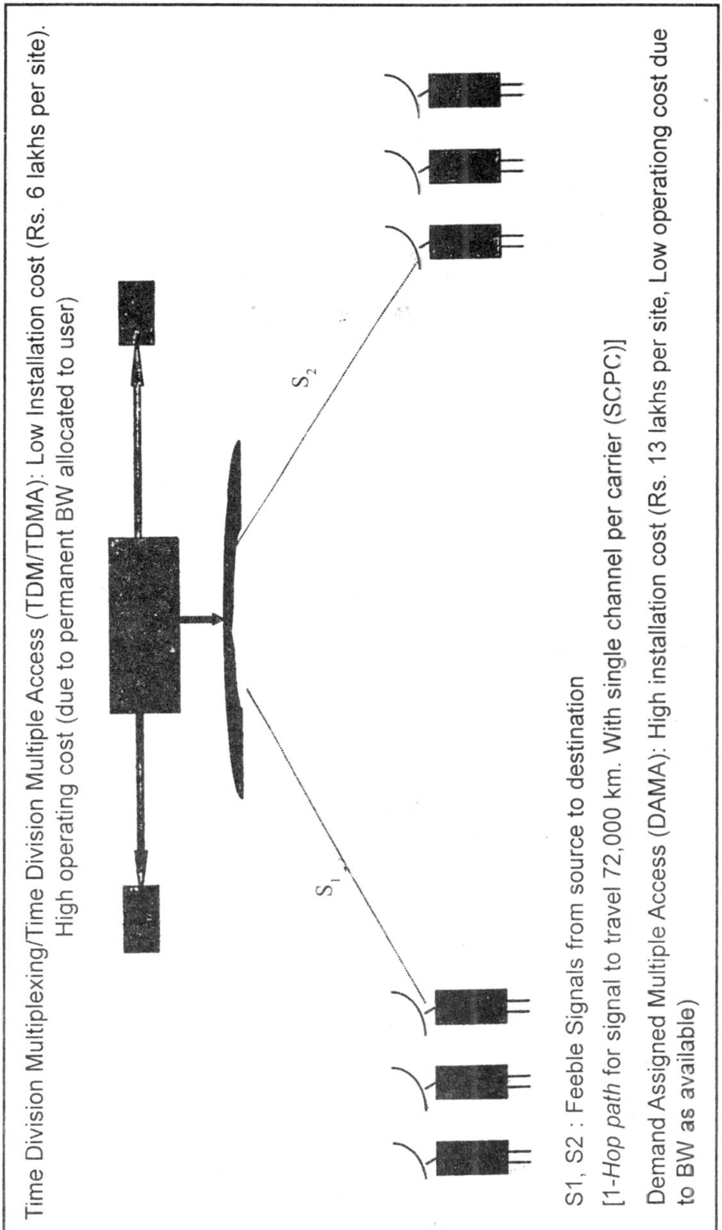

Fig. 3.2: TDM/TDMA and DAMA

With the large-scale availability of *Broadband* and with the widest possible range, the *Bandwidth* has ceased to be a problem. Public sector MTNL and VSVL, and private sector telecom organisations like Airtel now offer *broadband* services on a highly competitive basis, with user-organisations being decidedly gainers.

VSAT NETWORKS

Remote Area Business Management Network (RABMN), under the DOT, is one of the commercially most successful VSAT networks in India, utilised by user industry at large. While Hindustan Zinc, located at Udaipur, uses it, for instance, to connect to their raw-material procurement site at Vishakhapatnam, Tata Steel, situated at Jamshedpur, uses RABMN to get their sales and marketing data from the Tata Centre in Kolkata, within a distance of 300 km. On the other hand, there are industries which use RABMN in a group of nodes—hired by an individual organisation—and establish networks, like a group of 59 RABMN nodes forming the BHEL network in the public sector and a group of nodes by the century-old cigarette giant Godfrey Philips in the private sector. In the government services, the NIC uses a satellite networks based on VSAT technology for exclusive government use—connecting district headquarters, state government secretariats, Planning Commission and the state planning boards.

VSAT nodes can be connected to the existing terrestrial networks for establishing Wide Area Network (WAN) connectivity. Packet assembler/dissembler (PAD) switch is one such device, which facilitates linking local terrestrial networks with VSAT modes. For instance, an organisation may possess a number of PCs, minis and super-mini computers. These machines can be connected to the various ports of the PAD switch—with one of the ports connected to the VSAT controller. The port of the VSAT controller has X.25 compatibility, where the single physical channel can act as multiple logical channels. Interestingly, the computer can also be connected to the PAD switch, using dial-up modems The public telephone networks bandwidth in commercial VSAT network in RABMN has been, till recently, only 1.2 kbps and provides a simple transfer of text and data messages. With higher bandwidths under consideration, voice and image transmissions are also possible. As regards economic feasibility of VSAT networks, the tariff rate of the VSAT nodes has two components: fixed and variable. The use of VSAT network can be even more economical than the conventional method of postal letter or fax. A Hewlett-Packards survey for a 2-page electronic message reveals, for instance, approximately the following cost-details:

International	Fax: $ 1.66
	Letter: $ 0.51
	E-Mail: $ 0.22.
In-land	Fax: Rs. 30
	Letter: Rs. 20
	E-Mail: Rs. 10.

VSAT technology offers an excellent means of providing communication in remote areas: where terrestrial communication is either inadequate or unreliable. Indian industries are able to use this technology for information exchange and data communication among the following:

- Head offices and remote project-sites
- Raw material procurement centres in far-away townships
- Factories at rural areas
- Distribution stockyards or outside towns
- Erection-sites at remote areas
- Sub-contracting units near villages or towns.

E-MAIL SERVICES

As we have seen, e-mail messages are similar to postal messages used on addressed electronic envelopes. The text of the message is either typed from the keyboard or redirected from an existing computer file. The message travels across networks and eventually reaches the destination computer. International mails are sent today by connecting various networks through *gateway* switches. While an e-mail user needs his ID, mailbox and security, he avails of the service of an e-mail service provider—with the latter having his *hub* (server) and *backbone* (band).

A comparison among communication methods would show that, while landlines are expensive, time-consuming, often busy, needing tie-up among two parties and is usually one-to-one (except in tele-conferencing), e-mail is cost-effective, speedy, never too busy, without waiting for a tie-up and easily one-to-many. Further, while surface mail is slow and with long turn-around, e-mail is instantaneous and very flexible. Finally, while even courier service is relatively slow, expensive and inconvenient, e-mail is speedy, cost-effective and convenient. Indeed, e-mail has very rapidly become much more than just interpersonal messaging and found a niche in numerous industrial applications. To take only one instance in the sales and marketing area, its usages are: sales leads; sales reports; customer orders; price changes; announcements of special offers; market research; formulation of proposals; competitive information; and customer service.

Many e-mail service providers (ESPs) are operational today, out of whom Hotmail.Com (now purchased by Microsoft at an impressive price), Yahoo.Com and Rediffmail.Com are immensely popular. There were many early starters among ESPs like BI Access, RPG-Sprint, Dartmail, Gem 400, Elnet (on *leased line*), and Hughes Escort, HFCL, Telstra, Wipro RT (on *VSAT*)—some among whom have fallen by the wayside: due to the sheer competitive edge, wrong pricing or simply poor management. There have also been many new entrants, though it is highly doubtful how many of them would stick it out in the long run. There is also a good chance that many of the free ESPs will eventually convert into pay channels, as the advertisement revenue is not enough in the kitty for all to survive. This possibility is already looming on the horizon. As an interesting parallel, China has an eminently successful ESP on '263.Net' which is a pay channel, against their free ESPs like '163.Net', Sima.Com.Cn' and 'Sohu.Com.'

A typical example of information distribution and collection in a corporate e-mail network is illustrated at Figure 3.3.

SELECTION AND EVALUATION OF E-MAIL SERVICES

Against the above background, it may be a good idea to evolve a few selection and evaluation criteria for the ESPs by the end-users. Some of the suggested criteria are briefly enumerated below:

Compatibility with OS and GUIs

While Novell Network has been slipping down among networking software, Windows NT has been gaining ground. Among the GUIs, there are some back-end DBs tightly coupled with front-end GUIs, like UNIX-UNIFY; SQLBase-SQLWindow; Centura-Syface; and so on. A good choice is desirable.

Maximum Cities Connected by the ESP

This is especially true for the new players.

Local Call Facilities or STD

This will assume importance once the pay-channels idea gains ground.

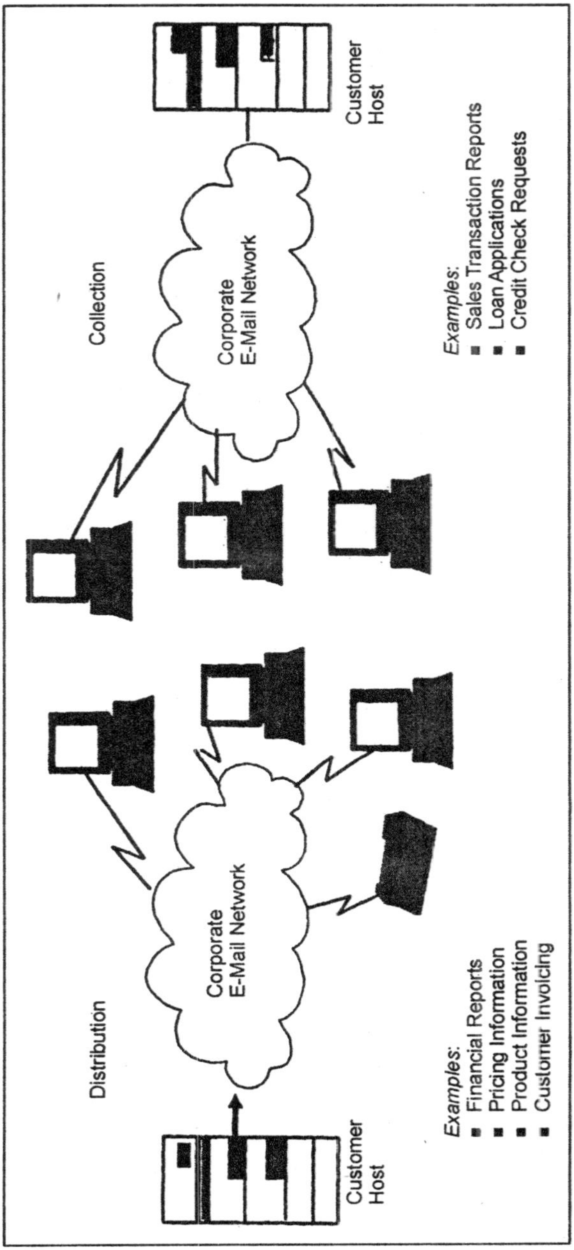

Fig. 3.3: Information Collection and Distribution in a Corporate Network

Facility for Fax

Both at national and international levels, Faxing facility is essential. There have been some interesting variations at the international front, based on volume, time of the day, and so on.

Facility for File Transfer

File Transfer Protocols (FTPs) improve by the day, but large-sized files are still taboo, except for *groupware* like Lotus Notes, operated by many client-server companies.

User-base and User Feedback

This can be a good base-level check-up for the newcomer ESPs.

Minimum Transmission Time

The time taken by the e-mail, from its source to its destination to reach, makes interesting observation. The more instantaneous it is, the better for the end-user organisation.

Internet Connectivity

Almost all ESPs have this facility, but a few are only domestic and do not connect to ESPs with servers abroad. International gateways are crucial for the end-user.

Minimum Cost

For bulk e-mail connectivity, there are many structured tariff-rates, which should be carefully compared.

Graphics Interface

Apart from the GUIs, good graphics capability—below or above 1 MB per user—needs be checked up.

Support Service

This is often a nightmarish problem, particularly when the ESP's server is either down or saturated. Frequencies of such ailments and their remedies should be checked up.

User-Friendliness

Most ESPs are user-friendly today. Still the end-user organisation should look into the matter.

CONCLUSION

While data processing is obtaining meaningful information by the recipient, communication processing is to make this information available to the right people at the right time. Inter-connecting VSAT nodes with the existing terrestrial networks can form enterprise-wide networks in large corporations, which are geographically dispersed. In such networks, the data processing outlets can be instantly fed to the communication processing system (with negligible delay of 540 milliseconds due to the 2-hop connectivity, as against 4 microseconds for point-to-point landlines) and rented to the end-user destinations: making the total information processing instantaneous, inexpensive and effective.

PART TWO

TECHNOLOGY REVIEW

4

MANAGEMENT OF CLIENT-SERVER TECHNOLOGY

INTRODUCTION

Managers working in a LAN environment do not get any processing support from the central hub. In client-server technology, such processing support is built in the servers, which exist to assist in application processing for user-managers. The latter operate as clients and receive the benefit of customised development from the server, which is a huge advantage.

FILE SERVER IN LOCAL AREA NETWORK (LAN)

In LAN, a *file server* is the central *node* (computer in the network), where the users can access it and use it for starting files. Typically, the file server in the LAN is also the central hub for sharing peripherals, such as, printers, print-queues, modems and software packages—given enough disk-storage space. Since any application running on a workstation in a LAN reads and writes files on its file server, all the files are often pumped across the network: on behalf of the operations taking place on LAN PCs. A file server plays no role in the processing of an application and does no more than storing application files that run on LAN PCs. A user, for instance, may have a personal DB on his LAN PC and can request information from a file on the server. The file server would send all or part of the data file across the LAN to the user's workstation, without participating in the DB management and DB Workstation. When the user saves the data file, he copies it back to the file server.

The above analysis will make it clear that the LAN file server, by the very design, cannot adequately provide service to the demands of multi-user and shared-data applications that mainframes can easily support. The two major bottlenecks are as follows:

1. The LAN file server does not deliver the *data concurrency* (simultaneous access to a single data-set by multiple users) that is required by multi-user applications. This is due to the fact that the file server operates files with large data sets and prevents the user from sharing the file when another user has locked it out.
2. Users who operate on the same data set are forced to wait to access the concerned file. File storage and file processing do not typically take place on the same computer. File processing occurs on workstations that store and retrieve files from a central file server. Therefore, if many workstations request and send files in a LAN, the network can become saturated with traffic—thus creating problems that degrade the overall system performance. A typical 4-node LAN is shown in Figure 4.1.

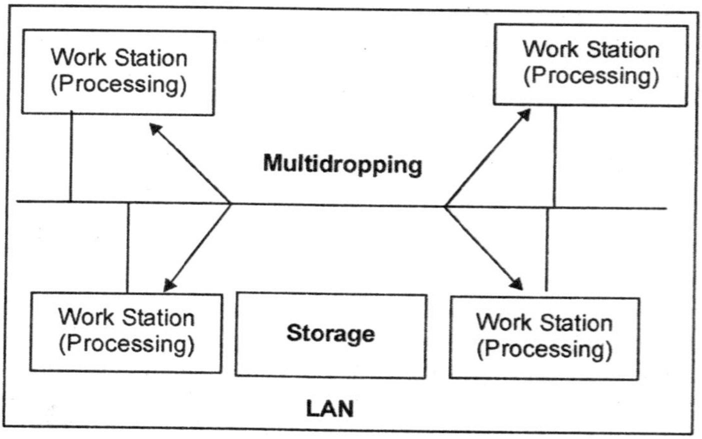

Fig. 4.1: A Typical 4-Node LAN

CLIENT-SERVER NETWORK

The above problems have led to the genesis of the client-server computing model. The latter is also known as *distributed application processing* and *cooperative application processing*—delivering the benefit of the network model, along with the shared data access and high performance characteristics of the host computer model. This is shown in Figure 4.2 for 4-node clients.

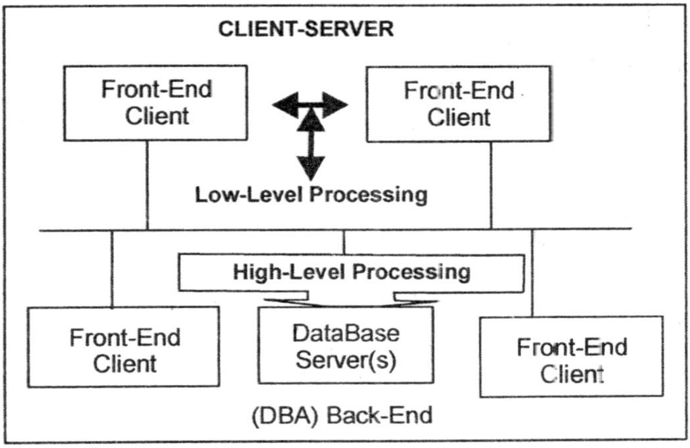

Fig. 4. 2: A 4-Node Client-Server Network

In a client-server network, the *back-end server* focuses on managing resources, such as, DB of information, efficiently. The server's primary job is to manage its resources optimally among *multiple clients* that concurrently request the same resource from the server. A client application in the *front-end* is that part of their system which users employ to interact with the data—ensuring global data integrity under the rules across all client applications. Summarised comparison between LAN and client-server network is provided in Table 4.1.

Table 4.1: Summary Comparison between LAN and Client-server Network

LAN	*Client-server Network*
• Central node for file server for storage, printers, modems and software packages;	• Back-end server/s for managing DB resource;
• No simultaneous access to a single data-set by multiple customers;	• Single DB serving among many concurrent user-clients;
• Centre hub not for processing, but only for storage; File processing done at workstations, in consequence, many retrievals can saturate network with traffic.	• Back-end server/s for high-level processing and optimal management for multiple clients; Back-end server splits processing load, hence distributed application processing.

MANAGEMENT NEEDS FOR CLIENT-SERVER TECHNOLOGY

For any developing, or developed, business organisation, big or small, client-server computing has become a must, because it provides near-instant access to information for *decision making*, facilitates communication, and reduces time, effort and cost for accomplishing tasks. Client-server computing satisfies the business needs by appropriately allocating the application processing between the *client* and the *server* processors. While the client requests service from the server, the server processes the request and returns the result to the client. From the user-manager point of view, the *client* is a single user workstation that provides presentation services and appropriate computing, connectivity and DB services—with interfaces relevant to the *business needs*. The server is the host machine with shared memory, providing computing, connectivity and DB services—with interfaces relevant to the *user requirements*.

Client-functions normally manage the user-interface, validate the data entered by the user, dispatch the requests to server programmes and occasionally also execute business logic. One of the key elements of a client workstation is the Graphical User Interface (*GUI*), to present the outcome. A server-function is dedicated to fulfil the client request by performing the task asked for and use the GUI for the presentation purposes on the network. The server functions as a *software engine* that manages shared resources, such as, DBs, printers, communication-links or high-powered processing. The roles of *clients* and *server* can be summarised as follows:

Role of Clients

- Solution specific;
- User interface to accomplish jobs;
- Managing presentation-logic (e.g., *pop-up* list on a data-entry form, *bar* graphics, etc.);
- Validating data entry;
- Requesting and receiving information from DB;
- Executing simple business logic.

Role of Server

- Development specific;
- Managing DB among concurrent users, with DB perceived to be *distributed* by the clients;

- Controlling DB access and protecting security;
- Receiving and executing client request;
- Executing complex business logic.

Beside the above, all the transactions and queries fired against the DB are managed by the *database administration* (DBA), maintaining consistency and integrity. The tasks of DBA can be summed up as follows:

- Accessing
- Updating
- Inserting
- Locking
- Security
- Logging
- Back-up
- Recovery.

BENEFITS OF CLIENT-SERVER TECHNOLOGY

Major benefits, arising out of client-server network applications, can be outlined as follows:

1. Transparent Data Sharing

Data, collected as part of the normal business process and maintained on a server, are easily and immediately made available to all authorised users. *Transparent* network services ensure that the same data are available, with the same *concurrency* to all authorised users.

2. Integrated Services

All information, that the client (user) is entitled to get, is available at his desktop. There is no need to change into *terminal mode*, or, log into another processor to access information.

3. Sharing Resource among Different Platforms

This is referred to as *open system computing*, which can combine the hardware and software products from different vendors: to form seamless information network aligned with available standards, independent of vendors. *Open systems* provide transparent access to all of the enterprise's applications and information, regardless of the platform on which they reside. Thus, users may obtain client services and transparent access to DB, communications and all the applications, which have been created and implemented. OS software and hardware remain independent of all such applications.

4. Data Interchangeability

Data can be accessed for high-volume operation and updation at multiple locations—without affecting DB integrity across the distributed locations. Network services help in providing transparent connectivity between clients and servers: local or remote.

5. Location Independence of Data

The user is not concerned about the location of the data or the technology of the processor involved. This transparency enables tables to be moved to other platforms and locations, without affecting the application code—making the application *portable* and giving birth to *open systems.*

6. Centralised Management

As processing steers away from the central data centre to the remote offices, monitoring and supporting systems from central locations take care of the network and the data, and "oversee" the operation of hardware and software on the network by experienced support personnel.

7. Rapid Application Development (RAD)

Using objects-oriented programming techniques can create applications and new client-server systems can be developed rapidly. The user can quickly create visual screens using *GUI* tools—without having to write the underlying code.

8. Effective Implementation of BPR

A client-server system can be easily modified to handle future modifications that may be made in any business processes through business process re-engineering (BPR)—to eliminate the shortcomings of the current system, and to modify the process for improvement and implementation of new technologies. Due to this feature, BPR can be effectively done: by automating tasks, enabling immediate information access and sharing, eliminating repetitive tasks and facilitating workflow.

Other perceived benefits are:

- Front-end and back-end tasks
- Scalability in keeping with the user-managers' needs
- Proof against failure (covered later)
- Access authority clear
- User friendliness.

OVERVIEW OF CLIENT-SERVER TECHNOLOGY

It should be clear from the above that the management as clients gets transparent, concurrent and efficient service from the client-server technology in an enterprise, which can be spread worldwide, with servers in one country and clients all over the globe. An overview can be provided in the following lines:

- Running on appropriate hardware and software platforms, based on *modular* concepts;
- Separation of large software into modules for easy development and good maintenance;
- 2-*tier architecture*: enabling clients talking direct to servers;
- 3-*tier architecture*: with an intermediate agency buffered in between clients and servers to prioritise requests, meter the execution and translate needs into appropriate instructional logic;
- Server's high intelligence: ensuring concurrency, minimizing networks traffic and improving performance.

2-tier and 3-tier architectures are illustrated in Figures 4.3 and 4.4 respectively. The 2-tier architecture is still predominantly used because of the following advantages:

1. Available with well-integrated PC-based tools like Visual Basic, Power Builder, MS Access and 4Gl tools;
2. Tools relatively inexpensive;
3. PC-based tools capable of *rapid application development* (RAD), i.e., applications developed in short time;
4. User having access to attractive GUIs.

However, the disadvantages of 2-tier architecture are as follows:

1. Maintenance of applications on client side ("Fat Client')
2. Increased network load

3. Applications loaded on individual PCs
4. Software distribution procedure complicated
5. PCs considered insecure.

In 3-tier architecture, most of the software is removed from the client ('Thin Client') with a DB-server tier responsible for data storage. Advantages are as follows:

1. Many users able to access a wide variety of server applications;
2. Business objects and data storage brought as close together as possible;
3. Change management easier and faster.

Management should opt for 3-tier architecture when different DBMSs are involved; application programmes are written in different languages; workload is heavy; and workload is expected to grow over time.

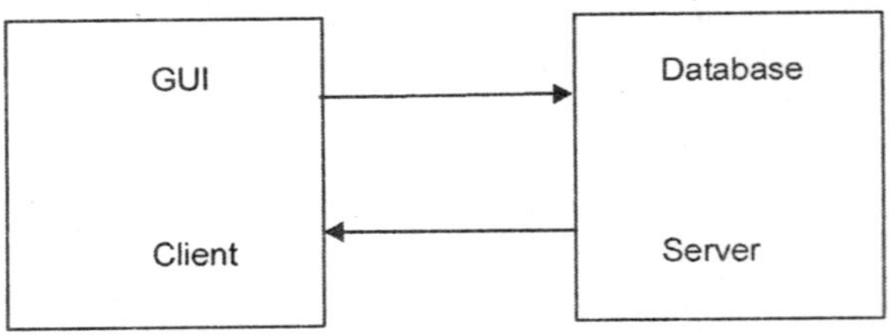

Fig. 4.3: 2-Tier Client-Server Architecture

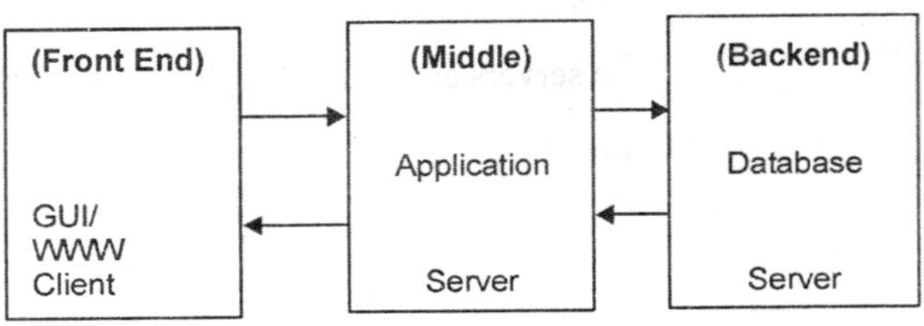

Fig. 4.4: 3-Tier Client-Server Architecture

SELECTION AND EVALUATION OF CLIENT-SERVER TECHNOLOGY

Due to the obvious technology advantages, the management of user-industry is prone to easily accepting and adopting this technology. There is, therefore, a need for a user assessment of the available vendors from the market and for making a judicious selection of the technology. This can be done on the following basis:

Server

(Available from multinationals like IBM with 15 per cent of the world market, Compaq 32 per cent, Hewlett Packard (HP) 13 per cent and Digital 5 per cent, *besides Sun Solaris and others)*

1. *Performance of CPU*

From among the available P-3, P-4, Celeron and Athlon chips with CPU speed from 300 MHz upwards to 1 GHz and beyond;

2. *OS Platform*

From among Novell Networks, Windows NT, Unix and Linux;

3. *Number of Users*

With considerations for the total number and the concurrent ones;

4. *Nature of Applications*

Covering development load and application support, with all their attendant features;

5. *Distributed Processing*

Load being borne by the servers and shared by the clients and servers;

6. Server Reliability and Data Criticality

Nature and extent of availability;

7. *Resource Sharing*

The volume to decide the configuration;

8. *Proof against Failure*

(i) Redundant CPU within the same server;
(ii) Hard disk *mirroring* (for dynamic on-line back-up in another HD of, say, 4.3 GB)

(iii) *Redundant array of inexpensive devices* (RAID) for, say, disks to have *'Hot Swap'* (on-line change of disk);
(iv) *Clustering* for stand-by server/s (especially for Railways or airlines reservations);
(v) *Automatic server recovery* (ASR) to affect *'fail-safe'*, i.e., automatic *re-booting* of another server against the impending *crash* of the existing server;
(vi) Redundant power supply (for CPU) as the *'Hot Plug'* facility;

9. *Scalability and Expandability*

For RAM, CPU and HD;

10. *Support Services*

Providing for RAM, CPU and HD, in the event of a *pre-failure warning;*

11. *Level of Enterprise*

Whether entry level, departmental level or enterprise level;

12. *Compatibility*

With other chips available in the market;

13. *Bus Speed for Data Communication*

In the range of 66 MHz, 100 MHz, etc.;

14. *Tight Integration and Manageability*

As a safeguard feature, among servers and clients, and among multiple servers;

15. *Solutions for Future Growth*

To protect fresh investment on a large scale, future solutions to be planned in advance;

16. Cost of Ownership

With obvious concern for the price—to be as reasonably priced as possible.

Clients

(*Available from multinationals like IBM, Compaq and HP from semi-multinationals of India like HCL and Wipro; and locally assembled products*)

1. *Choice between 'Dumb' and 'Smart' Clients*

If 'dumb'—monitor resolution; if 'smart'—connectivity with server: local with RS232C cabling and remote with Ethernet; 'dumb' being gradually replaced by 'smart';

2. *Performance*

From among P-3, P-4 and other chips, with CPU speed from 300 MHz upwards;

3. *OS Platform*

From among Windows NT, Novell Netware, etc.;

4. *RAM and Hard Disk*

RAM varying from 4-16 MB to 128-256 MB, depending on the nature of client, hard disk configuration to be decided on application-load; available latest hard disk (like USB) to be preferred;

5. *Needed Accessories*

Client requirement of CD-ROM drive, Fixed DD, modems and printers;

6. *Compatibility*

With the entire client-server technology being considered;

7. *Quality and Warranty*

In favour of licensed brands (other than cost which is in favour of locals); building and tuning with OS;

8. *Cost of ownership*

In favour of locals.

BROWSER-SERVER TECHNOLOGY

The Web-based technologies are indeed the natural choice for large data warehouses. The Internet can be used as a virtually free resource, which provides a universal connectivity within and between companies and among individuals. The start-up advantages of the Web facilities are:

- Ease of performing complex administrative tasks of managing distributed environments; and
- Storage and management of both data and applications by companies on *servers* that can be centrally managed and updated—thus eliminating problems with software and data accuracy.

The emerging third-generation Websites offer Web-based *application server* (by replacing HTML gateways). These third-generation *servers* provide users with all the capabilities of existing decision-support applications through the *Web browser*, without the need to load any *client* software. Several vendors of decision-support applications (for query and reporting) have converted their tools to work on the Web, adopting the following approaches:

- *HTML Publishing*: Transforming a query output into the HTML page that can be downloaded into *browser* (without interactive access to data or reports);
- *Helper Application*: Residing within a *browser*—as in 'fat' client where data can be downloaded and used for analysis by the user;
- *Plug-ins*: like helper application, downloaded from the *server* prior to their use. Plug-ins are *browser*-specific and may not run on all platforms or with all *browsers;*
- *Server-centric Components*: Created by vendor as a new server component (or re-building a desktop) that can be integrated with the Web (via a Web gateway); and
- *Java and Active X applications*: Re-developed by vendor in Java or Active X—as a true 'thin' client. This is the most promising and flexible approach.

APPLICATIONS OF CLIENT-SERVER TECHNOLOGY

As a major backbone in today's IT, client-server technology is becoming important by the day. Some of the advanced applications are described below:

1. Expert System

Essentially, an expert system (ES) captures the rules that describe an expert's logical response to a business situation. ES provides a great opportunity for higher productivity and reduced costs, as these *stored rules* in a *knowledge-base* (KB) are consistently applied to business. ES is currently used mainly in government and financial organisations. In new applications of ES in *network management*, a rule-based analyser is used to help diagnose the problems in the network. The rules-base can also help a regular preventive maintenance of the network.

The retail sector has begun to use ES for real-time management. ES audits the reality against the expectations in real-time. This allows unexpected results (good or bad) to be detected early and lets the manager concentrate on existing customers or new marketing programmes, by using client-server technology.

2. Point-of-Service (POS)

Traditional point-of-sale technologies are now re-named point-of-service (POS) technology: found in most restaurants, supermarkets, department stores and auto-service stations—for pricing, managing staff, accounting, distributing products and controlling inventory. Moreover, integration of technology, business process and management information in POS has made it an important player in the era of client-server applications—making POS a very widely used area of client-server technology.

3. Geographic Information System (GIS)

GIS is a technology that provides the use of a wide scattering of over-ground and under-ground resources and facilities, and building meaningful overview of information to meet business needs, GIS can cover the topology of a landscape, including features like roads, buildings, sewers, electric cables, minerals and soil content. When integrated with ES, GIS can play a useful role in business processes.

To start with, GIS—in a powerful package like *Arc-Explorer*—enables to store virtually unlimited global information as a series of layers. With GIS, a customer can design *customised* printed map to fulfill a particular need—by simply using the software to select the relevant layers. GIS also helps architects and designers to specify attributes wanted by the users and associate them with an objective map. Such attributes may be descriptive data, detailed measurements of any kind or legal description. GIS can today be based on client-server technology—with the latter's *connectivity* permitting shared access to the layers maintained by diverse departments. In fact, the GIS information base, if combined with attributes stored in other DBs, can provide important value-addition. Location of a polling booth with easy access for all, for instance, can be done by combining voters' list with the street-maps—so that no natural or artificial barriers block the way.

CONCLUSION

Client-server technology has emerged today as a powerful tool to satisfy user needs at the client's front-end, by providing both application and development support by the servers at the back-end. There is a clear focus on the customised development of application solutions, which clients can use, and even enhance with their own GIS tools.

5

MANAGEMENT USE OF DATA WAREHOUSE AND DATA MINING

INTRODUCTION

In asserting the prerogative of decision-making, based on data and information, today's end-users have the advantage of falling back not only on the immediately available past data, but also on the distant past data that could have remained otherwise as an inscrutable legacy system: perhaps made of much-patched COBOL programmes. Tools and techniques are now available that can aggregate such data—glossing over their disparate nature—and glean meaningful inferences from the massive accumulation of data. We look here into the overall philosophy of such aggregation into massive data warehouse and then apply the scalpel of analysis: to derive the much-needed management insight.

REQUIREMENTS OF THE END-USER

Many end-users need to use *decision support systems* and raise a large number of queries, seeking responses. Some end-users need specific query tools so that they can build their queries themselves, while others seek specific information. In addition, trend analysis tools and others, which can show the outcome in a graphical way, are often used for statistical analysis. With historical data in *decision support system*, one can carry out all these types of applications to support management in its decision support function.

Recent experience shows that it is difficult to build effective *decision support systems*, since the techniques available are unable to satisfy the end-users. The latter would, in fact, like to have *all* the techniques at their disposal, such as, Graphical User Interfaces (GUIs), statistical techniques, windowing mechanisms, and visualisation tools that they can use for accessing the data, etc. This implies that a great deal of local computer power is needed at each workstation and the *client-server technology* is the best solution. This technology involves dispersing the software over several computers and creating an environment for the end-user so that each appears to be working on one single system. The heavy load of GUIs and other visual techniques can be processed on the local machine, while the specific database server handles all the DB tasks. Thus, the database server can be completely optimised for the DB. There are several *client-server tools* that are able to copy DB records from the operational DB to specific DB servers that may eventually store the data warehouse.

DATA WAREHOUSE: GENERAL

In its simplest form, a data warehouse is no more than the collection of key informations: used to manage and direct the business for its most profitable outcome. This could be anything from deciding the level of stock on a supermarket-shelf in focused targeting of customers for loyalty schemes and direct mail promotion, to strategic decision-making on major market segments and company profitability. More rigorously speaking, data warehouse is a repository of subjectively selected and adapted operational data—which can answer any ad-hoc, complex, statistical or analytical analysis.

Data warehouse is usually situated at the centre of a *decision support system* of a company and contains integrated, historical data (both summarised and detailed information)—common to the entire organisation. It has become essential today for effective business intelligence, business strategy formulation and implementation in a globally competitive environment: where larger and larger amount of data (doubling every 18 months!) are required to be processed faster and faster (in a few seconds!), for comprehension of their real meaning and impact. Data warehouse enables the following: easy organisation and maintenance of large data; fast retrieval and analysis in depth, required from time to time.

Data can be classified into 3 categories: *reference and transaction data; derived data;* and *de-normalised data. Reference and transaction data* come from operational sources and are normally kept in a conventional DB. It is loaded periodically for refreshing the data warehouse. When purged from the source, it is archived into the data warehouse and cannot be modified any further. *Derived data*, on the other hand, is only derived from the above, based on certain rules and computations. These can be derived afresh or suitably modified. *De-normalised data,* which are the basis for on-line analytical processing

(OLAP) tools, are prepared from time to time and directly based on the details of reference and transaction data.

DATA MARTS

Data flows from a data warehouse to different departments for their customized DSS usage. These departmental components, called *data marts,* are bodies of DSS data for an individual department that have the same architecture of a data warehouse. As a subset of a data warehouse, a data mart is comparatively quite popular. This is because a central data warehouse keeps growing and becomes gradually more complex and perhaps even unwieldy—and harder to customise. While the analyst of DSS can easily customise, summarise and analyse results from a small data warehouse, he finds the cost of processing going up as the volume of data increases. The available software to access the large quantity of data is not as elegant as the processing of small data from the data mart.

For the above reasons, the analyst finds the data mart very useful for fast and easy analysis—as also the end-users of the results of such analysis. The individual department can easily summarise, sort, select or structure the data by way of customisation. The historical data can be kept limited and the processing load restricted to the department's own requirements. This is, of course, besides the fact that, at times, the *data marts* for individual departments may become incompatible with each other, since usually there is no necessity to look into each other's requirements.

The *data marts* can be categorised as *multidimensional OLAP* and *relational OLAP*. In the former, the numerical data—which is basically multidimensional in nature—can be *sliced* and *diced* in a free fashion, i.e., free from the data-modelling constraints of a DBMS. For example, data of commodity-prices can be presented commodity-wise, region-wise, outlet-wise, etc. In the relational OLAP, the data mart can contain both text and numerical data—and are supported by RDBMS. These data marts are used for general-purpose DSS analysis with many indices: for ad-hoc queries, or, for pre-structured queries.

Metadata (or data about data) describes the details about the data in a data warehouse or in a data mart. Such descriptions may be in terms of the contents and sources of data that flow into the data warehouse or a data mart, as follows:

- Description of sources of data;
- Description about the tables, attributes and relationships of the data mart;

- Definitions of all types; and
- Description of customisation, as the data passes from data warehouse into data mart.

The linkages and relationships between metadata of a data warehouse and metadata of a data mart have to be well established and understood by the manager using the metadata. With such linkages, the manager using data mart metadata can easily find the heritage data in data warehouse. Further, the DSS analyst also needs to know how calculations are made and how data are selected for the data mart environment.

The *data* in a data mart can be of detailed level, summary level, ad-hoc level or pre-structured level. They may contain a lot of summary data and processed data. The *software for access and analysis*, which is unique for a data warehouse application, allows the end-user to search through the data mart: to find, compute or derive the data required by him. *Tables* in the data mart may include: summary tables, detailed tables, reference tables, historical tables and analytical (spreadsheet) tables, among others.

DATA WAREHOUSE: DESIGN AND IMPLEMENTATION

While technical design and its implementation is beyond our scope, the following design issues are important for end-users to observe:

- *Homogeneity of data sources*, requiring efforts in data conversion and quality levels of data integrity, reliability and authenticity;
- *Old and historical data*, to be relevant and useful, and not too old;
- *Tendency to grow very large*, with discretion to limit the size of the workhouse.

The *performance* considerations for a well-designed data warehouse are as follows:

1. It should be completely business-driven and based on specific business-goals through OLAP;
2. Although OLAP applications on a data warehouse do not call for *stringent, real-time responses* as in OLAP itself, the interactions of the user should be *on-line* and interactive, with good enough speed and rapid query processing.

The *implementation* of the design should keep in view the following steps:

- To collect and analyse business requirements
- To create a data model and physical design after hardware platform selection
- To define data sources
- To choose the DBMS and software platform
- To extract data from operational data sources (legacy or DB), transfer, clean-up and load
- To choose DB access and reporting tools
- To choose DB connectivity software
- To choose the data analysis (OLAP) tools and presentation software (client GUI)
- To keep refreshing the data warehouse periodically.

One of the best statistical *access tools* for data warehouse, available in the market, is SAS (from SAS Institute), which has its own DB engine: capable to impart data warehouse and data source. This facility of SAS is not available with most other data warehouse and OLAP vendors.

SELECTION AND EVALUATION OF DATA WAREHOUSE

For a judicious selection by end-user of the data warehouse, the following can be the major considerations:

I. Hardware Platform Criteria

Hardware should be with substantial data storage facilities—with disk-space 2 to 3 times the data for processing and indexing. Both mainframes and non-mainframes (with UNIX or Windows NT or LINUX) are in order. Among non-mainframes, high-end series like Sun Solaris, should be reliable.

II. DBMS Criteria

- Scalability
- Performance in high-volume storage
- Processing and throughput of high order.

IBM (DB2), Oracle, Sybase, etc., are good DBMS candidates. They should support parallel DB processing.

III. Communication and Networking Criteria

- Security
- Privacy
- Accessibility.

It should be Internet (Web)-enabled. If Intranet is operational within the user organisation, it should be intranet-enabled.

DATA WAREHOUSE: BENEFITS AND DELIVERY

The following obvious benefits should accrue to the end-user from a data warehouse:

- Tapping wealth of information and knowledge created over time;
- Integrating multiple versions of an application (like finance data, to be combined if a transition accounting system to the US accounting system or the British accounting system, on merger or acquisition);
- Merging data from current multiple applications;
- Combining the data from the old system (mainframe-based, custom-built, legacy) with the new system (standard business application); and
- Allowing for year-on-year analysis, even though basic operations are changed. *Open Business Solution Interacting* (OBSI) provides a good protocol.

The above benefits can be derived from an *enterprise* data warehouse about many aspects of information being used by the business (probably larger than sectoral databases or data marts, used beyond a single functional need). For such enterprise information, the delivery process should appear as in Figure 5.1.

DATA WAREHOUSE TO DATA MINING

A data warehouse, as we have seen, is a subject-oriented, integrated and time-variant collection of data and aims at supporting business decisions. Data mining, on the other hand, is the process of extracting previously unknown, valid and actionable information from large data (as transaction data, DB or data ware house) and then using the information so derived to make crucial business secessions. Thus, data mining process is a natural and logical continuation to data warehouse. An organisation that has already invested in a data warehouse may definitely like to reap the benefits from it by going ahead for data mining process, as shown in Figure 5.2.

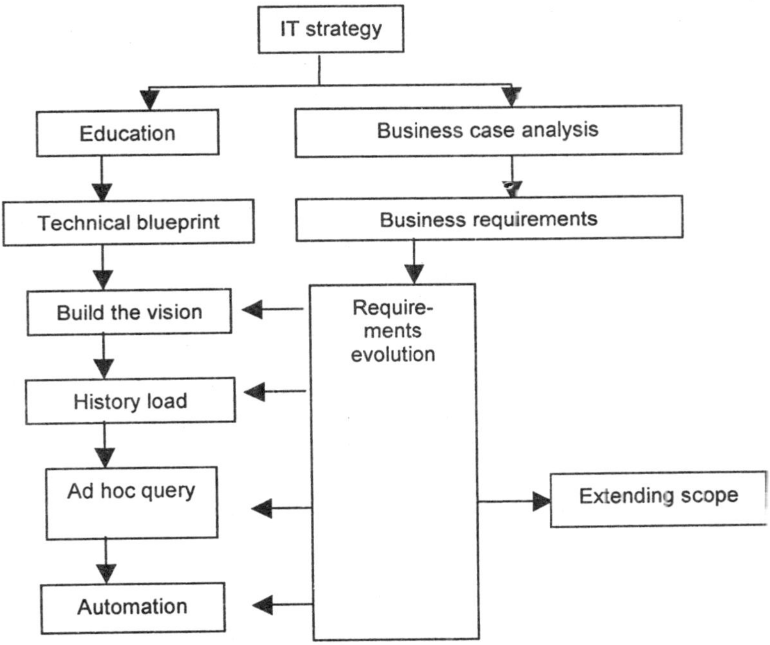

Fig. 5.1: The Data Warehouse Delivery Process

Presentation to Users for Decision-making

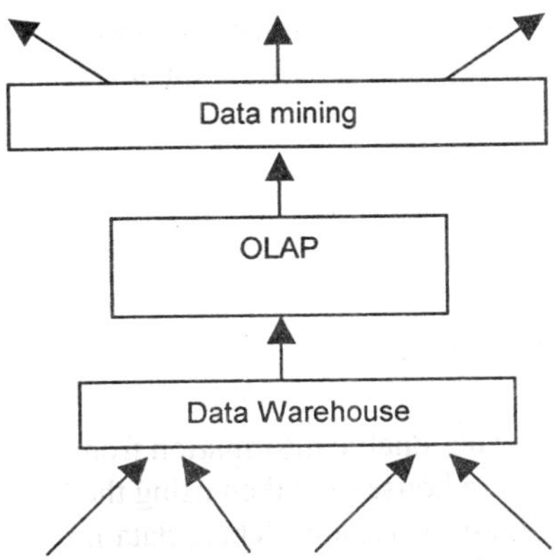

Fig. 5.2: Data Mining Hierarchy for Business Decision Support

It is, however, possible that an organisation can take up data mining even without having built a data warehouse: by using either direct transactional data records or, a single DB of data transactions on which mining is required to be done. The *growth process* of data mining has been as follows:

1. In the 1980's, all major organisations built infrastructural DBs, containing data about their clients, components and products. These DBs have been potential goldmines: containing gigabytes of data with much *hidden* information that cannot be easily traced using SQL or other shallow query facilities. Data mining algorithms can find *optimal clusters* or *interesting regularities* in a DB. While SQL is a query language helping to find data, data mining algorithms typically zoom into interesting sub-parts of the DB—using many, repeated SQL queries and other intermediate results.
2. As the use of *networks* grow, it is increasingly easy to connect DBs. Thus, by connecting a client's file to another file with demographic data, one may get unexpected views on the spending patterns of certain population groups.
3. Over the recent years, machine-learning techniques have expanded: with *neural networks, genetic algorithms* and other learning techniques. These can find interesting connections between DBs and knowledge-bases (KBs), as can be seen in Table 5.1, in building expert systems with and without machine-learning tools.
4. The *client-server technology* gives the individual *knowledge worker* access to central information systems, from his/her own terminal. Policy-makers and marketing analysts also wish to avail themselves of these technical possibilities.

Practical problems faced for transitions from data warehouse to data mining are the following:

- Lack of long-term vision
- All files not up-to-date
- Struggle between departments
- Poor cooperation from the computer departments
- Legal and privacy restrictions
- Files hard to connect for technical reasons
- Time-delay problem and
- Interpretation problems.

Table 5.1: Expert System with or without Machine-Learning Tools

Expert Systems and Applications	*Machine-Learning Tool Used for Development*	*Number of Rules (Reflecting Complexity)*	*Development Man-Years*	*Maintenance Man-Years (Required Per Year)*
MYCIN (Medical diagnosis)	None	400	100	Not available
XCON VAX (Computer configuration)	None	8,000	180	30
GASOIL (Gas-oil separator system configuration)	Expert Ease, Extran 7	2,800	1	0.1
BMT (Configuration of fire-protection equipment in building)	1st class, Rule Master	30,000	9	2

DATA MINING FOR CUSTOMER RELATIONSHIP

The impetus for data mining tools and techniques has come in the wake of the Internet: which is transforming the competitive landscape at a breakneck pace. With the Internet's phenomenal growth, Web-enabled customer interactions are bound to increase. The World Wide Web (WWW) is fuelling the growth of e-commerce, which has grown by more than 100 per cent in the 5 years preceding 1999. Also, according to the US Department of Commerce, the Internet sales have already surpassed $300 billion by 2002, comprising both business-to-businesses (B2B) and business to consumer (B2C) e-commerce.

Amazon.com, located at Seattle, Washington, provides an example of building customer relationship. Amazon.com opened its original doors in July 1995 with a mission to use the Internet to transform book buying into the fastest, easiest and most enjoyable shopping experience possible. Since its inception, the company has grown to be a leading on-line shopping-site, selling to over 20 million people in more than 160 countries and offering millions of titles in books, audio, video, games and electronic categories. Whenever a Web-customer conducts a search of the Amazon.com, the latter offers additional products, using enticing phrases like "customers who bought this book also bought". For each search, several additional selections or authors are offered. Amazon.com offers a clear example of *customer acquisition profiling* that does not require advanced algorithms of data mining. Far more precise or specialised applications can be developed with today's technology and applied to e-commerce.

Data mining techniques, as surveyed later can, in fact, be applied throughout the complete customer relationship life-cycle, as practised in customer relation management (CRM), exemplified in Table 5.2.

Table 5.2: Data Mining for Customer Relationship

CRM Life-Cycle Stage	*Activities*	*Data Mining Example*
Finding	Lead generation	• Customer acquisition profiling • Web-mining for prospects • Targeting market
Reaching	Marketing programmes	• Customer acquisition profiling
Selling	Contact selling	• Customer acquisition profiling • On-line shopping • Scenario notification • Customer-centric selling
Satisfying	• Product performance • Service performance • Customer service	• Customer retention profiling • Scenario notification • Staffing level prediction • Enquiry routing
Retaining	Customer retention	• Customer retention profiling • Scenario notification • Individual customer profiles

The capabilities of data mining, usefully built and enhanced CRM performance, should include the following:

- Learning patterns that allow rapid routing of customer enquiries;
- Learning customer buying habits to suggest likely products of interest;
- Categorising customers for focussed attention;
- Automating intelligence responses to customer 'events' (e.g., enquiry, purchase, etc.);
- Providing productive models to reduce cost and allow more competitive pricing (e.g., waste control); and
- Assisting buyers to select customer-centric inventory.

STEPS OF DATA MINING

Data mining steps can be outlined in general terms as follows:

Step 1: Dividing Business Objectives

The business objectives, which drive the data mining process, are to be decided in advance. The minimum requirement is prescribed as a business problem or opportunity, along with some sponsorship by the organisation.

Step 2: Data Preparation

- *Data Selection*: identification of the sources of data and extraction of data;
- *Data Pre-processing*: data sampling and quality assurance; and
- *Data Transformation*: data conversion into an analytical model: representing a consolidated, integrated and time-dependent re-structuring of the data selected and pre-processed from various operational sources.

Step 3: Data Mining

Applying the data mining algorithms (calculations) to pre-processed data interactively combined with the analysis of results, as seen below. The actual process here may result in some database segmentation or a cyclic predictive process, depending on the algorithms used.

Step 4: Analysis of Results

This is the most important step in the entire cycle of data mining—using visualisation aids and tools analysis of data, done by a highly skilled data analyst together with a business analyst.

Step 5: Assimilation of Knowledge

This step comprises action taken to implement the knowledge gained by data mining and utilised in the business cycle. The final benefit may be recognition of a new focus on the importance of data as an asset to the organisation.

DATA MINING ALGORITHMS: EXAMPLES

These apply different techniques to identify previously unknown characteristics, patterns and associations in the data, illustrated below:

Technique 1: Database Segmentation

This involves partitioning a database into segment of similar data (or records of data). Data records that share similar properties are considered as *homogeneous* and are made into *clusters*. The data elements within the same cluster have more similarity than those in other clusters.

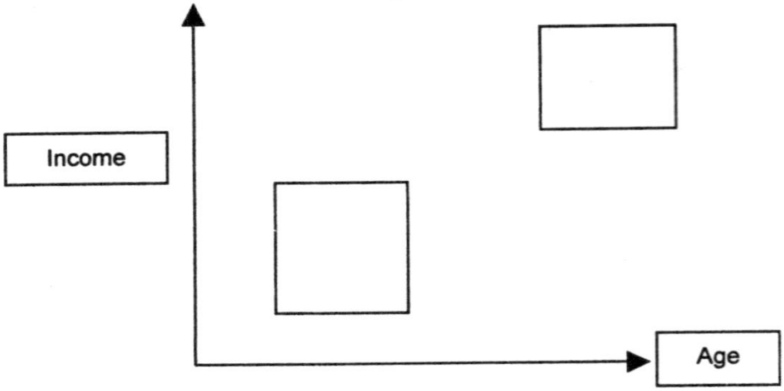

Fig. 5.3: Example of Data Clustering

Technique 2: Predictive Modelling

This analyses the existing DBs to determine some essential characteristics about the data helpful to make fairly accurate predictions. It can be, for instance, some *If-then-else* clauses of decision-logic applied to analyse the behaviour-pattern of a government organisation where the resignation data have been analysed to reveal a particular pattern.

Technique 3: Link Analysis

This seeks links among data records, or sets of data records, through *association* discovery, *sequential pattern* discovery and *similar time sequence* discovery. These are familiar methods used in statistics.

Technique 4: Deviation Detection

This also uses statistics and visualisation techniques: to identify interesting patterns of, say, customer behaviour at different time sequences and drawing conclusions for product promotion techniques.

BENEFITS OF DATA MINING

Data mining has tools and techniques capable of discovering meaningful correlations among various aspects of business. They also bring out hidden patterns and trends. The overall benefits emerging from applying these tools can be summarised as follows:

- To assess the project scope, using business objectives and processes as context;
- To assess and pull together data from practically any source;
- To analyse the meaning and insight of data, with statistical tools, free diagrams, neural networks and data visualisation tools;
- To act for making decisions fast and share clear reports with customers via the Internet, intranet or e-mail; and
- To automate repetitive analysis and share analytical capabilities by building customer interface.

The anatomy of data mining can be visualised, as shown in Figure 5.4. It gives an integrated picture from the legacy RDBMS and data files to the point of GUI for the end-users.

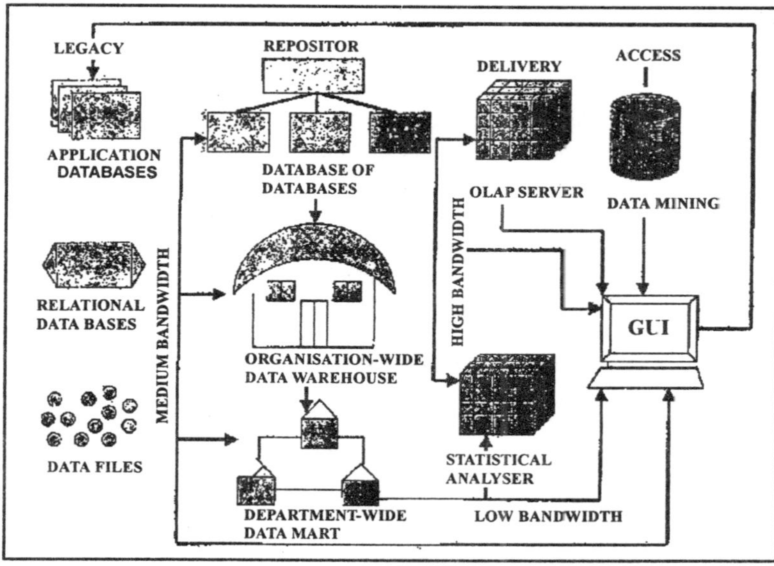

Fig. 5.4: Anatomy of Data Mining

STATISTICAL TOOLS FOR DATA MINING

The commercial tools for data mining are illustrated as follows:

1. IBM-*Intelligent Miner*
2. SAS Institute—*SAS System*
3. Silicon graphics—*Mineset*

4. Data Mind Corporation—*Data Crusher*
5. Integrate Solution—*Clementine*

Besides the above, the popular *Statistical Package of Social Sciences* (SPSS) has come out with their data mining version. SPSS has been long known for proficiency in tests of correlation, association and significance; regression analysis (bi-variate and multi-variate); factor analysis; canonical analysis; and so on. Now, SPSS 7.5 package is available to provide a range of data mining tools to assist decision making in: sales and marketing, business planning, finance, manufacturing, quality control, human resources, and so on.

The tool kit of SPSS 7.5 for statistical discovery and verification methods include the following:

- *SPSS CHAID*: From *regression* and *cluster analysis,* locating important variables and their interactions through *tree diagrams*, e.g., market research, direct marketing and Credit risk analysis;
- *SPSS Neural Connections*: Exploring data with neural networks and building adaptive models for non-linear, complex and fuzzy data, e.g., forecasting, predictions and segmentations; and
- *SPSS Diamond*: Exploring data in innovative ways, using data visualisation tools and transforming complex data and statistics into vivid graphics, to illustrate relationships.

DATA WAREHOUSE AND DATA MINING: APPLICATIONS IN INDIA

The data warehouse and data mining technologies have extensive potential for application in the large industrial sector and in government, which carry elaborate data archives. Prabhu outlines (CSR Prabhu, *Data Wave housing: Concepts, techniques, Products and Applications,* Prentice-Hall of India, ND, 2002) several Central government sectors like agriculture, rural development, health and energy. Some of these sectors are outlined below:

1. Census Data

The Registrar General and Census Commissioner of India decennially compiles information of all individuals, villages, population groups, etc. This information is very wide-based, such as, the *individual slip*: a compilation of information of individual households, of which a DB of 5 per cent sample is maintained for analysis. A data warehouse can be built from this DB, upon which OLAP techniques can be applied. Data mining can also be performed for analysis and knowledge discovery.

2. Prices of Essential Commodities

The Ministry of Food and Civil Supplies compiles daily data on weekly basis for about 300 observation centres throughout India on the prices of essential commodities like rice, edible oil, sugar, kerosene, etc. These data, compiled at district level, are transmitted to the ministry for aggregation and storage. A data warehouse can be built for these data and OLAP techniques applied for its analysis. A data mining and forecasting technique can be applied for advanced forecasting of the actual commodity prices—strengthened try taking into account external variables like rainfall, population-growth and inflation.

3. Agriculture Data

The agricultural census, performed by the Ministry of Agriculture, compiles a large number of agricultural parameters at the national level. District-wise agricultural production, area and field of crops are also compiled. These data can be built into a data warehouse for analysis, mining and forecasting.

4. Rural Development Data

The Below-Poverty-Line (BPL) survey data on individuals can be built into a data warehouse. From the Drinking Water Mission, drinking water census data can be effectively utilised by OLAP and data mining techniques.

5. Health Data

Community needs assessment data, immunization data and data from national programmes on controlling blindness, leprosy and malaria can all be used for data warehouse implementation, OLAP and data mining applications.

6. Foreign Trade Data

Data on India's imports and exports, available with the Director General of Foreign Trade, Ministry of Commerce, can be analysed and converted into a data warehouse. Now that India is a member of the World Trade

Organisation (WTO), a monitoring system on world price can be made to perform better using data warehouse and data mining techniques.

CONCLUSION

Many laudable attempts have been made already in India for building data warehouse and data mining, such as, General Information Service Terminal of National Informatics Centre (GISTNIC) at Hyderabad for the 1991 census, essential commodity prices study in Tamil Nadu (using SAS tools), and so on. But, as visualised above, much more can be done: even in sectors like tourism, programme implementation revenue, economic affords and government accounts. Data warehouse and data mining technologies could lead to the most significant advancements in the government functioning.

6

MANAGEMENT OF ON-LINE INTEGRATED SYSTEMS

INTRODUCTION

While batch-processing (using such a 3rd generation language as COBOL) has been in vogue for long, the advent of relational DBs and 4th generation data manipulation and data accessing languages have made it easy to look at individual transactions and to process them expeditiously, so as to obtain results on on-line, real-time basis. Many large organisations like NTPC have been at the vanguard of such processing, using centralised as well as distributed servers. The concept has now taken off globally: with real-time, integrated infrastructure services taking off across continents.

BATCH PROCESSING (BP) TO TIME-SHARING

In *batch processing* (BP) applications of the earlier years, it was enjoined on the end-user to submit his application needs to the computer and, at the end of processing, collect the output from the workstation. This was typically in a third-generation environment where languages like COBOL were used for commercial data processing activities. In on-line transaction processing (OLTP), the essential difference that has arisen is that the computing power is made available to the end-user on a computer terminal where the job is submitted and the results of processing are also available from the same terminal. The facility is so pronounced and taken so much for granted that many users have not seen the central computing facility or even know where it is located! It remains for the software professionals to make available the processing service and, having done that, maintain it in a full-proof fashion. Since this facility is collectively available to all end-users, they virtually share computer time among themselves from the central computing facility. The arrangement, therefore, is known as *time-sharing*.

The environmental change from BP to OLTP (in time-sharing fashion) is so fundamental that one can draw the popular analogy that if, earlier, it was Mohammed (end-user) going to the mountain (computer), now it is the mountain (the computing power) that comes to Mohammed (end-user)!

ON-LINE TRANSACTION PROCESSING (OLTP)

As we have noted, the BP mode visualised inputting of an entire batch to the machine, awaiting its processing ant finally collecting the output. The individual transactions were not differentiated from the end-user point of view. In OLTP, each individual transaction goes through the entire process: from input through processing to the ultimate output: before the next transaction is triggered in. A classic example is the IMPRESS System of the Indian Railways, developed initially by the CMC, and maintained and updated by the Centre for Railway Information System (CRIS), running the On-Line Passenger Reservation System. Every request for a passenger reservation—with its details for the train no., date, class and category of reservation, starting and destination stations, passenger name, age and preferences, and other relevant particulars—are fed in a specific terminal as part of a single transaction and processed, till the output—in the form of a Railway ticket—is printed out, before the next transaction can be fed in the same terminal. The On-Line Passenger Reservation System of the Indian Airlines follows a very similar approach.

The salient features of OLTP can be outlined as follows:

1. It is a proven method for providing on-line, real-time data processing capabilities from a corporate DB.
2. It is a mode of application where messages received from terminal devices are processed.
3. It provides information matching the flow of business operations—transaction by transaction.
4. In consequence, there is a vital link between the OLTP information and the money generated in the business operation, the On-Line Railway Reservation System being a case in point.
5. It provides corporate executives greater control over business information, enabling them to take better and faster decisions.
6. It gives corporate personnel instant access to crucial decision parameters.

As we shall see, On-Line Application Processing (OLAP) takes the last two points forward—by being concerned with the complete application and not just individual transactions, and, to that extent, helping even further in the decision-making.

ROLE OF DATABASE

In OLTP, a data model is presumed to exist for the entire organisation, as a *repository* of all critical records and files that make its business run. Thereafter, all the data and information, relevant to managing day-to-day business operations are presumed to be forthcoming for performing qualitative decision implementation. Within this framework, OLTP has seen the following six incarnations from DP at one end to the high-profile executive computing (see Chapter 12 for details), through their attendant phases of computerization:

1. *Data Processing of Routine Tasks:* For example, pay roll, stores ledger, etc.
2. *On-Line Query and Response:* With re-structured applications and involving data communications, for example, on-line inventory control, on-line goods inwards inspection, etc., which are generally non-real time systems to save cost.
3. *OLTP*: Rationalising computer techniques and using administrative database for processing transactions, where an organisation looks upon itself as an integrated entity and uses Centralised access as well as Centralised processing.
4. *Distributed Access and Centralised Processing (Centralised OLTP)*: With one central database being maintained, for example, on-line production and inventory control system.
5. *Distributed Access and Distributed Processing (Distributed OLTP)*: With on-line distributed databases being maintained with linkages to the central database, for example, on-line production planning and control system, integrated with other corporate functions like sales, purchase, finance, etc.
6. *Executive Computing*: With freedom to organise corporate executive's own information, for example, finance manager looking into production, collecting forecasts from the integrated DB, drawing inferences from the same integrated DB and chalking out own cash flow forecasts.

The last phase is the most important development that has taken place in the domain of OLTP, where corporate host computer supports end-users either through local terminals (or workstations) or through remote terminals (or workstations).

ON-LINE REAL TIME AND ON-LINE NON-REAL TIME

Many organisations still attempt a mix of BP and OLTP applications, with a judicious distribution of jobs to be done at each end. This is an ideal way to achieve overall economy, since BP is far more economic (though laborious) than OLTP, which tends to be expensive, though very convenient—provided such a mix suits the corporate applications environment.

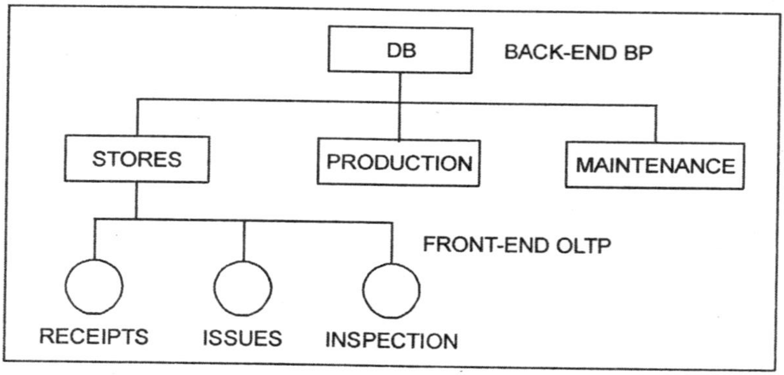

Fig. 6.1: On-Line Real Time & On-Line Non-Real Time (A Mix of Batch Processing (BP) & On-line Transaction Processing (OLTP))

As illustrated in Figure 6.1, this user-organisation performs *back-end DP* at the Mainframe that carries out the updating of the host DB possibly at night when there is least demand on its processing time otherwise. The Mini, on the other hand, is located at the Stores section and earmarked for *front-end OLTP* during the working day: for handling all the transactions on the daily receipts (including returns from production), issues (fresh and renewed) and inspection. While the day-to-day data on these transactions are available on an immediate basis, the updating of the master files is done only at night, from where the overall information for decision-making (on raising indents and placing fresh orders based on EOQ/EBQ, demand analysis on slow-moving and non-moving items, etc.) is obtained every morning—for taking action on material procurement, disposal and so on. The cost saving is enormous and at any given point, the organisation can switch over to the more convenient (though costly) option of *back-end OLTP*.

CENTRALISED AND DISTRIBUTED OLTP

For organisations in a confined territory, it may be worthwhile to operate *Centralised OLTP* with the following characteristics, as outlined in Figure 6.2:

- Corporate host computer (generally Mainframe) used as one central processing complex;
- One central filing system used on the host computer;
- Transaction processing workload handled under Centralised environment; and
- Terminals/workstations: either local or remote.

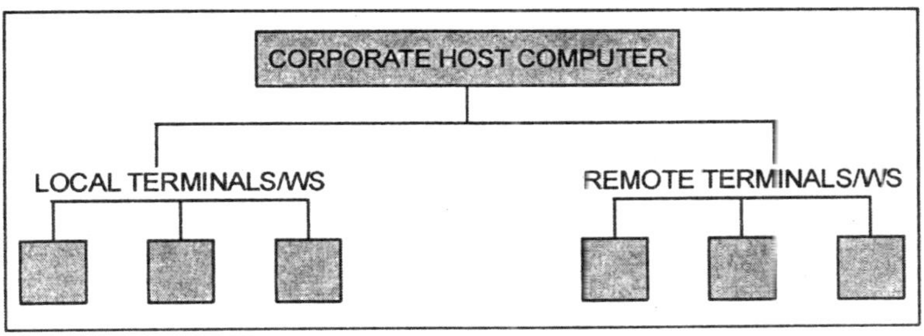

Fig. 6.2: Centralised OLTP

In contrast, for organisations spread over a vast geographical expanse, it is customary to go for *Distributed OLTP* with the following characteristics, as outlined in Figure 6.3:

- Multiple computer systems distributed throughout the organisation;
- Multiple computer systems located in different departments;
- Transaction processing offloaded from host system to satellite computers; and
- Central filing system keeping track of the distributed filing.

An illustration of an on-line, integrated system for Centralised OLTP can be drawn for a research-oriented user-organisation, where personnel information, accounts-related functions, purchase and stores-related functions, administrative-related functions and security-related functions can have their own workstations, along with own sets of terminals—connected to the host database at the corporate office. This configuration

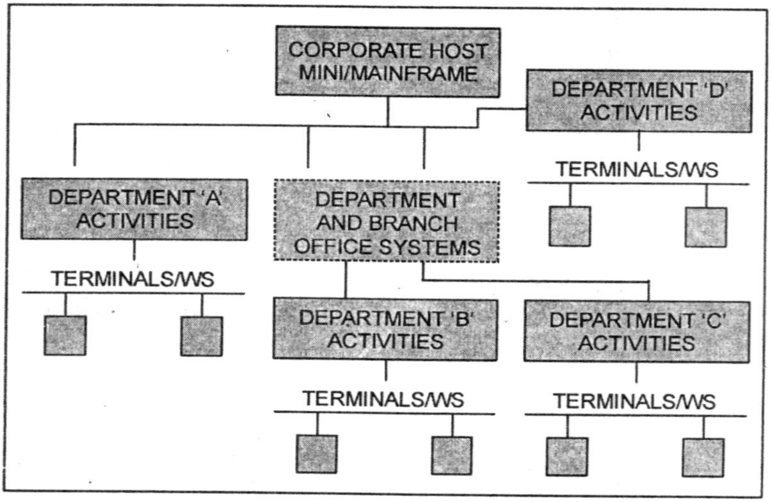

Fig. 6.3: Distributed OLTP

is generally characterised by LAN, with user login facility. Personnel information, for instance, may comprise Employee Record DB, with employee salary data (salary, Income tax, Provident Fund, etc.); Library data (books issue and return); Visitors data ((in-out time); Employee-based purchase data (employees, items, suppliers); Leave data (for employee leave transactions); and so on. Relational DBs can be Oracle 8i, with front-end GUIs like Developer 2000 and 4-GL tools like Accell. The system support facilities can comprise Disk Mirroring, Disk Strapping, etc Data integrity is fully ensured in the OLTP network and parent-to-child relationship among the records at host and local/remote locations is strictly maintained. Such a configuration is shown in Figure 6.4.

On the other hand, an illustration of an on-line, integrated system for distributed OLTP can be drawn for a manufacturing enterprise—with production planning functions, production scheduling functions, raw material-related functions, order processing functions, indenting and purchase functions, inventory-related functions and finance-related functions. While administration and finance can still remain centrally linked to the corporate DB doing Centralised OLTP due to organizational efficiency, the functions—like Production planning and control. Inventory, Purchase and Stores—can be at widely dispersed territory, occupied by the enterprise, with their own single or multiple Minis and connected

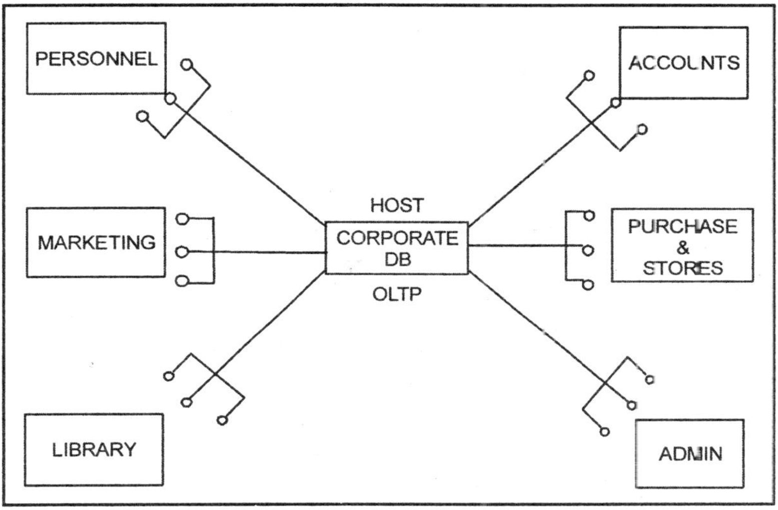

Fig. 6.4: Centralised OLTP (for R&D Organisation)

workstations, and own set of terminals. Production planning, for instance, may comprise Sales and marketing data, to start with, leading to Production scheduling data (for various departments); Production control data (for product records); Inventory control data (for indenting and cost); Purchase data (raw material); and Finance data (for financial requirements of various departments). The DB considerations and support facilities remain the same as for on-line integrated systems with Centralised OLTP. Such a configuration is shown in Figure 6.5.

SOFTWARE FEATURES OF ON-LINE INTEGRATED SYSTEMS

From the end-user point of view, the user interfaces are uniform, since all systems use the same package. This leads to the following three advantages:

1. It eases requirement of user training
2. New systems can be easily implemented
3. General user manual applies for all systems.

For updates, transactions need locking of data: for ensuring confidentiality especially in Finance and Personal data. This needs provision of *Read Lock* and *Write Lock,* which are automatically applied whenever updatings are invoked. Sharing of data remains totally user-

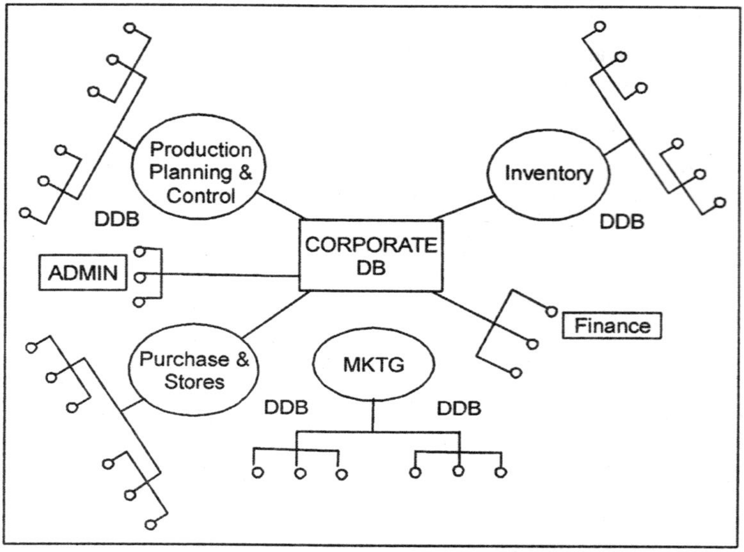

Fig. 6.5: Distributed OLTP (for Manufacturing Enterprise)

transparent. Modular designs are adopted for integration of databases, with the following advantages:

1. New systems are added with ease
2. Software modules for new systems are transparent to other users.

Implementation of computerised systems is possible in a phased manner. Common data, used by multiple systems, can be easily checked and verified. On-line nature of systems reduces the load arising otherwise in batch mode: with transactions entered as and when available—with no waiting involved for processing time.

Two major disadvantages can arise for on-line, integrated systems as follows:

- If the system crashes, it may result in total failure of the OLTP. This may be avoided with on-line disk back-up and substantial UPS;
- The system, with the host DB, tends to be too unwieldy for very large enterprises. The way out is to change the technology: to Materials Requirement Planning (MRP); MRP-II and Enterprise Resource Planning (ERP)—covered in Chapter 9.

ON-LINE APPLICATION PROCESSING (OLAP)

On-line Application Processing (OLAP) is a recent extension of OLTP, where the same principle is followed: of processing one complete application at a time (rather than one transaction at a time as in OLTP), before taking up the next application. Obviously, the computing resources have to be large, doing justice to the heavy processing needs. Special packages have been developed for this purpose by software vendors.

A quick comparison between OLTP and OLAP is given in Table 6.1.

Table 6.1: OLTP and OLAP

OLTP	*OLAP*
Current data	Current & Historical Data
Small DB	Large DB
Fast in execution	Slow in execution
Little time to receive output	Long time to receive output
No analysis capability	Good analysis

CASE: IT IN POWER SECTOR—FROM 'BP' TO 'OLTP'

The following account provides a complete case study of planned transition from BP to OLTP for National Thermal Power Corporation (NTPC), a major power utility in the public sector, established in 1975, for sustainable power development with around 11 thermal power and 7 gas-based power projects. Generating over one-fourth of India's total thermal power, it has an installed capacity of over 21,000 MW. The work on computerisation has remained an on-going process: aiming at maximizing efficiency, availability and control of power in the following area, among others:

- *Plant Maintenance Management*: covering preventive maintenance, condition monitoring, lubrication schedule and equipment history;
- *Materials Management*: covering purchase, stores and risk management;
- *Financial Management*: covering accounts, budgets, purchase accounting, payroll and provident fund; and
- *Bond Management*: covering maintenance of bond registers, issue of interest warrants and bank reconciliation.

In the case, the stages of automation were as follows:

I. All the above applications were developed in-house in COBOL and in BP mode, in 1983 onwards. The base modules of BP were completed in 1995 and implemented first in NTPC's Dadri plant.

II. In order to provide up-to-date information to the end-users at their desk, the need was felt to recast the BP applications into on-line, integrated mode, in the 1990's. Initially, standard business packages were looked for, but found inadequate to meet NTPC's requirements, as they were very generic in nature and would have involved a lot of customisation. Taking advantage of the available RDBMS tools and UNIX servers, applications in on-line environment were designed and implemented in-house and more modules, functionality and features were added to meet the users requirements in the following manner:

(1) *On-Line Plant Maintenance Management* operated a set of 12 integrated modules to provide equipment information to: work specification, preventive maintenance, breakdown maintenance, job planning, work permits, condition monitoring, shut-down maintenance and MIS reports. It was then integrated to *Materials Management,* to help in checking availability of inventory, raising of *Store Issue Vouchers* (SIVs) and cost of spare parts on each work order.

(2) *On-Line Materials Management* was found essential to cover procurement, stores, risk management and MIS reports, as *Power Plant Operation Management* is critically dependent on availability of spare parts and consumables for efficient and effective plant operation. The system was implemented in Dadri plant in 1995 and captured total data inflow in procurement and storage. *On-line procurement module* had facility for generating requisitions for non-stock items, as well as stock item requisitions when stock went below re-order level. Procurement activities covered requests for quotations; bid evaluation; purchase order placement (including amendments) and generating follow-up report. *On-line stores module* covered all stages: from the feeding of requisitions into the system up to the creation of unloading reports; MIS-cum-SRV (stores receipt voucher); SIV (store issue voucher); MRN (material

requisition note); MTN (material transport note) and adjustment vouchers: directly on-line. Vouchers were priced on-line as soon as they were generated, leaving scope for re-pricing at the time of PSL (priced store generation).

(3) *On-line Risk Management* covered declaration register, printing of claims and updation of claim settlement, apart from MIS reports.

(4) *On-line Financial Management* was first developed and implemented at Singrauli Plant in 1997, covering books accounting (cash/bank, general ledger, trial balance, business graphics, etc.); purchase accounting (verification of invoices with purchase orders and SRV, creation of JV. Payment to suppliers and transporters, and release of retentions); establishment and payroll accounting (integrated with loan ledgers, advances ledgers, interest on loans and pay slips); and funds management.

In order to improve plant performance, efficiency and profitability in the power generation, the following in-house/outsourced applications were developed and implemented later in a distributed OLTP mode:

1. *On-line Operation and Performance Monitoring*: used on-line by plant personnel to enter operational data (generation, load, outage, fuel stock/consumed/received, etc.); store at the plant's server DB, as replicated at the corporate's host DB; and access to prepare MIS reports and performance analysis for management at all the plants and Corporate Centre.

2. *On-line Environment Monitoring*: used at all the plants and Corporate Centre to prepare MIS reports, related to environment, for management and government.

3. *On-line Commercial Management*: used, as an integrated system, by regional HQ and Corporate Centre to collect on-line information on status and realisation of billing, for improving cash flow.

4. *On-line Wear and Debris Analysis*: used, as a Web-based system, by R&D department to store, analyse, report and disseminate information on lubricating oil and locomotive engines across the entire NTPC, for predicting failure of locomotives and taking corrective action.

5. *On-line Executive Information System*: used, as a Web-based system, to cover all strategic areas at Corporate Centre for strategic planning and decision making. KRAs (key result areas) are: plant operation; finance, personnel, projects, commercial and general

information. The system extracts data from plant and corporate DBs and flat files, and make information available on-line to end-users with GUIs.

6. *On-line Human Resource Management*: used, as an ERP module of Peoplesoft package (see Chapter 9) at Corporate Centre, to help in workflow and automation of all activities in the personnel department—with personnel information available at their desktop and having interfaces with financial system.

DISTRIBUTED OLTP ON AN INTRANET

Distributed OLTP can be configured on an Intranet within a widely spread industrial enterprise. Such an Intranet, for instance, has been developed within the NTPC network: to meet the objective of effective, on-line communications on corporate policies, circulars, salary details, employee claims and sundry other details. The NTPC Intranet is available anywhere within the NTPC. *Key Areas* covered are: organisation chart, human resources, finance, IT, common forms and links to Intranets in sister projects.

DISTRIBUTED OLTP ON AN INTERNET

A recent phenomenon has been *Infrastructure Management Services* (IMS), where customers are supported across continents on incident management, problem management and change management IMS provides the customer real-time snapshot of the overall 'health' of the entire organisation: across spread-out verticals and departments. Host servers and corporate DBs in India are able to monitor, maintain, manage and protect the entire IT infrastructure and storage systems of hundreds of small, medium and large global companies: remotely and on-line. Instances of such global companies are: a $50 billion service provider in the UK; a large utility firm in the USA; a medium retail chain in West Asia; and a small telecommunications infrastructure company in southern Europe. In effect, IMS covers the entire gamut of services from monitoring and managing to enhancing the performance of a client's infrastructure backbone. Since up to 70 per cent of the whole IT functions of an enterprise can be outsourced to third-party service providers, according to industry experts, IMS is likely to become a mainstream sector in real-time outsourcing by 2007.

CONCLUSION

In summary, OLTP has provided a very welcome transition for large organisations in their transaction processing: from time-consuming, error-prone BP (although surprisingly economic!) to OLTP, and further to OLAP, where the end-users can receive thorough insight into their applications on an on-line, real-time basis. As we saw for the Indian power-giant NTPC, the concept has been pushed to their remote project-sites within the Indian subcontinent, without losing control by the corporate office—with its host DBs—on the site operations. Stretching the concept across continents, new on-line, integrated areas are coming up everyday.

7

NATIONAL RESOURCE MANAGEMENT WITH SATELLITE TECHNOLOGY

INTRODUCTION

Over the last three decades, Indian space programmes have taken major strides, with two operational space systems: *Indian National Satellite* (INSAT) system for telecommunication, television, educational broadcasting and meteorological services; and *Indian Remote Sensing Satellite* (IRS) for national resource monitoring and data management. As we shall see, space application programmes have, with the participation of multiple user-agencies, enabled their benefits to reach the grassroots-level society. Right from the dwindling snow-cover on the high Himalayas, up to the forest degradation, ground-water observations on flood and drought, cropping patterns, mining resources, underground water detection, ocean-fishing and ocean-bed minerals—the watchful eye from the space makes its observations handy for multifarious applications to a large range of users. While research in space science has contributed towards enhanced knowledge and understanding of several scientific phenomena, the capabilities built under space programmes are used for commercial gains through international marketing of space hardware and services.

STRUCTURE AND TECHNOLOGY OF 'INSAT'

The INSAT is already one of the largest domestic communication and meteorological satellite systems in the Asia-Pacific region. Fulfilling an uncanny prediction made by the science savant Arthur C Clark way back in 1930, these satellites are placed 36,000 km above the earth's surface—on its equatorial plane—into a Geo-synchronous Transfer Orbit (GTO) where they appear outwardly motionless, balancing the earth's gravity-pull against their tangential thrust away from gravity. The Geo-synchronous Satellite Launch Vehicle (GSLV) for launching them. Happily, the cryogenic engine that used the Russian technology earlier is now almost entirely indigenous: capable of launching 2,500-kg class of satellites into GTO. The Indian GSLV (with 3-stage fuelling, viz., liquid, solid and cryogenic) is now well poised to enter into the international market and become a powerful dollar-earner! Antrix Corporation, the marketing with Indian Space Research Organisation (ISRO), has already a sales turnover exceeding Rs. 3 billion, with more than 75 per cent of the earning in foreign exchange!

The *first generation* INSATs were: 1A (1982) with service for 6 months; 1B (1983) for 7 years; 1C (1988) for 1.5 years; and 1D (1990) for above 10 years. All these were multi-purpose communication and meteorological satellites. The *second generation* INSAT-2 series was: 2A (1992); 2B (1993); 2C (1995); 2D (1997—mission failed); and 2E (1999). Deployed with long orbital life, 2A and 2B provided mobile satellite service and business communication, whereas 2C extended TV outreach beyond Indian boundaries. 2E is still operational, with footprints in Central Europe and Australia, leasing 11 out of 17 C-band transponders for regional and global coverage.

Among the *third generation* satellites, INSAT-3 series is entirely operational, comprising: 3A (2003—12 normal C-band and 6 Ku-band transponders); 3B (2000—12 extended C-band and 3 Ku-band transponders); 3C (2002—24 normal C-band, 6 extended C-band and 2 S-band transponders); and 3E (2003—24 normal C-band and 12 extended C-band transponders). While KALPANA-1 (2002) has been an exclusive meteorological satellite, 3D (planned in 2007-08) will also carry state-of-the-art meteorological payloads. GSAT-2 (2003—4 normal C-band and 2 Ku-band transponders) is carrying mobile satellite services (MSS) payload, similar to 3B and 3C. The *fourth generation* satellites are on the planning board, such as, 4A (planned early 2005—12C-band and 12 Ku-band transponders); 4B (mid-2005—also 12 C-band and 12 Ku-band transponders); 4CR (Sept. 2007—12 Ku-band transponders); and 4D (planned in 2006-08—24 C-band transponders), besides experimental GSAT-4 (planned in 2006-08—8 Ka-band transponders). Indeed, with Insat 4CR, India has now 212 transponders in the sky.

END-USE OF INSAT

With the growth of the Internet, the INSAT system has been a major contributor to corporate business communication and other value-added services like e-commerce (see Chapter 8), satellite RV channels, launch of direct-to-home (DTH) service (specifically planned under INSAT-4A) with high-definition (HD) picture and CD-quality sound; commercial remote sensing applications; basic telephony and Internet telephony. Some generic services, important for end-users, are as follows:

1. Educational Service

EDUSAT (2004, 6 C-band transponders covering entire country, I Ku-band transponder for mainland and 5 Ku-band transponders for intended regional coverage) is India's exclusive satellite for serving the educational sector. Its *pilot project phase* is already using INSAT 3B in Karnataka (Visveswaraya Technological University), Maharashtra (YB Chavan Open University) and MP (Rajiv Gandhi Technical University)—providing lessons in regular classes for primary, secondary and engineering education. The *semi-operational phase* will use 20 uplinks and about 1000 classrooms in the country, covering all the 5 regional beams. The *third phase* will see the 1950-kg EDUSAT as fully operational—with manufactures and service providers replicating the systems with funding by the end-users and techno-managerial support from Department of Space (DOS). In the meanwhile, there is an on-going educational TV service for primary school children in Tamil, Marathi, Oriya, Telugu and Hindi languages, apart from '*Gyandarshan*-I' (2000) educational programme for IGNOU and '*Gyandrshan*-III' (2003) channel for technical education originating from IITs.

2. Village Resource Service

Communication satellites like INSAT, together with earth observation satellites like IRS (covered later on), are being used to launch Village Resource Centres (VRCs): aimed at aiding traditional rural societies. VRCs are to provide access to spatial information like land use/land cover; soil and ground water prospects; and on-line interaction between agricultural scientists and local farmers on farming systems; action plans based on weather; community-specific advice on soil and water conservation, and critical information of fishermen's interest like sea-status and wave heights. In addition, VRCs are to act as community-centric learning centres at village level and connectivity to speciality hospitals in order to bring the expert doctors' advice to the villages. The first cluster of VRCs and their interactive VSAT networks have started at 4 nodes in Tamil Nadu, with the central node at MS Swaminathan Research Foundation in Chennai, and being extended to Kerala (with 25 nodes) and other states.

3. Village Communications Service

GRAMSAT programme is another initiative to provide communications networks at the state level—connecting the state capital to districts and blocks—for computer connectivity; TV broadcasting; e-governance; development information; tele-conferencing; disaster management; tele-medicine; and distance education. The pilot projects are operational in Gujarat, Karnataka, MP, Orissa and Rajasthan—being extended to the Northeast, Andaman and Nicobar (VSAT-based) and Kerala.

4. Tele-medicine Service

Beginning in 2001, INSAT communications are being used to connect remote hospitals and health centres to the speciality hospitals located in the cities. By 2005, the tele-medicine network has connected 100 hospitals: 78 remote/rural/district hospitals/health centres to 22 super-speciality hospitals located in major cities like Delhi, Kolkata, Mumbai, Bangalore, Chennai and Lucknow.

5. Television and Radio Service

INSAT has been a catalyst to expansion of TV coverage in India: 65 per cent of landmass and 90 per cent of total population, with 22 Doordarshan and 32 private TV channels. Radio networking through INSAT provides high-fidelity programme channels for 213 All India Radio (AIR) stations.

6. News Gathering Service

INSAT system has enabled on-the-spot, real-time news coverage to cover different locations for transmission and dissemination.

7. Telecommunications Service

A total of 704 telecommunication terminals (excluding NICNET) are operational in INSAT telecommunications network, providing 7,163 two-way speech circuits. These include 31 *Closed User Group* (CUG) VSATs; 287 *multi-channel-per-carrier* VSATs; 45 high-speed VSATs; and 160 VSATs under *Remote Area Business Management Network* (RABMN).

8. Mobile Satellite Service (MSS)

Developed with the participation of Indian industries, this enables a small, portable satellite terminal, which works with INSAT-C and GSAT-2, for voice communication from any location in India during an emergency.

9. Satellite-aided Search and Rescue Service

INSAT-3A, placed at 93.5deg. East, is equipped with search and rescue payload that picks up and relays alert signals: originating from the distress beacons of maritime, aviation and land users: under the international COSPAS-SARSAT programme, complemented by the *low-earth* orbit search and rescue (LEOSAR) system. The *local user terminals* (LUT), established at Lucknow and Bangalore, provide coverage to the bulk of Indian Ocean region, rendering distress-alert to Bangladesh, Bhutan, Maldives, Nepal, Seychelles, Sri Lanka and Tanzania.

10. Satellite Navigation Service

Satellite-based, regional *GPS and GEO Augmentation* system for navigation, given the acronym GAGAN, is under implementation by 2006-08 for the entire Asia-Pacific region.

STRUCTURE AND TECHNOLOGY OF 'IRS'

Indian earth observation system began with the launch of *remote sensing satellites* from 1988 onwards: with the largest constellation of IRS satellites in operation today. They fall into two broad categories: satellites in *medium-earth* orbits at 10,000 km above the earth (in a sun-synchronous polar orbit) carrying payload up to 1500 kg; and satellites in *low-earth* polar orbit with payloads 1000-1200 kg or less. The Polar Satellite Launch Vehicle (PSLV) is now India's workhorse launcher, which has launched its ninth IRS satellite CARTOSAT-1 (2005) for cartographic applications and a small satellite HAMSAT, as piggyback, for amateur radio (HAM) operators.

First generation IRS satellites were: IRS-1A (1988), IRS-1B (1991) and IRS-1C (1995) and IRS-1D (1997). *Second generation* IRS satellites on Indian PSLV were: IRS-P2 (1994), IRS-1D (1997) and IRS-P4, P5 and P6 (between 1999 and 2002). Apart from them, OCEANSAT-1 (1999) was launched to study physical and biological aspects of oceans; and Technology Experiment Satellite (TES, 2001) launched to validate technologies for use in future cartographic missions. IRS-P6, called RESOURCESAT-1 (2003), was a multi-institutional endeavour at 620 km high, with academic and private participation from IIT Chennai, M/S Godrej, Bharat Aluminium Company (BALCO), Hindustan Aeronautics (HAL) and Bharat Electronics (BEL) for making data products available to end-users. Out of them, IRS-1C, IRS-1D, IRS-P3, OCEANSAT-1, RESOURCESAT-1 and TES are still in operation.

Earlier IRS satellites 1C and 1D carried, typically, three cameras: panchromatic camera (PAN); medium-resolution, linear-imaging, self-scanner (LISS-3) operating with 3 spectral bands in visible and *very near infrared regions* (VNIR); and *wide-field sensor* (WiFS) operating with 3 spectral bands in VNIR and *short-wave infrared* (SWIR) regions. Later, P6 carried a fourth camera, viz., high-resolution, *linear-imaging, self-scanner* (LISS-4), operating with 3 spectral bands in VNIR region. The two components of a camera are: *spatial resolution* (pixel), defining the fineness of a picture, and *swathe* (picture-frame). This is illustrated in Figure 7.1, for both *visible* and *infrared* sections of the total spectrum.

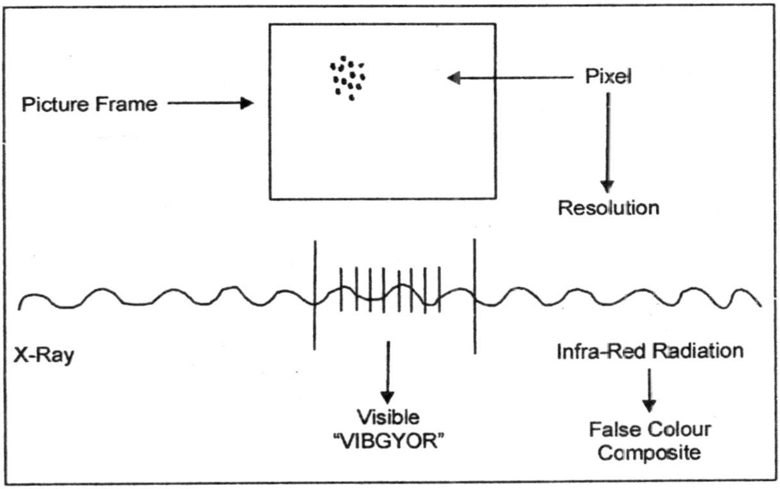

Fig. 7.1: Remote Sensing

For instance, IRS-1C and IRS-1D have *spatial resolution* (in metres) as: 5.8 for PAN; 23.5 for LISS-3 VNIR; 70.5 for LISS-3 SWIR; and 188 for WiFS. They have *swathe* (in km) as: 70 for Pan; 142 for LISS-3 VNIR; 148 for LISS-3 SWIR; and 810 for WiFS. Other parameters are as follows:

- Frequency for total mapping = 22 days, for getting the complete country coverage;
- Sun-synchronous orbit-plane rotation = Earth's rotation around the sun;

- Altitude = 900 km;
- Single orbit across the poles = 1 hour 45 minutes, i.e., 14 orbits in a day; and
- Ground brightness = constant (equator at 19-30 morning every time).

The following comparative analysis for international RS satellites in low-earth orbit, SPOT and LANDSAT, as being used commercially all over South and Southeast Asia, is useful to appreciate the huge cost-effectiveness of the Indian IRS system:

Table 7.1: Comparative Analysis for International Remote Sensing Satellites

	SPOT (FRANCE)	*LANDSAT (USA)*
Spatial Resolution	10 m B&W	30 m B&W
	20m False colour composite 80 m	False colour composite
Swathe	60 km × 60 km	180 km × 180 km
Frequency	26 days	16 days
Cost	$2500 for 10m × 20m	$1500 for 30 m × 80 m

Nepal, for example, has made good use of these international images for their deforestation survey and forestation programmes, while Bangladesh has used them for their upstream river water analyses—for political parleys with India.

The most recent development in IRS was a unique launch of a Polar Satellite Launch Vehicle (PSLV) from the launch-pad of Sriharikota in January 2007, carrying four satellites at one go. The single PSLV-C7 included one satellite that would return to earth to set the stage for India's manned space mission. The Space Capsule Recovery Experiment (SRE-1) weighing 550 kg is designed to test re-entry technology. The capsule would orbit earth for 13-30 days. The other three satellites in the PSLV payload were: (i) the remote-sensing Cartosat; (ii) Lapan-Tubsat from Indonesia; and Argentina-built Pehuensat-1. In particular, CARTOSAT-2 was India's 12th IRS satellite. With nearly one-metre resolution, it would provide scene-specific spot-imagery, with end-use designed for mapping purposes. Incidentally, this was India's 11th successful rocket-launch since 1982, when the first satellite in INSAT series was put into orbit. Earlier, Indian satellites were put into orbit aboard European Ariane rockets from the European Union's space centre in British Guiana. ISRO is working on launch of unmanned mission to orbit moon in 2008, followed by manned mission by 2020.

END-USE OF 'IRS'

With the launch of IRS satellites, the earth observation applications are overseen by *National Natural Resource Management System* (NNRMS). The latter aims at deriving optimum utilization of India's natural resources by systematic inventory: using earth observation data in conjunction with conventional techniques like *census* data, *national sample survey* data and *field-survey* data. NNRMS applications of remote sensing are grouped under: (i) agriculture and soils, (ii) bio-resources, (iii) geology and mineral resources, (iv) water resources, (v) ocean resources, (vi) cartography and mapping, (vii) urban management, (viii) rural development, (ix) technology and training, and (x) meteorology.

NNRMS projects of direct relevance to national, regional and local level are carried out under the aegis of National Remote Sensing Agency (NRSA), Hyderabad, Space Application Centre (SAC), Ahmedabad and North-east Space Application Centre (NE-SAC), Shillong—with the help of the following *regional remote-sensing service centres* (RRSSCs):

- *South*—Geological Survey of India, Bangalore
- *Central*—National Bureau of Soil Survey and Land-Use Planning, Nagpur
- *West*—Central Arid Zone Research Institute, Jodhpur
- *East*—IIT, Kharagpur
- *North*—Indian Institute of Remote Sensing, Dehradun.

The end-users are involved in these projects for which the following application areas have been especially relevant over the years:

VISIBLE SPECTRUM (B&W)

1. Glacial Resources

- For snow-fed rivers leading to flood/drought
- For snow-covered areas of Beas and Chenab catchments

2. Cropping Resources

- Three-crop inventory parameters: acreage, conditions, yield
- Intermediate crop potential
- Incidence of insect/pest/fungus

3. Forest Resources

- Categories: closed, degraded, open
- Deforestation: scope, coverage, period
- Forestation and regeneration of forests

4. Surface Water Resources

- Water-levels in tanks, bunds, reservoirs, dams, lakes
- Water-levels in rivers: flood, drought
- Flood occurrence: scope, coverage, frequencies, period

INVISIBLE SPECTRUM (FALSE COLUR COMPOSITE)

5. Underground Water Resources

- Rise/fall of water-levels (dry/wet season)
- Occurrence of "Aquifers" (e.g., Thar Desert)
- Drinking water and related water management

6. Underground Mineral Resources

- Occurrence: scope, coverage, exploitation potential

7. Underwater Fishing Resources

- Movement of "schools" of fish (e.g., Tuna in shallow sea during the afternoon hours)
- Fishermen's cooperatives and fishing trawlers industry

8. Sea-bed/Ocean-bed Mineral Resources

- Occurrence of iron, manganese, tin, cobalt, nickel
- Secret "scooping" of existing minerals by submarines from developed countries.

The following taskforces have been operating since the inception of IRS satellite launching:

(i) *Taskforce on Agriculture*: Activities related to crops, soil, water-resources, climate and local weather;
(ii) *Taskforce on Forestry*: Timber harvest, forest regeneration on damage assessment due to forest fire and animal degradation;

(iii) *Taskforce on Geology*: All geo-scientific resources and cartography;
(iv) *Taskforce on Oceanography, Marine Resources and Coastal Studies*: Fish, ocean-bed minerals, coastal formation and degradation;
(v) *Taskforce on Soils and Land-use*: Special survey on salinity, waterlogging, ravines, shifting cultivation, wasteland and fallow land;
(vi) *Taskforce on Urban and Rural Studies*: Non-agricultural rural planning, urban land-use, National Capital region (NCR), environment monitoring; and
(vii) *Taskforce on Water Resources (Ground and surface Water)*: Precipitation, coastal morphology, wet land.

The more recent end-uses of IRS are illustrated below:

A *RESOURCESAT data utilisation:* GIS applications and studies on snow-cover discrimination, multiple crop discrimination, crop condition assessment, surface wetness zoning, soil mapping and salinity units, analysis of forest types, regional-level land-use and land-cover, and updating of regional geological mapping.

B. *Groundwater Prospects Mapping:* Groundwater prospects mapping, under Rajiv Gandhi National Drinking Water Mission, in AP, Karnataka, Kerala, MP, Chattisgarh and Rajasthan.

C. *Crop Acreage and Productivity Estimate (CAPE)*: A countrywide project for principal crops, such as, wheat, rice, sorghum, cotton, mustard and groundnut for evaluation of yield models, using trend and weather data.

D. *Forecasting Agricultural Output through Agro-meteorological and Land-based Observations (FASAL):* Phased implementation of CAPE through the state agencies.

E. *Agro-climatic Planning and Information:* for 27 major horticultural crops in six districts of Karnataka.

F. *Horticultural Mission in North-eastern States*: For passion fruit area, cashew plantations, pineapple and 'Jhum' cultivation.

G. *Precision Farming*: For optimising agricultural inputs towards maximizing productivity and maximum return for farmers, including yield-gap analysis and decision support system for site-specific crop management.

H. *Potential Fishery Zone*: By monitoring ocean colour and sea-surface temperature, for tri-weekly dissemination to end-users like fishery departments, government agencies, corporates and fishermen groups.

I. *Biodiversity Characterisation*: Covering four rich biodiversity regions of the country—NE region, Western Himalayas, Western Ghats, and Andaman and Nicobar Islands.
J. *Desertification Monitoring*: For both hot and cold desert areas of Rajasthan, Haryana, UP, AP, Tamil Nadu, J&K, and HP, under integrated resources information system for desert areas (IRIS-DA).
K. Snow-melt Run-off Forecast: For Sutlej river basin up to Bhakra reservoir.
L. *River Bank Erosion Mapping*: For Brahmaputra River.
M. *National Wasteland Inventory*: For preparing a wasteland atlas and delineating new areas under wastelands, For a GIS database.
N. *Route Alignment Study of Transmission Lines*: For Sikkim and West Bengal under GIS environment.
O. *Land Use and Land Cover for Pipeline Routes*: For the Consortium of Petroleum Industries in West Bengal, Orissa, AP, Tamil Nadu, Gujarat and Maharashtra.
P. *Site Identification for Thermal Power Stations*: Along east and west coast of India.
Q. *Watershed Development and Monitoring*: For micro watersheds in Karnataka and changes in watersheds in 12 states.
R. *Salinity and Waterlogged Area Monitoring*: For 'command' areas in Rajasthan, Haryana, Bihar, Jharkhand, Karnataka and Goa.
S. *Reservoir Capacity Survey*: For capacity estimation and storage loss in 20 reservoirs, namely, Hallali, Bhadar, Palitana, Jakhan, Parbati, Ramsagar, Isapur, Kyredemkulai, Nizamsagar, Panam, Sondur, Dudhwa, Narayanpur, Mayurakshi, Gumti, Umiam, Idukki, Damanganga and Kallada.
U. *Ocean Applications*: For sea-surface height observations and in-situ temperature profiles.

NATIONAL RESOURCES INFORMATION SYSTEM (NRIS)

NRIS involves decentralizing the process and practice of national planning at district level—by highlighting inter-and intra-regional disparities in development; pinpointing natural resources; capturing location-specific data; and integrating with area-specific data over space and time. These are natural resources data characterised by: large diversity; multi-user and multi-purpose nature; inter-linked and integrated parameters; qualities of reliability, consistency and sensitivity; involvement of many organisations for data collection; standardisation and compatibility of procedures; and use of data for decision-making.

NRIS databases for 30 district nodes in 17 states have been created, with work still progressing in 14 states, e.g., Mizoram, Sikkim, Gujarat, Haryana, Karnataka, MP, Maharashtra, Orissa, Punjab, UP, J&K, HP, Rajasthan and Tamil Nadu. Some major projects under NRIS are illustrated as follow

1. *Large-scale Mapping*: This has been initiated for the entire country, using high-resolution satellite images, to generate cartographic maps on 1: 10,000 scales. Prototype work for about 5,000-sq km area is on for urban, rural and disaster-prone areas.

2. *Aerial Remote Sensing Survey*: This included the recent aerial photography of 28 towns in and around Kolkata; aeromagnetic survey over Indo-Gangetic plains; and mapping of 4 Andhra towns for services to the poor. Interestingly, NRIS is poised to provide satellite imaging and aerial survey for digital mapping of Maldives and setting up of a Remote Sensing Unit in its capital Male.

3. *Natural Resources Census*: Initiated for the whole country, this includes natural resources like land, water, soils, forests, etc.—to be conducted with repeat cycles for depicting changes and to provide a snap-shot of the country's natural resources. Using IRS images, NNRMS has begun the census of land-use and land-cover of the entire country. The broad cycle of for the census for geomorphology and soils would be 20 years, while it would be 5 years for land degradation, land use, land cover, vegetation cover, wet lands, snow cover and glaciers. Prototype studies are completed in 11 states. Developmental planning for Chattisgarh state has been taken up: using remote sensing and GIS data for its 20,400 villages spread over 135,000 sq km.

4. *National Resources Repository*: Based mainly on remote sensing data, this repository comprises information generated through user-funded projects and others on standard formats.

DISASTER MANAGEMENT SUPPORT

These services aim at creating digital databases for facilitating hazard zoning, damage assessment, monitoring of major natural disasters: using satellite and aerial data. While space-based systems have no capacity to detect and forecast natural calamities like earthquakes and Tsunamis, the satellites do provide the means of assessing damage, enabling relief operations planning and establishing emergency communication systems. Some specific applications have been as follows:

- Drought assessment: design response;
- Land-slide hazards: zoning;
- Early warning systems: fish catch, cyclone prediction (based on formation of low-pressure nucleus over the high seas) and land communications, for West Bengal and Kerala
- Prediction of appearance of fractures inside the water mass of a dam;
- Near real-time flood monitoring and assessment, applicable to a total of 40 flood maps in 2003, prepared for Assam, Bihar, Orissa, Uttar Pradesh and West Bengal;
- Forest-fire monitoring, using synergistic satellite data, for the 53 per cent forest cover of the country, which is prone to low-to-high severity of burning; and
- Decision support systems for sectoral planning, district industrial centres, veterinary centres and health facilities, including yield-gap analysis and spatial decision support system (SDSS) for site-specific crop management.

In the aftermath of the *Tsunami disaster* in December 2004, one flyaway VSAT, ten INMARSAT (see later) telephones and four Mobile Satellite Service (MSS) phones were airlifted to Port Blair for augmenting the telecommunications links. VSAT-based video conferencing facilities were set up to connect seven taluk headquarters to the secretariat. As a specific user-interface, the Andaman and Nicobar administration used the VSATs at critical locations like Diglipur, Mayabandar, Rangpur, Hut Bay, Car Nicobar, Camorta, Cambel Bay and Port Blair for relief and rehabilitation: through regular video conferencing facility between these locations and the main land via INSAT-3E.

END-USER INTERFACE—NOW AND FUTURE

User-specific applications-survey in a single year of 2004 brought out the following:

- 21 major flood events were monitored in Assam, Bihar, Orissa and West Bengal using satellite data;
- Using multi-date satellite data, the condition of Pereechu Lake in Tibet was monitored on daily basis to assess the possible threat

of flash floods in Sutlej basin. The information was critical for planning precautionary measures, including evacuation of villagers downstream and operations of Himachal Pradesh power plant;

- Using the vegetation index, derived from satellite data, drought conditions were regularly monitored at district and taluk levels for 14 states, viz., Andhra Pradesh, Bihar, Chattisgarh, Gujarat, Haryana, Jharkhand, Karnataka, Maharashtra, Madhya Pradesh, Orissa, Rajasthan, Tamil Nadu, Uttaranchal and Uttar Pradesh;
- As a follow-up to the major landslides in Uttar Kashi, airborne survey was carried out and detailed land-slide hazard-zoning maps were prepared for administration, along with suggestions for management measures.
- Using IRS data, landslide hazard-zoning maps were taken up for the Dehra dun-Yamunotri route. Similar study was initiated for Shillong-Silchar-Aizal highway corridor;
- State-of-the-art ground water prospects are under preparation: to narrow down target—zones for selection of appropriate sites for drilling of wells and planning for recharge structures, based on further ground hydrological and geophysical surveys;
- Satellite-based performance evaluation was done for the Hirakund project-irrigated command area—to augment the irrigation utilisation and agricultural productivity for the cultivable land in the project. Satellite data for the *rabi* season of 1992-93, 1994-95 and 2001-02 was analysed to map the extent of water logging and surface salinity in the command area;
- Capability of space technology in addressing the information needs for the interlinking of rivers was demonstrated, including a focussed study on the link between Godavari and Krishna river basins;
- Cyclonic and anti-cyclonic eddies were studied, using sea surface height observations and chlorophyll concentrations;
- Since marine fisheries provide support to millions of fishermen, remote sensing technology was used to study potential fishing zones: for successful fishing by reducing time, fuel and manpower, because of their symbiotic coverage. *Potential Fishing Zone* (PFZ) forecast is provided to the fishing community in near real-time since 1992, which has revealed 3-4 fold increases in fish catch along Visakhapatnam coast alone;
- High-resolution data from IRS-1A on five different dates were analysed to study the water quality in and around the Mumbai

port area, including parameters like colour, transparency, suspended sediment concentration, sewage outflow, floating material (oil and petroleum products), water dynamics and port activity: to monitor the pollution levels;

- Parts of Western Ghats in Karnataka were studied by IRS P6 for rapid assessment of national forest cover, retrieval of biophysical parameters, forest phrenology and forest fire monitoring;
- In the crop studies in Krishna, Kurnool and Guntur districts in AP, sensitivity of reflected radiation in IRS-P6 to leaf water content was found useful to discriminate crops and to assess their conditions; and
- Digital databases are under completion for 56 districts in five states (Assam-21, Bihar-10, Orissa-10, West Bengal-6 and AP-10), for land-use and infrastructure.

On the whole, India's 6 remote sensing (polar orbiting) satellites—the largest such constellation in the world—are under use for: providing education in far-flung areas; connecting fishermen out on the seas to price their catch before they reach port; enabling taxi-operators to locate their cars using GSM, to give only three instances at random. India's 7 communication (geo-stationary) satellites—the biggest civilian constellation in the Asia-Pacific region—are under use for: providing television coverage to 90 per cent of the population; extending remote healthcare services; and carrying education to the rural poor. India is perhaps the only country whose satellite technology meets societal needs in a cost-effective manner and delivers the services to the needy.

On the economic front, CARTOSAT-1 (2005), mentioned earlier, has a resolution of 1 m and is equipped to keep vigil on the borders, thus saving money spent on buying from the US satellite images of military standard at very high cost. Indian launch vehicles (GSLV and PSLV) have carried satellites of Belgium and Germany—in preference to European Space Agency—making for good economics. India's space technology is already a competitive money-earner in selling infrared images to countries, including the US, where they are used for mapping. Several countries in Asia and Africa are users of India's satellite services: among the best, considering both cost and quality.

Regarding future planning, India is poised to launch the world's biggest satellites by 2007-08, as part of the Mark 3 programme. India's launch of the ambitious *'Chandrayan-1'* for orbiting the moon, due within half a decade, is for 'blue-sky research', namely, to create 3-D maps of the moon's surface at a record resolution of 5-10 metres and to map the distribution of limonite, a mineral that traps helium-3, as a possible source of energy for future bases on the moon. This, too, has potential to become a money-spinner in the decades to come.

COORDINATION WITH INTERNATIONAL SATELLITES

After the *Tsunami* disaster, December 2004 saw the activation of the International charter on Disaster, when satellite data from foreign satellites, such as, SPOT, RADARSET and ENVISAT were over Port Blair, Car Nicobar, Cuddalore and Nagapattinam were shared with the Indian imagery.

On a continued basis, India participated—as a founding member—in the formation of the International Maritime Satellite organisation (INMARSAT): with headquarters in London and with 72 countries as member-countries. INMARSAT supports services of direct-dial telephones, Telex, FAX, e-mail and data communications among mobile users around the world, connected to the international PSTN. It also provides two-way messaging and data communications on 'store-and-forward' basis. Indian VSNL commissioned a land earth station (LES) at Arvi, Pune, in 1992 for Rs. 23 crores, and continues to hold the investment on behalf of India.

Since multiple destinations are linked in the INMARSAT network, it makes possible to break the news of natural disasters and provide life-saving communications at a moment's notice. Some of the important users are mobile industries, which may comprise surveyors at remote points, pipeline teams, task forces for road construction and maintenance at high altitudes, drilling mining crews on the high seas who need to keep in touch with the port-base and international business communications managers who need to link with remote project sites. *Mobile earth stations* (MES) can be located on jeeps and can be used to roam villages and towns far inland, in order to:

- Disseminate location-specific and time-specific information on crops, seeds, fertilizers, irrigation, pesticides, harvesting, marketing and a myriad other aspects;
- Transmit queries to the subject-area specialists and block development officers (BDOs) at a moment's notice;
- Permit mutual consultations in health, industry and other sectors.

CONCLUSION

NRIS is well geared today to provide user support with a range of software products that includes: GEO-encoded resource integrated system (GRIDS) as a computerised monitoring software for evaluation of development programmes; district information system (DIS) as a menu-driven software on natural resources, socio-economic sectors and application modules at district level; geographical information system (GIS) as a software for spatial data input (vector and raster), analysis and digital image analysis and graphic display; and land information system (LIS) as a land records software for *Jamabandi* (record of rights), *Khasra* (record of land-use and crops, *Intekal* (mutation), *Lal Kitab* (village statistics) and *Sirm* (maps). NRIS priority application areas, of potential interest particularly to agriculture and agro-industry, are: resource mapping for local areas; integrated watershed management; evaluation of land potential and land use (including water body, swamps, forests, pastures, mines and built-up areas); irrigation scheduling, waterlogging and salinity; and minor irrigation structures (including groundwater, rain potential and water harvesting). The overwhelming user-advantages that NRIS offers—by combining IT and satellite technology—are: standard databases; structured data for decision-making; fast retrieval of inference; rapid updating; compatibility in space-time distribution of data; rapid aggregation of facts and figures; and, eventually, rational planning.

PART THREE

MANAGEMENT ENTERPRISES

8

E-COMMERCE AND ELECTRONIC DATA EXCHANGE

INTRODUCTION

Electronic Commerce (E-Commerce), associated with buying and selling of information, products and services over computer communication networks, has ushered in a new opening to the traditional mode of commerce—with information, at the heart of any commercial activity, being transmitted from computer to computer in an automated way. In its essence, E-Commerce is the paperless exchange of business information: using Electronic Data Interchange, Electronic Mail, Electronic Bulletin Boards, Electronic Funds Transfer and other network-based technologies. Beyond automating manual processes and paper-based transactions, it fundamentally helps to move organisations towards a fully electronic environment and change essentially the way they had operated in the past.

E-Commerce got its boost with the US government announcing that all federal purchases would be made paperless. Australia and Singapore governments, too, have opted in favour of electronic transactions, instead of paper-based ones, for all international trade. In the recent years, the Internet has given another fillip to E-Commerce: as a low-cost alternative to the proprietary networks. While E-Commerce standards are still under development, the more common Electronic Data Interchange (EDI)—the inter-organisational exchange of business documentation in structured, machine-processable form over computer networks—is still the dominant part of E-Commerce, covering nearly 80 per cent of its global volume.

CLASSIFICATION OF E-COMMERCE

From communication perspective, E-Commerce is procurement and delivery of information, products and services and their payments via telephone lines, networks or other means of computer-to-computer connectivity. These networks can be of Post, cable TV, leased line or wireless kind. From on-line perspective, E-Commerce is the capability of buying and selling of information, products and services on the Internet, EDI and other on-line services.

From business process perspective, E-Commerce is an application of technology for automation of business transaction and workflow. From service perspective, E-Commerce is a tool applied to firms, consumers and management: to cut cost of service; to improve quality of goods; and to increase speed of delivery. The overall goals are: cost-effective performance for quality; customer satisfaction through informed decision-making; economic efficiency by reduced costs; rapid exchange of business information; and high-speed accelerated or on-line transition of activities.

The bedrock of E-Commerce is IT-based transactions. These transactions are exchanges taking place, when one economic entity buys or sells information, product or service to another economic entity. The transfer of information, product or service occurs across a technology-linked interface, such as, *consumer-client* or *producer-server* interface.

HISTORY OF E-COMMERCE

The phenomenal development of E-Commerce has not been overnight, but an evolutionary process over the years. Its gradual growth, spanning the recent decades, can be broadly traced as follows:

Early 1970's: This was the period of *Electronic Funds Transfer* (EFT) covering banks over secure private networks. The computer-to-computer connectivity under Society for Worldwide Inter-bank Financial Telecommunications (SWIFT) was a prime example operating among major global banks, using mainframes, to provide such services like debit cards at *points-of-service* (POS). Direct deposits across the international banking networks amounted to $4.0 trillion, equivalent to 55 per cent of the US federal payments.

Late 70's to early 80's: This period saw the emergence of *electronic messaging* technologies, e.g., bar codes, e-mails and value-added networks, leading to EDI, all of which helped to reduce paper work and increase automation. These services enabled the following business practices to grow:

- Technical data interchange in the engineering area;
- Demand-driven mass customisation in manufacturing area;
- 'Virtual' and team-based enterprises;
- Desktop video conferencing;
- Document work flow system;

- Electronic mail for easy communication; and
- Electronic data interchange, beginning in the procurement area.

Late 80's to early 90's: This was the time when 'electronic messaging' got consolidated. By the mid 80's, social interactions became evident through file transfers, 'knowledge' sharing and 'virtual' community. From the mid 90's onwards, World-wide Web on the Internet opened the cheap way to do business and achieve economies of scale.

TOOLS OF E-COMMERCE: EDI

Electronic Data Interchange

Electronic Data Interchange (EDI) is the electronic exchange of business documents in a standard, compute-processable universally accepted format between trading partner. EDI is quite different from sending 'electronic mail' messages or sharing files through a network, or a bulletin board. In EDI, the computer applications of both the sender and the receiver, referred to as *trading partners* (TPs), have to agree upon the format of the business document: sent as a data file over an electronic messaging service. EDU cannot be undertaken unilaterally, but requires the cooperation—and active participation—of TPs. TPs normally consist of an organization's principal suppliers and wholesale customers. The components of EDI system are illustrated in Figure 8.1.

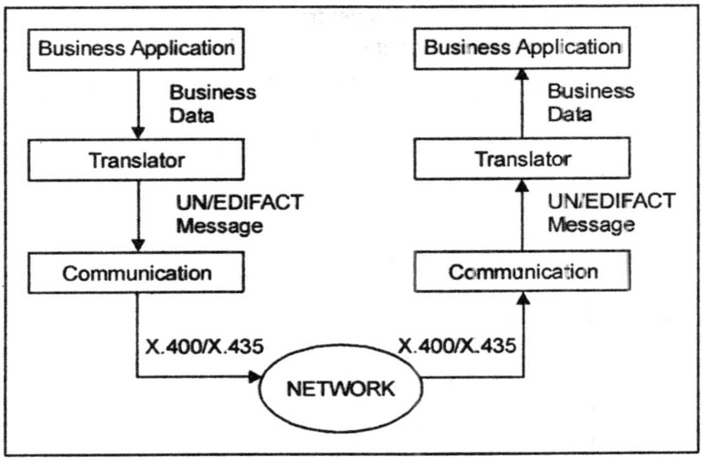

Fig. 8.1: The Components of EDI System

Incidentally, since data are exchanged in pre-defined formats, business documents can be exchanged irrespective of the nature of computer application at either end of communication. For instance the supplier's 'accounts receivable' application for raising invoice can still be in a file using COBOL, while the customer's 'accounts payable' may be based on an RDBMS, such as, ORACLE.

Historically, EDI came into force to transmit electronically documents, such as, purchase orders, invoices, shipping notices, receiving advices and other standard business transactions between trading partners. EDI also began to be used to transmit financial information and related payments in electronic form. When used for payments, it is known as financial EDI and is synonymous to *electronic fund transfer* (EFT), referred to earlier. Obviously, EDI is much more than merely substituting electronic transaction for paper ones. It is a means to streamline procedures, improve efficiency and enhance productivity. In fact, EDI allows a new look at the 'processes' within the organisation, with a view to re-engineering them, if necessary, through 'business process reengineering' (see Chapter 9).

Where EDI is implemented, computers electronically exchange business documents with each other—ideally without human intervention. Hence arise both direct and long-term benefits as follows:

Direct Benefits

- Since computer-to-computer transfer of information is automatic, data are entered only at source;
- Processing-cost of EDI documents is smaller than processing paper documents;
- Customer service is improved: by quick transfer of business documents and decreased errors;
- Information is managed more effectively; and
- There is more job satisfaction among DE operators, when re-deployed in more creative jobs.

Strategic Benefits

- Customer relations are improved through better quality and speed of service;
- Competitive edge is enhanced;
- Reduction in product costs can be achieved;
- Business relations with TPs are improved; and

- Better sales forecasting and business planning is feasible for improved information availability.

It will be evident that EDI is most suited in areas where the following factors exist:

1. Large volume of repetitive standard actions.
2. Tight operative margins.
3. Strong competitive requirements.
4. Operative time constraints.
5. Trading partners' request for paperless exchange of documents.

COMPONETS OF EDI

The major components of EDI are: standards, software and third-party (value-added) communications networks.

EDI Standards

UN/EDIFACT (*EDI for Administration, Commerce and Transport*) is a widely used standard today, announced as 'standard' by the United Nations in 1987. EDIFACT standard has since been promoted by the UN for all international trade, as looked after by the International Standards Organisation (ISO) for developing syntax rules and data dictionary; and the United Nations Economic Commission (UN/ECE) for the use, promotion and standardization of EDIFACT messages. The six EDIFACT boards of the world comprise: Western Europe, Eastern Europe, Pan America, Australia/New Zealand, Asia and Africa. India is one of the members of the Asia EDIFACT board.

EDI Software

EDI software has the functions of data conversion, data formatting and message communication. EDI software comprises computer instructions that translate information from unstructured, company-specific format to the structured EDI format, and then communicate the EDI message. At the other end, EDI software receives the message and translates it from standard format to the company-specific one.

Since electronic trading began before UN/EDIFACT came into vogue, there was a number of national standard software, including ANSI ASC X12 in North America and TRADACOMOS in the UK. As these are well-entrenched software among their user communities, it could take sometime before these users migrate to UN/EDIFACT. In particular, as the US user-community is still to shift wholesale to the UN-EDIFACT, this is a little problematic for transacting global trade with UN-EDIFACT, but the American migration is well on its way.

Looking closely at UN-EDIFACT, it sets out a new 'language' for document interchange and has, like any other language, a syntax or grammar—as specified in ISO 9735 standard. An UN/EDIFACT 'message' is a collection of data relating to business documents and transactions between the TPs engaged in EDI—with about 150 standard message-formats covering different business documents in the following areas:

1. *Administration:* Customs declaration, legal administration request/response, job application, etc.
2. *Commerce:* Tender, purchase order, invoice, payment order, remittance advice, etc.
3. *Transport:* Transport booking, multi-modal status report, arrival notice, etc.

Third Party (Value Added) Communications Network)

Value Added Networks (VANs) provide users with a single-point interface to the trading community, freeing the user from the hassles of handling different communication protocols, time-zones and availability of computer systems at the other end. There can be more than one VAN—offering EDI services, with a range of service options to choose from. The management strategy for selecting a VAN would be the following:

- Level of customer service;
- Extent to which there is inter-connection with other service-provider's VANs;
- Commitment to handle forthcoming changes in standards: inevitable for enhancing future functions; and
- Costs.

IMPLEMENTATION STRATEGY OF EDI

Three basic options available to an organisation for implementing EDI are as follows:

EDI Service Bureaus: Used for all the EDI messaging requirements of an organisation, without implementing EDI in-house;

Stand-alone EDI: Using electronic forms, instead of paper, in a non-integrated approach on a stand-alone PC and sent over a communications network;

Integrated EDI: Using mainframe business system for EDI, or, a front-end processor to take care of translation and communication. The latter can be connected over a LAN to the organisation's application computer.

PRIME TOOL OF E-COMMERCE: THE INTERNET

Electronic mail is used on the Internet as a means to transmitting EDI messages. A special IETF protocol is used to 'envelope' the EDI data within the e-mail message. Under this arrangement, the Internet replaces the proprietary VANs for exchange of EDI messages between trading partners. While the cost of transmitting EDI message is substantially reduced here, problems of security and integrity of transactions, as well as reliability of the Internet, remain dominant issues.

On the whole, EDI over the Internet is quite flexible, since there is no need for a prior network connection. Thus, many small firms can exchange EDI messages Typically, it is a *hub-and-spoke* arrangement: with a big buyer set up as the hub and dictating to its suppliers at the spokes. Virtual trading communities get formed over the Internet, since the Internet replaces much of this limited network with its own links. Filtering software, based on rules, processes the EDI messages and issues an auto-response for information about setting up a TP relationship.

The commercial Websites of business organisations have Web documents that offer useful product information, interactive brochures, news, reviews, etc User organisations can create e-brochures in colour with graphics, audio and video, and can reach customers worldwide at very little cost—instead of the expensive alternative of printing brochures and mailing them The electronic malls, shops and stores on the Web enable one to see and order merchandise by using a forms-interface: an *electronic form* that contain blank boxes for the user to enter information on product codes, credit card numbers, etc.

Outside the digital technology, electronic commerce is conducted in the following sectors, among others:

- Financial Services
- Travel
- Music
- Retailing
- Books
- Cars
- Advertising and Marketing
- Pornography (!)

TOOLS FOR E-COMMERCE: EXTRANETS

Extranet is a collaborative Internet connection from an *intranet* (closely held Internet-like protocols among a select group of organisations) with outside organisations and business partners. It can also be an extension of an *intranet* which makes the latter accessible to other organisations or individuals with or without an intranet. Parts of an *Intranet* are made available to customers or business partners for specific applications.

The key characteristic of an *extranet* is that it extends the Intranet from one location to another across the Internet by secured data flows: using cryptography and authorisation procedures—to another intranet of a business partner. Thus, intranets of business partners, material suppliers, financial services, distributors, customers, etc., are connected to extranets by an agreement between collaborative parties. In fact, business-to-business e-commerce is growing on *extranets*. Organisations gain comparative advantage through speedier transactions and access to new markets, as also by simplified and faster distribution of information, products and services. In addition to simple commercial transactions of sale-purchase, collaborative activities between business partners on the extranet are enabled at much lower costs than over VANs.

TOOLS FOR E-COMMERCE: EFT

Electronic Fund Transfer (EFT), as stated earlier, is the electronic mechanism for exchanging banking instruments across the world: without frontiers. Today the medium for such bank-to-bank operations is the Society for Worldwide Inter-bank Financial Telecommunications (SWIFT), mentioned in Chapter 3. Set-up by a consortium of US and European banks in 1973: for data processing and computing standards of inter-bank communications, SWIFT now operates the largest cross-border system in the world for exchanging banking messages—besides fund transfers—electronically over its proprietary network. Operating in 130 countries, an estimated $ 2.5 trillion worth of payment is processed every day among 5,000 banks and other co-owners of SWIFT.

SWIFT developed its own electronic 'envelope' and content formats for sending banking messages and instruments with a high degree of reliability and security. With the introduction of UN/EDIFACT in many banks and financial institutions, EDIFACT messages are also sent nowadays within SWIFT 'envelopes' to their destinations. Though SWIFT is primarily operational for EFT, new standards and solutions are introduced from time to time to enable extension of this service—comprising EFT and financial EDI—to the consumer segment as well.

In India, the first five participative banks that were selected for the Indian Bans Association's pilot project on implementation of EDI for banks are: State Bank of India, Union bank of India, Punjab National Bank, Corporation Bank and Bank of India. The documents selected for exchange among branches of participative banks—for both inter-bank and intra-bank requirements—were: demand draft advice, caution advice, cancellation advice and inter-city credit advice. These messages were converted into EDIFACT and routed through NICNET's EDI mailboxes to the destination branch.

FRAMEWORK OF E-COMMERCE

As seen above, E-Commerce was, until recently, realised through EDI and EFT (over proprietary networks), with 80 per cent of global E-Commerce still conducted through EDI. The emergence of the Internet and the Web has now revolutionised commerce and created new paradigms. The new business paradigm is that of the 'virtual' corporation, which has come into being through the combination of intranets and extranets. Universal access to information—enabled by the Internet, intranets and extranets—is at the heart of new business models for E-Commerce. An enterprise today is an inter-networking organisation, conducting E-Commerce supported by networks, in the following manner: Consumer-to-Business; Business-to-Business and Intranet Procurement.

Consumer-to-Business E-Commerce

Electronic mall, shops and virtual storefronts allow individual consumers to browse for products and services using credit cards. This form of E-commerce is Internet-based, with unrestricted access to consumers. Credit-card payments have to be secured against unauthorized access by intruders on public networks.

Business-to-Business E-Commerce

At present, this comprises bulk of the E-Commerce conducted over networks: as much as 70 per cent. Business-to-business procurements and fulfillment, including financial transactions, has traditionally been conducted over such private/proprietary networks as SWIFT. Today, this segment of E-Commerce is driving Extranets—restricted to business partners and subject to security procedures based on 'firewalls'.

Intranet Procurement

Business transactions, which are internal to an enterprise, accounts departments and subsidiaries, also come under E-commerce. Internet sales, order processing, intra-company charging and billing, fund transfers (for accounting), etc., can be very large and voluminous and can even span the entire globe in caser of multinationals. Intranets respond to this challenge.

It should be noted that E-commerce on the Internet, Intranets and Extranets use the common standards TCP/IP on the inter-networked information infrastructure—when they span the entire world. EDI capabilities are added to the existing business applications of organisations, and enabled on the Web for E-Commerce.

In the light of the above discussion, a diagrammatic representation of E-Commerce is indicated at Figure 8.2 and its framework is illustrated in Table 8.1.

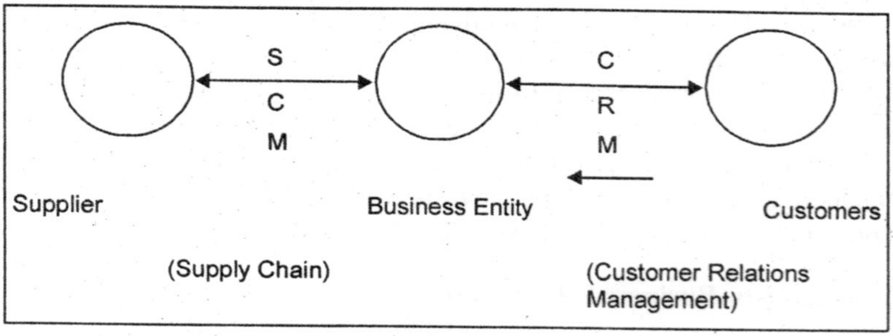

Fig. 8.2: E-Commerce

Table 8.1: Framework of E-commerce

Policy Issue	*E-Commerce Application*	*Standards & Protocol*
Public Policy, Legal Policy and Privacy Issues	Video-On-Demand • Promote banking • Procurement and Purchasing • On-Line Market and Advance • Home Shopping	Technical Standards for documents, security and net-works protocols

MANAGENET STRATEGY FOR E-COMMERCE

Management of the information distribution involves many key-issues, such as, infrastructure (EDI, E-Mail, Hyper-Text in the Internet and information transfer protocols); Multi-Media (content creation and network publishing infrastructure, such as, HTML, JAVA and WWW); and Network Infrastructure (Telecommunications, Cable and Wireless, Internet).

KEY QUESTINONS FOR MANAGEMENT

- What is driving changes in business? How quick are these changes? What are the competitive pressures?
- Who are the key competitors? What is the external threat?
- How easy or difficult it is to incorporate changes into business? How much 're-engineering' is needed?
- How does one design a new organisational structure? How does one assess its efficiency?

NEW MARKETS FOR E-COMMERCE

New markets in E-Commerce have taken the following directions:

- Innovative virtual business, market and trading communities;
- Virtual distributors, outsourcing, physical warehousing and goods to couriers (logistics);
- Buyer intermediaries, sellers forming industry-specific Internet markets, e.g., Real estates, automobile parts, construction, etc;
- Global manufacturing industries (aerospace, computers, automobiles) integrating their supply chins.

First Generation E-Commerce

Merchants used the Internet as a marketing medium (e.g., Web presence, advertising, products, disseminating organisational information).

Second Generation E-Commerce

Electronic ordering began to enhance market research and increase revenue.

Third Generation E-Commerce

Delivery of enhanced market and increased revenue started flowing in, while increasing customer service and reducing operating cost.

Generic new markets are summarised in Table 8.2.

Table 8.2: Generic New Markets

Commerce	*Transport*	*Government*
Trade & Industry Munufacturing	Road, Rail, Air & Sea Forwarding Dispatch	Customer & Exercise Natural and International Trade
Finance & Banking Tourism and Travel	Warehousing Logistics management	

COST-BENEFIT ANALYSIS OF E-COMMERCE

These elements are summarised as follows:

Cost

- Research and business planning cost;
- Investment cost in IT equipment, software, personnel and telecommunication for EDI;
- Implementation cost for EDI technology, messages, etc.;
- Re-engineering cost in infrastructure; and
- Indirect costs of management, legal audit, security, etc.

Benefits

- Multi-media presentation of business products and services;
- Large volume of repetitive and standardised data;
- Improved access to global 'electronic' customers;
- Ability to re-think existing customers with improved efficiency;
- Strong competitive edge;
- Improved corporate trading relations;
- Improved service—fast and accurate information, and novel ways of working together;
- Reduced cost of trade in computer-to-computer mode;
- Faster trading cycle;
- Ability to adopt new business process (e.g., 'TIT' in manufacturing technology);
- Effective organisation management (with re-alignment of functions/resources).

E-COMMERCE IN INDIA

India's Ministry of Commerce established an EDI Council to introduce EDI in this country in 1994. The Council formally adopted UN/EDIFACT messages as the standard for conducting EDI in India. The India EDIFACT Committee set up a number of Message development Groups to promote use of EDIFACT standards in exchanging international trade transactions. The following agencies in the areas of international trade were identified as 'key government regulatory organisations', with the four-fold mandate of: (1) Making them EDI-capable; (2) Accepting trade documentation electronically in the EDIFACT format; (3) Processing the same in their respective automated computer systems; and (4) Responding electronically to the user:

- *Customs Department*: for processing customs declarations;
- *Directorate General of Foreign Trade*: for processing import-export licence applications;
- *Apparel Export Promotion Council*: for processing quota certificate release;
- *Reserve Bank of India*: for receiving data about customs declarations electronically from the Customs;
- *Banks*: for duty payment and drawback release in the context of international trade and introducing EDI in normal banking transactions;
- *Port Authorities*: for interface with the shipping lines, shipping agents and freight forwarders;
- *Airport Authority of India*: for interface with airlines and air-cargo agents;
- *Container Corporation of India*: for the Inland Container Depots;
- *Railways*: for integrating freight operations as part of EDIin the transport sector; and
- *ECGC* and other trae-related agencies.

A number of private sector EDI service providers have come up in the country as follows:

1. Mahindra Network Services
2. Global Telecom Services
3. Satyam Infoway Network
4. Manipal Control Data
5. IBM Global Network Services
6. GE Information Services.

INTELLECTUAL PROPERTY RIGHTS IN E-COMMERCE

In the whole world, the IT sector **patents** now form 10 per cent of the total patents granted, figuring hi-tech names like IBM, Canon, Hitachi, Lucent, Microsoft and Oracle. With the rise of the Internet, their patent portfolios are in the areas like business methods and E-commerce. Special attention is being given globally to the copyright of computer software, Web-publishing and multi-media contents posted on the Internet. On-line Intellectual Property Rights (IPR) are involved in group computing, business alliances, e-enabled services, etc., where copyright and related rights in respect of protection of data and DBs are essential components. Many countries have either legislated, or are legislating, on category of software to be provided patent protection so as to comply with TRIPs Agreement. Thus, copyrights are providing major contribution to their GNP and economy—calling for new measures like Digital Rights Management.

A new global concern is about corporates a watch over their trademarks and domain names, to prevent infringement. IBM, for instance, makes nearly $2 billions from patent infringement litigations as pure profit. Even in India, by licensing of popular tunes to the cell phones @ Rs. 1.50 per year, Indian Performance Rights Society (IPRS), Mumbai, collects revenue of over $3.5 million per year.

IT projects, due to sheer obsolescence, need IPR valuation, particularly in the absence of standards. The valuation is also needed for assessing true returns on efforts and investments made to launch, protect and nurture 'brands' for gaining market shares. Here is an imperative need for valuating acquired and homegrown IPRs for future decisions on whether to invest or obtain licences form abroad, alongside licence-terms and associated payments.

CONCLUSION—SECURITY CONCERNS

Any discussion on E-commerce is incomplete without dwelling upon the crucial security issues as follows:

Confidentiality: Information should be protected from unauthorised internal users, external hackers and immune from interception during transmission on communications networks.

Integrity: On retrieval, or receipt at the other end of a communications network, information should appear exactly as was stores or sent.

Availability: Information, stored or transmitted across

communications network, should be available wherever required and to whatever extent desired.

Authenticity: It should be possible to prevent any person or object from masquerading as some other person or object.

Non-Reputability: After sending/authorizing a message, the sender should not be able later to deny having done so.

Auditability: Audit data must be recorded in such a way that all specified confidentiality and integrity requirements are met.

Implementation of E-commerce solutions for all the security services listed above is based on *Cryptographic* techniques, comprising *Encryption* (process of making information unintelligible to the unauthorised reader) and *Decryption* ((reversing encryption to make the information intelligible once again). Two types of basic Cryptographic Systems are: *Symmetric* and *Asymmetric.*

SYMMETRIC CRYPTOSYSTEMS

Begun in 1977, major commercial use of symmetric cryptosystems adopt *data encryption standard* (DES), as adopted as a US Federal standard. In the DES cryptosystem, data is encrypted in 64-bit blocks, using a 56-bit 'key'. This is illustrated in Figure 8.3.

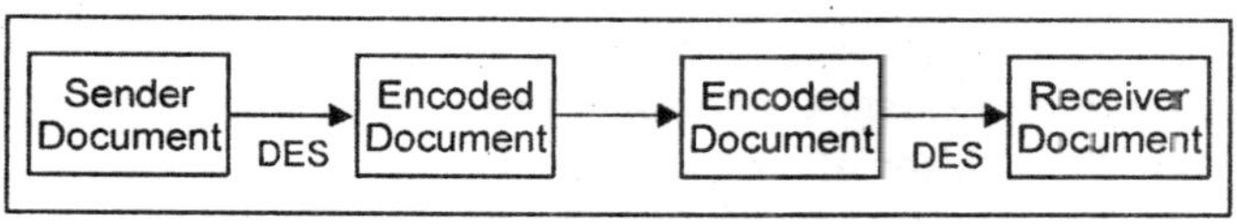

Fig. 8.3: Symmetric Keys

ASYMMETRIC CRYPTOSYSTEMS

Asymmetric or 'Public Key' cryptosystems are built around the possession of a pair of keys—a *public key* and a *private key*—by each entity wishing to engage in secure communication. While everyone knows the 'public key', only the owner knows the 'private key'. The algorithm to generate these keys is such that if either of the keys is used to encrypt a message, only the other key—out of the key-pair—can decrypt it. Let us say that user B (with public key PKB and private key SKB) is to receive an authenticated message from user A (with public key PKA and private key PKA). Then the encryption and decryption procedure can be illustrated as in Figure 8.4.

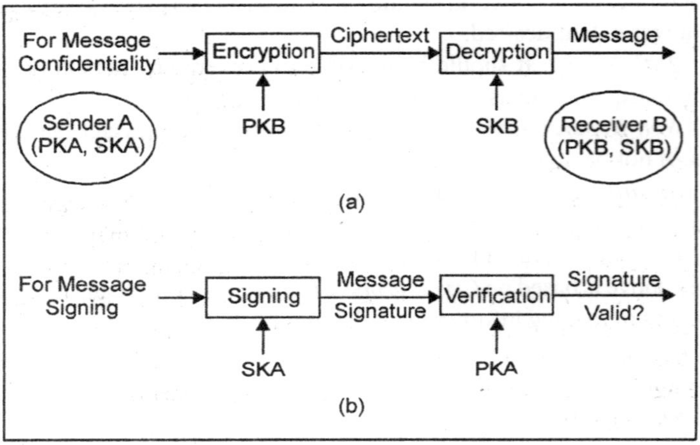

Fig. 8.4: Asymmetric Algorithm

Using the RSA cryptosystem, a message is encrypted with the sender's private key to generate 'digital signature'. The message is then sent to the destination along with this 'signature'. The recipient decrypts the 'signature' using the sender's public key and if the result matches the message-copy received, the recipient can be sure that the claimed originator sent the message. Although this is a highly secure way of digitally signing messages, it no doubt entails large processing overhead. 'Message digests' are often used to reduce processing overheads for implementing 'digital signatures'. Technology solutions have been developed to implement different security services: with varying degrees of strength and reliability. All the solutions depend on encryption using 'keys'—either single or in pairs.

For further reading, reference may be made to *E-Commerce: The Cutting Edge of Business,* Kamlesh K. Bajaj and Debjani Nag (Tata Megraw-Hill, New Delhi, 2003).

9

ENTERPRISE RESOURCE PLANNING AND BUSINESS PROCESS RE-ENGINEERING

INTRODUCTION

We examine here the characteristics of the fast-emerging scenario of Enterprise Resource Planning (ERP) tools and their potential applicability in the corporate sector. ERP is successful provided the existing deficiencies and inequities have been removed through Business Process Re-engineering (BRP), either sequentially or in tandem. Special attention is devoted to the selection and implementation strategies and their pitfalls, as ERP is a costly step to be taken for the business house and it would also be imperative to ensure that an advanced computer culture exists already. Above all, the top management's mindset should be favourable to provide leadership to the substantial changes going to be necessary.

ETERPRISE RESOURCE PLANNING (ERP) AND BUSINESS PROCESS RE-ENGINEERING (BPR)

There has been consistently two schools of thought about whether the corporate organisation should implement Enterprise Resource Planning (ERP) first without carrying out Business Process Re-engineering (BPR), or, whether BPR should have precedence over ERP. Some experts feel that once ERP is planned for, the management should go ahead to implement it, even while computerising all existing inefficiencies. The argument is that the inadequacies and inefficiencies would become visible soon enough under the ERP and can then be taken care of. The opposite school considers that if customisation is limited to the bare necessities and if better business practices—inherent in the ERP package—were to be adopted, then the implementation process itself would call for re-engineering. In that event, ERP system could serve as the bedrock and effectiveness of BPR done alongside BPR and would be fully effective.

Most authorities veer today to the second option and do not recommend the existing inefficiencies to linger on.

BUSINESS PROCESS RE-ENGINEERING (BPR)

For our purpose, a 'business process' can be viewed as an amalgam of people, energy, materials, equipment and procedures that are designed to produce a specific end-result as a work-product. A set of such processes forms a coherent 'business system'. The latter is a pointer to the way in which a business unit (or a collection of units) caries out the business. BPR is then a sequential in-depth analysis of all the existing processes that should lead to the way in which the inefficient and unwanted process 9 or Processes) can be eliminated so that the overall productivity can improve.

Varied examples of 'business processes' are: development of a new product; ordering of goods from a supplier; creation of a marketing or distribution plan; negotiating a diversification proposal for new or related products; processing or paying insurance claim; and writing an elaborate proposal for an eventual government contract

Re-designing 'business processes' with IT is a common place phenomenon today. The development of business vision and process parameters are: identification of the process to be re-designed; understanding and measuring of the existing process; identifying IT capabilities; design and building of prototype of the new process; etc. It is to be understood that BPR is not work simplification of the industrial engineering kind, but has such clear-cut objectives as *time reduction* (with increased automation); *cost reduction* (without affecting quality);—after identifying the process to be re-designed through detailed approach or high-impact approach. Also, while understanding and measuring the existing process is done, problems must be understood and not repeated, and accurate measurement must serve as a base-line for improvement of *time* versus *cost*. IT remains a *key-enabler* for the following:

- Analysis of the existing process
- As tool for re-designing business process
- Creating a new process-design option, and not just a supportive role.

ENTERPRISE RESOURCE PLANNING (ERP)

ERP is a set (or suite) of software concerning major 'business processes', such as, Finance, Material, Manufacturing Distribution. Logistics and Sale. This suite has its components as tightly integrated with one another, so that any business activity recorded at one place will be reflected everywhere. The neatest analogy is the tight coupling of goods wagons in a railway transportation system—driven by a locomotive—where any activity like sudden starting/speeding/stopping will be immediately sensed everywhere in the entire train.

In a typical ERP application, there is so much data consistency and integrity that data are entered only once and there is no duplication. For instance, material received/accepted in Stores Department results automatically in the following chain of actions:

- Stock immediately upgraded;
- Pending purchase order cleared;
- Purchase indent of Stores Department updated as 'serviced';
- Supplier of material credited by the supplier value; and
- Company's 'Outstanding' Payable increased by the same value.

ERP-DRIVEN SUPPLY CHAIN MANAGEMENT

For an ERP-driven business organisation, its suppliers tend to be chosen not only based on their ability to deliver *quality* and *price-choice,* but also on their ability to work with the buying company's processes. The advantages of such a supply chain management are as follows:

1. Just-in-time delivery;
2. Payment on usage, and not delivery;
3. Random checking with punitive clauses; Supplier quality certification;
4. EDI linkage; and
5. Compatible IT systems, e.g., Yahoo!, gmail, etc.

In such an ERP-driven supply chain management, material supply is not sourced from a large number of suppliers on 'best prices'. There is a reduction in the number of suppliers—leading to building 'Win-Win' situations and relationships with them.

ERP-DRIVEN DAY-TO-DAY OPERATIONS AND SYSTEMS

In such systems, there is a constant recording of '3-Ps': *Policies, Procedures* and *Practices.* '3-Ps' are watched at various levels: System, Facility, Supplier, Warehouse. Product; Buyer, Commodity, Purchase requisition, Contact, and Sale. '3-Ps' are recorded for monitoring status, watching deviations, and noting need for change—with external stimuli; analysis of delays and problems caused by the existing '3-Ps'.

MATERIALS REQUIREMENT PLANNING (MRP)

Materials Requirement Planning (MRP or MRP-I) is a useful stepping stone before the ERP is introduced. MRP tests the sinews of the user-organisation like food, pharmaceutical or automotive industry. For repetitive supplier scheduling. This segment can be gradually coupled with master production scheduling, automatic updating of purchase requirements; tightly coupled with marketing, supplier, production, inventory and finance; and finally to be coupled with production planning and control at the MRP-II stage.

Manufacturing Resource Planning (MRP-II) is often a useful intermediary step, especially for large manufacturing organisations, before a costly business launch is made for the comprehensive suite of ERP software. MRP, MRP-II and ERP are treated in that order of evolution—by gradually testing the IT muscle of the organisation and gently initiating it into more and more demanding IT applications, when the lower ends have proved sufficiently integrated, accepted and assimilated by management echelons and benefits made visible to all concerned.

ERP IN PROCESS INDUSTRY: A STUDY

An example of process industry, based on the conceptual premises of the jeans-market giant Levi's is shown in Figure 9.1, where output from any previous process is assumed as input for the next process.

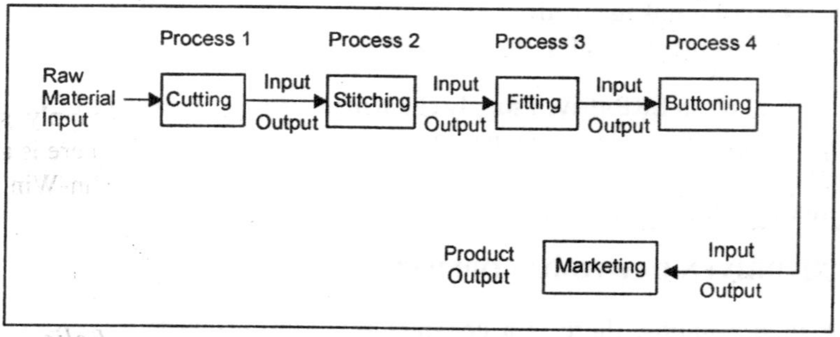

Fig. 9.1: Example of Process Industry (Levi's Jeans)

In a process industry, to start with, the potency of the system for tracking inventory of active items (in terms of physical quantity/value) is very high. The unit of measurement conventionally is any measure of a given resource in receipt, planning or production stage. Often, lot-sizes are used for products like textiles or food-grades. Forward and backward resource-movements need also be traceable. For achieving status control, grades are assigned to lots. This is especially true for the process industry like pharmaceuticals, food, textiles and paper. 'What-If' information is gathered to experiment with formula-compositions, for instance, to decide on whether to develop or enhance potency in pharmaceuticals; or what would Levi do if too much material rejection happens in their textile products. For determining formula cost, and new product cost and marketability, on line analysis is undertaken.

The other areas of ERP in a process industry are as follows:

- Co-products and by-products (e.g., handkerchiefs and caps as Jeans products)—with resources and formulas;
- Waste and Recycles—to control amount and cost;
- Process costing—ability to capture receipt, usage and shipment (e.g., delivery of Jeans products);
- Master scheduling—high-level plan for planned orders and production (e.g., Jeans material used and wasted); and
- Resource planning—consumption of critical production resources (e.g., chemicals for Jeans).

Benefits of ERP in process industry would accrue on the following lines:

1. Improved process yields,
2. Improved capacity utilisation,
3. Uncovered 'hidden' capacities,
4. Better cost identification and management,
5. Quicker response to market changes,
6. More efficient use of warehouse space, and
7. Improved operations flexibility.

ERP IN SMALL AND MEDIUM-SCALE INDUSTRY (SME)

ERP solutions cannot be 'shrink-wrapped' (i.e., fitted as though tailor-made) as a 'plug-and-play' package (i.e., automatically downloaded and applied). Careful implementation chores—in terms of jobs and activities—will be needed to select specific components of ERP: to reach the promised goals. Many authorities feel that a comprehensive ERP approach may not suit the SME sector and there may be some truth in that view.

BENEFITS OF ERP

These can be summarised into the following streams, covering all business processes:

Tangible

- Reduction in inventory in the range 15 per cent to 35 per cent;
- Significant reduction in 'stock-outs', i.e., zero finished-goods inventory or zero raw-goods inventory;
- Diversion of specified manpower from routine activities to profitable ones; and
- 'Payback' period as low as one year.

Intangible

- Executive time available for more productive causes;
- Elimination of unproductive reconciliation activities;
- Availability of timely/appropriate information;
- Improved process times; and
- Feasibility of control of operations.

Overall

- Global custody
- Money market
- Accounts processing

Foreign exchange
Funds transfer
Commercial lending

Spin-off

- Business data across spread-sheets;
- Publishing tools for client reporting; and
- Information consolidation from diverse systems.

SELECTION AND IMPLEMENTATION STRATEGY FOR ERP

Selection Strategy

Selection of ERP packages is dependent on a large number of factors including the following:

1. Base level of IT support for key business processes for 5-10 years;
2. Opportunity for business process-owners and key-users to build better commitment;
3. Understanding of strengths and weaknesses of key business people;
4. Structured implementation plan, covering costs, resources and benefits.

Planned Criteria

The above selection pre-supposes availability of the following planned criteria:

1. Machine resources,
2. People resources,
3. Realistic time-frames,
4. Incorporation of best practices during BPR, by removing defects and inefficiencies, and
5. Improving architecture.

Pitfalls in Planning

Two common pitfalls that are generally accosted are, either, no prior comparative analysis ERP packages has been done, or, customised functionality persists in the legacy systems. By a proper study of functionality, architecture and paradigms of ERP packages, such pitfalls can be avoided.

Implementation Strategy

This can be divided phase-wise as follows:

Pre-requisite Phase

- Top management commitment
- Measurable targets for business benefits
- Establishing linkages to team-performance
- Investment in education and training
- Focus on operational discipline
- Adherence to defined processes

- Elimination of parallel records during BPR.

Assessment and Planning Phase

- Detailed work plan
- Identifying project-team
- Assessment of existing business processes
- Broad areas of process re-engineering

Business Design and Prototype Phase

- Conducting a 'pilot'
- Process re-engineering for improved processes

Delivery and Assimilation Phase

- Setting up of package parameters
- Preparing new procedures
- User training for new procedures
- Testing of new systems/procedures
- Data-transfer from legacy system to new ERP system.

Key Success Factors for Implementation

These can be enumerated as follows:

1. Measurable business objectives;
2. Strong project management;
3. Business process re-engineering;
4. Thorough product knowledge;
5. Proven implementation technology;
6. Effective change management;
7. User ownership of system; and
8. Implementation experience in Indian/international environment.

INDIAN ERP MARKET: A CASE STUDY

A millennium-end study showed that the growth rate of Indian ERP market was 100 per cent between Rs. 575 million in 1996-97 and Rs. 1150 million in 1997-98. It anticipated further that the high growth would continue unabated in the next two decades. Its findings on the distribution of ERP sites in India are presented in Table 9.1.

Table 9.1: Millennium-end ERP Sites in India

BPCS	9 %
AVALON	10 %
RAMCO	17 %
ORACLE Financial	13 %
SAP R/S	16 %
MFS/PRO	10 %
Others	25 %

MAJOR GLOBAL PLAYERS IN ERP

The following is a summary from the millennium-end analysis according to end-user perceptions for the major market-players of ERP:

SAP (Systems Applications and Products) (German)

- Foremost ERP with the latest technology
- Store-house for open information
- Symbiotic relation with technology leaders
- Long-term partnership with customers
- Adaptable to customers' hardware (including 'Legacy').

Strengths

- Known to take care of total integration
- Best business practices
- Embodying business solutions for competitive-edge
- Helping to enhance customer service level
- Flexible and modular
- Potential of synergy among business, process and industry solutions
- Internet-enabled
- Integrating people and common tasks.

Key Success Factors

- Strong executive support
- Good project team
- Rapid decision-making
- Accessibility of implementation team
- Sound project monitoring
- Technical support from ERP
- Adaptability to old systems.

Weaknesses

- Complexity of vertical initiatives
- Little world-class functionality
- High cost.

Oracle (American)

- Applications on Windows NT
- Possessing 35 products:
  - Solutions for financial management
  - Supply chain management
  - Manufacturing project system
  - Human resources management
  - Sales force automation, etc.

Strengths

- Clustering technologies:
  - Oracle parallel server
  - Oracle fail-safe
- Scalability for fast-growing companies
- Global functionality
- Great flexibility
- Rapid implementation.

Weakness

- Very strong in financial segment, but not so much in the manufacturing segment.

Baan (Dutch)

- Bundling work-group server (WGS)
- Products: BAAN 4.0, Triton 2 and 3.

Strengths

- First-line support for DBMS, with a single point of contact
- Competitive pricing combining application and DBMS licence
- Readily acceptable for middle market
- Low cost.

Weaknesses

- Not preferred by large users
- Limited number of operational DBMSs (most likely Oracle).

Peoplesoft (British)

- Peoplesoft-6: SQL server of 32-bit architecture
- Peoplesoft-7: 2-and 3-tier architectures, Internet-enabled with Web extensions, Java client, OLAP tools available
- Peoplesoft-7.5: AS/400 and 11 new products.

Strengths

- Strong in HR product, 80 per cent business-share of market
- Europe's adaptation enhanced, with more language translation
- More European functionality (VAT handling, multiple currency, conversions to 'Euro').

Weakness

- Financial services coverage up to only 20 per cent.

IFS (Industrial and Financial System) (Swedish)

- Complete business system to bundle: engineering, manufacturing, distribution and maintenance
- IFS Applications: also financial, human resource management
- Customers: large and medium-sized corporates, within manufacturing/processing industry.

Strengths

- Long-term relations with customers
- Responsible for implementing maintenance
- Product development in cooperation with client
- Revenues from consultancy/licensing aiming at 59 per cent of sales
- Based on 'object-oriented' technology and possessing 40 independent modules
- Due to 'open architecture', IFS applications may compliment business system from other vendors for specific purposes

- IFS product development with development tools from Oracle, Microsoft, Ventura, Sunsoft: technical solutions rapidly implemented.

BPCS (Business Planning and Control System) (USA)

- 30 integrated products for manufacturing/process industries, distribution logistics and financial applications
- AS/SET: Fully integrated CASE (Computer-assisted Software Engineering) products.

Strengths

- EOI product-line: customers/vendors through EOI network: short cycle-time, high quality
- Based on advanced AS/400 and intelligent 'work stations'
- Over 5500 users worldwide, featuring prominently among 'Fortune-300' companies
- 'Computer-integrated' manufacturing (CIM)
- Applications/integration with trading partners.

Ramco Marshal (India—Chennai)

- Launched in medium/small industry
- Spent Rs. 1000 million to develop 'Marshal 3.0'
- Launched by Madras Cements' cumulative sales far above Rs. 1999 millions since launch in 1994
- 44 sites in India at the millennium-end, with 100 companies worldwide.

CONCLUSION

The above survey, though carried out at the millennium-end, is a valuable adjunct to the user-perception of the world's major ERP vendors. Though approximate at this date, it retains its validity as a comparative analysis tool. What emerges is that the ERP vendors have specific product-advantage in given sectors and clearly-defined deliverables. The end-user should remain alive to these specificities.

PART FOUR

MANAGEMENT STRATEGY

10

IT PLANNING STRATEGY: AN INDUSTRY EXPERIENCE

INTRODUCTION

An across-the-board survey of the Indian user-industry at the millennium-end revealed that it was embracing IT haltingly at the small-scale level. The pace of the IT adoption quickened at the size of industry got upscale from small to medium and speeded up at the large-scale-industrial front.

CASE I—SMALL-SCALE INDUSTRY

At the small-scale level with a turnover up to 100 crores, the IT applications were mainly data-processing, i.e., handling routine accounting and accounting-oriented activities. A rapid survey of a few key small-scale industry brought out some important restrictive factors as follows:

- Lack of resources
- Lack of investment capacity to update IT
- Fear among clerical staff to lose jobs
- Lack of appropriate know-how.

In fact, there was a lack of conceptual thinking at the top management, for whom IT was a single tool to convert legacy routines into data processing jobs—on a one-to-one basis. Very few MIS applications were evident, on the plea of dearth of skilled IT manpower and expenditure not being affordable.

The usage criteria considered for the survey included factors like the organisation's hardware, and telecom needs, its choice of application modules and before and after integration, and education choices in IT.

CASE 2—MEDIUM-SCALE INDUSTRY

The canvas at the medium-scale industry level (with turnover from Rs. 100

to 500 crores) reached a different colour and texture. The IT usage here showed the emergence of On-Line Query (applications which included capture, processing and transmission of fast or instantaneous data and screened information) and Graduated Automation (of large, repetitive jobs like record keeping). On-line and automated clerical and shop-floor systems were observed—with payoffs to management still simple and immediate. The IT usage transited from data processing to MIS for most of them. The illustrative cases were as follows:

Swadeshi Polytex

It was a joint-sector enterprise, base at Ghaziabad (with a turnover around Rs. 100 crore), which produced polyester staple fibre and a by-product of methanol' Its Galaxy-21 system produced about 100 batch-processing reports covering financial accounts, share accounts, payroll, fixed deposits and so on.

Its phased action plan for IT included upgradation from 486's at the EDP centre and separate workstations at department levels to monitor production parameters, top management level for planning and control, and local communication facilities between EDP center and relevant departments in Phase 1, and LAN for inter-departmental connectivity and managers' workstations for 20 nodes with multi-user FoxPro, in Phase 2. Other plans include inter-office (including Ghaziabad and Delhi) and inter-branch connectivity, along with email, supported by modems and telephone lines, and hierarchical training. The total cost estimated up to the end of phase 2 is Rs. 16.5 lakh. However, the limiting factors observed here resemble uncannily to those at the small-scale units. But the largest impediment has been frequent changes in leadership and IT plans not being implemented mostly due to lack of top management support.

Godfrey Phillips

The Rs. 400-crore company, which has been operational since last century, produces cigarettes and tea. All the six locations of the company used IT for financial accounts, sales accounts, purchase accounts, payroll and pilot software on MIS. The major IT characteristics were 486s and Pentium for divisional managers and above; 'notebooks' for managers, and PCs for secretaries; with LAN at Mumbai and Ghaziabad factories, email with VSATs; and Remote Area Business Management Network (RABMN) in 11 locations.

Similar to other medium industries, this old firm was low on MIS. On the whole, there seemed to be little return to management on IT.

Rollatainers

The survey found innovative thinking, investment capacity with little constraint, planned use of training, constant look-out for new technical know-how, and enhancements in MIS at the world's largest manufacturer of cartons with a turn-over of Rs. 300 crore. Till recently, its hardware was Unix-driven, HCL minis with dumb terminals at Faridabad and Sonepat, and applications were totally data-processing kind, without any MIS or use for decision-making. The current strategy established a 50-node client-server LAN with networking software and databases with graphic-user interfaces—covering inventory with purchase function; financial accounts, production monitoring, marketing, and dispatch information, employee services (with provident fund and TDS) and others, with quite a few MIS and DSS applications at the corporate office. It also had multi-user LANs at other units, with a minimum of 10 nodes, with e-mails as well as Internet connectivity available at all locations. In the next phase, as seen in Figure 10.1, the action plan included upgrading development and application servers at the corporate office, Pentium servers and upgraded software at all locations, conversion of SQL base to Oracle, integrated production planning and control with completed and 'clubbed' Bills Of Material (BOM), fixed parameters, and machine capacities, scheduling for 'bunching' jobs for optimum production, firm, verbal, and assumed sales orders, and all logic updation for any re-scheduling, several MIS analyses for production, sales, and BOM, WAN, and ERP.

Overall, it was observed that, leadership provided by top management, which remained a clear determinant of IT effectiveness, was poor. Interestingly, the financial investment in IT tools seemed to veer around 0.5 per cent to 1 per cent of the total turnover, which was clearly on the low side.

IT-Usage Trends

- Ad hoc-ism prevalent in general
- Unclear use of PCs
- Preponderance of data-processing applications
- Unclear software choices

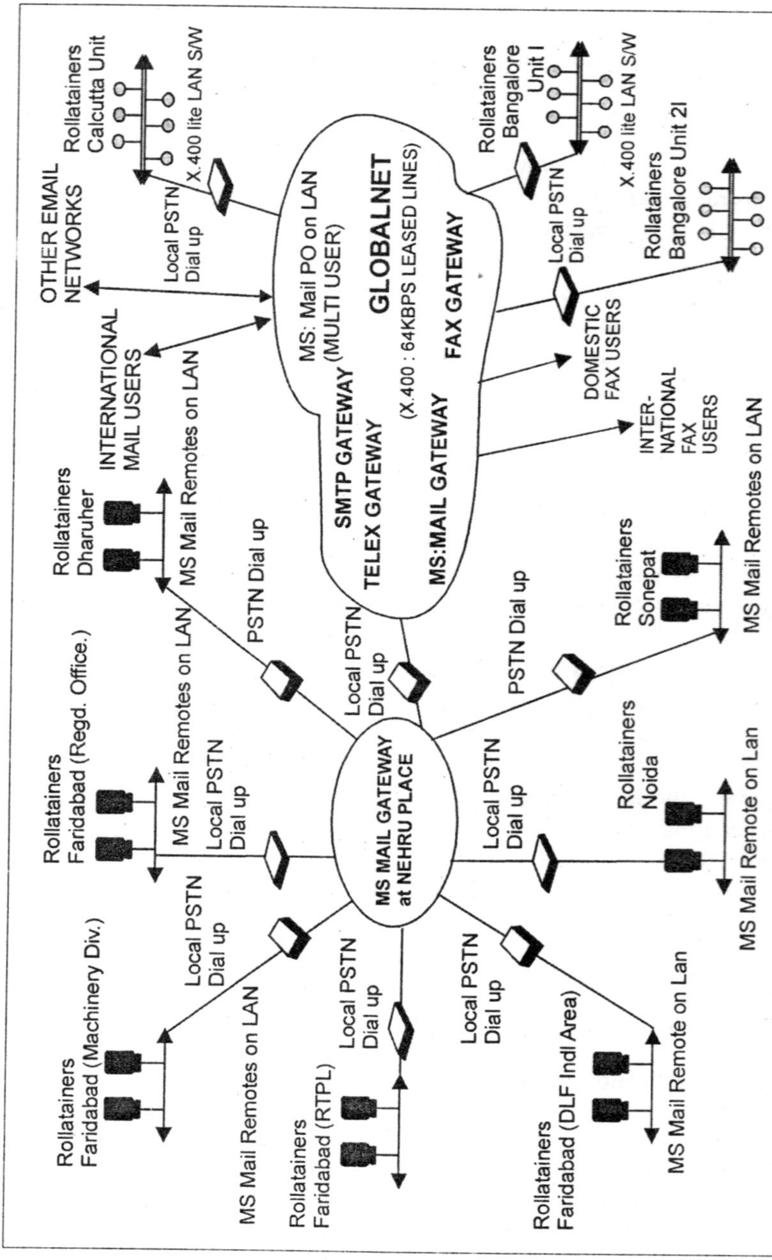

Fig. 10.1: Proposed E-Mail Network for Rollatainers

- Under-utilized LANs
- E-mail and VSATs not fully assessed
- MIS tools not fully available
- Integrated planning and control not yet within reach
- Poor user education.

CASE 3—LARGE-SCALE INDUSTRY

The canvas for the large-scale industry (turnover above Rs 500 crore), on the contrary, showed a clearly advanced phase comprising MIS with structured and periodic management, Decision Support System (DSS) with semi or unstructured management interaction on ad hoc basis, decision-making activities with 'screened' and 'filtered' information, and automation of random, discretionary, non-repetitive activities. This phase made for knowledge-worker systems at managerial levels. Payoffs were subtle and complex, often with wide ramifications.

National Thermal Power Corporation

One of the 'Nava Ratna' PSUs in the country, NTPC (with turnover of around Rs. 6400 crores) virtually covered the entire country and fed power to the regional grids. Its WAN application areas included process plant operation, plant maintenance and material management.

As seen in Figure 10.2, the phased action plan for IT was to ensure that there integrated management, EDI and the Internet connectivity so that decision-making was based on the latest data. There was also to be enterprise-wide availability of current information, which would lead to team-spirit, better resource-sharing and ultimately lead to 'office is, where you are' stage.

TATA Steel

As illustrated in Figures 10.3 to 10.6, Tata Steel (with a turnover of around Rs. 6500 crores) was a pioneers of IT usage since the 1960's—evolving from the second-generation to the Unisys machines and then stepping into IBM 9000, Sun Micros and VAX, with a large network of 'Token Ring' and process control computers: linked to the network. There was a separate sales network, not linked to 'Token Ring', comprising minis and desktops, called TISCONET.

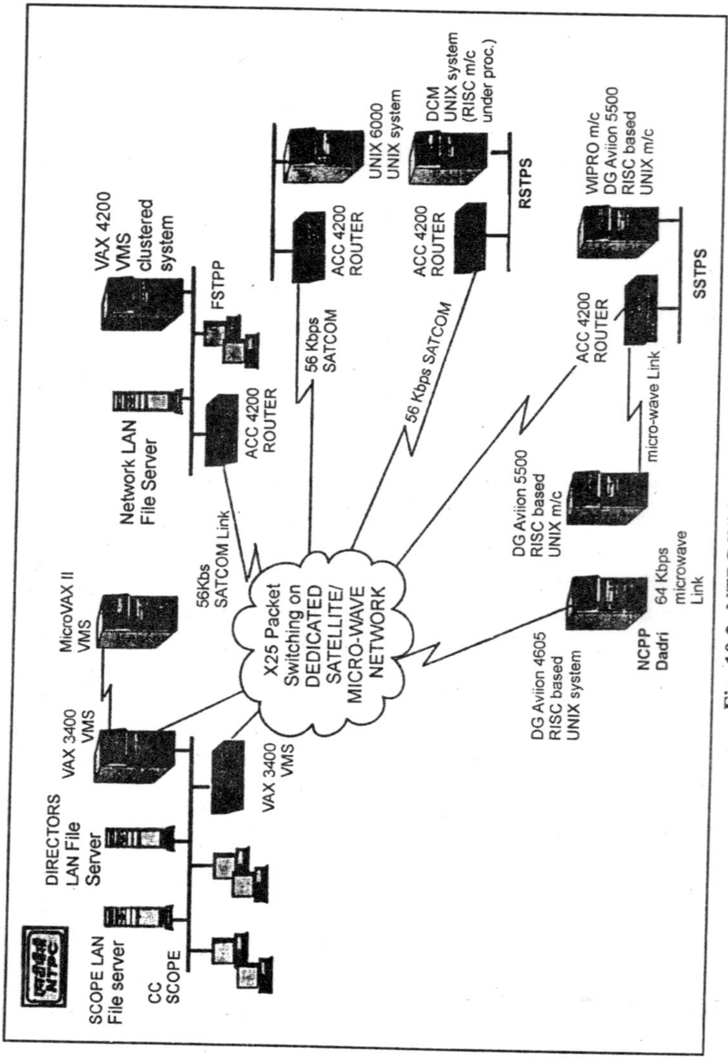

Fig. 10.2: NTPC Wide Area Network

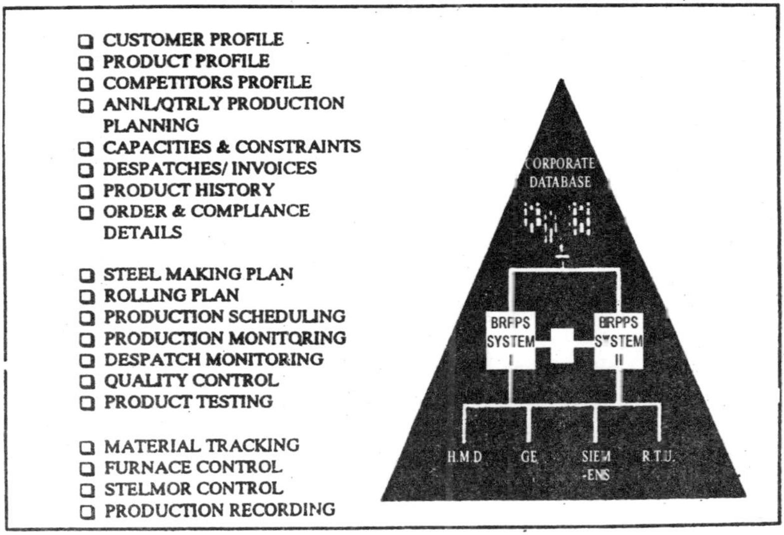

Fig. 10.3: Tata Steels Process Control Network

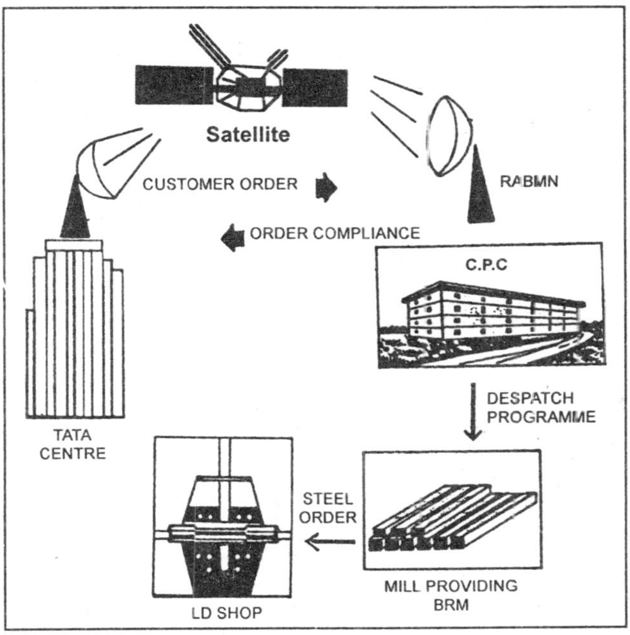

Fig. 10.4: Tata Steels Jamshedpur Linkage Overview

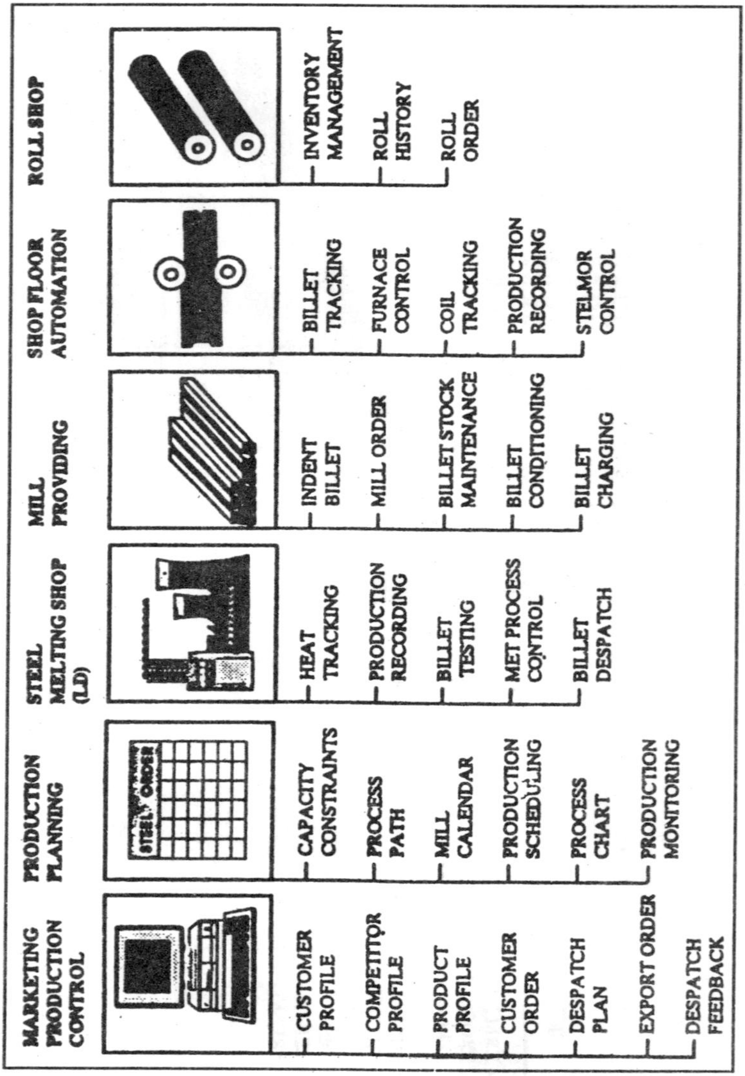

Fig. 10.5: Tata Steel's Phased Plan for IT-I

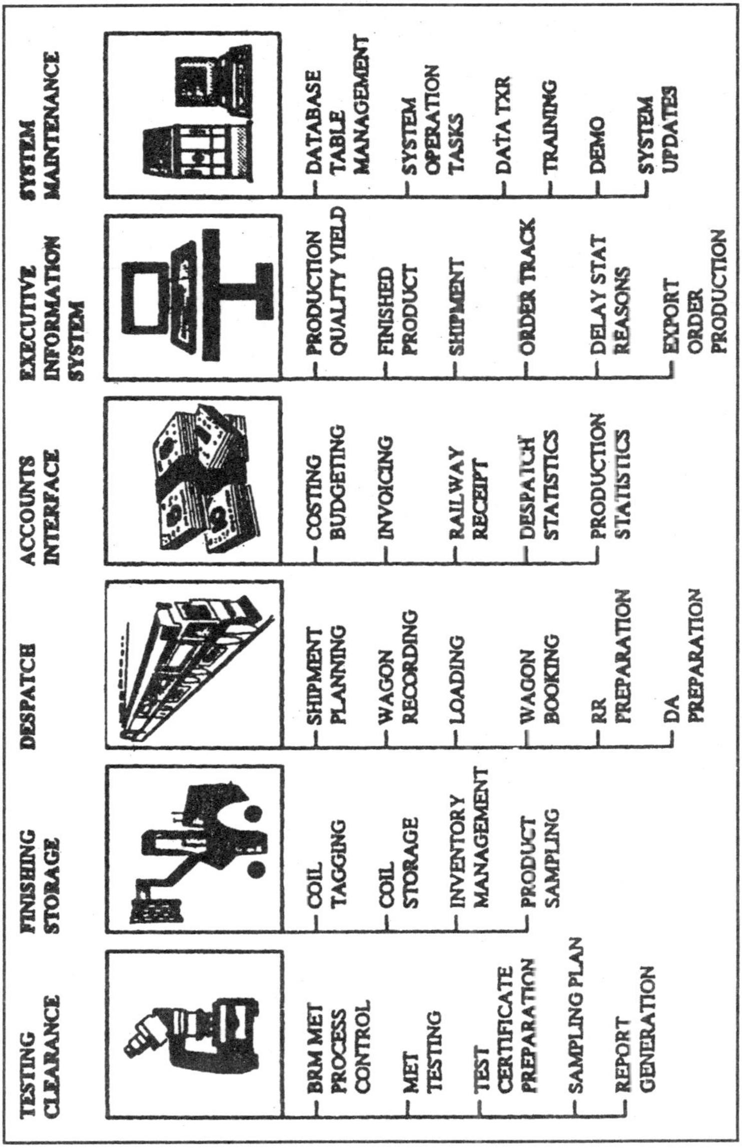

Fig. 10.6: Tata Steel's Phased Plan for IT-II

Survey Findings

The survey revealed that there was clear evidence of strongly supportive top leadership at large enterprises using IT. In the case of Tata Steel, for instance, the IBM expert-team from Australia was involved for IT modernisation strategy at the senior management initiative. The investment in IT tools was estimated between 1 percent and 1.5 per cent of the turnover, which is above the medium industry norm and seemed adequate.

It seemed at the time of the survey that the Indian industry had not reached the mature stage of IT—a phase where IT was to be used to change the very process of management, where jobs were not only automated, but also performed in essentially new ways, where IT was a strategic as well as policy tool for business. As would be seen in the next Chapter, IT as a strategic tool must catch up with strategic outlook of top executives—yet be sophisticated enough to match with their complex needs in daily routines, and by changing and influencing the very 'mission' or 'purpose' of the enterprise. Similarly, when IT was policy tool for business, it would speed up changes and perceptions in corporate policies: by instant capture of information from work-force, distant customers and company branches, and present key facts in any form to make decision-making and policy-making smooth. Policy changes could be tested ahead of time: by quickly running and comparing a dozen of 'what ifs' and 'if then, else' issues.

IT-Usage Trends

- Systematic planning evident
- Copious use of PCs
- Phasing from data-processing applications to automation of discretionary and once-off activities
- Well-planned software choices
- MIS and DSS tools increasingly used to aid decision-making
- Integrated planning and control introduced
- User-education hierarchy levels defined.

CONCLUSION

On the whole, the canvas for the large-scale industry shows a clear advanced phase over its medium and small-scale counterparts—

comprising MIS with structured and periodic management and DSS with semi-or unstructured management interaction on ad-hoc basis. Since payoff can be vast and far-reaching in the mature phase of IT, one looks forward to such an imaginative thrust from the Indian industry-houses in the present millennium—where IT would pervade the top executives and stakeholder systems.

Published in *DATAQUEST* in Vol. XVI, No.12, 30 June 1998.

11

IT AS A BUSINESS STRATEGY TOOL

INTRODUCTION

As has been seen earlier, IT has assumed today a new role, with all-round openness, accountability, networking and accessibility. As a strategic resource for business and industry, it has changed emphasis from technical and efficiency issues to strategy, management and policy issues. IT has given rise to managers who deal with information as 'knowledge workers' who are primarily end-users. There has been a paradigm shift: from the first phase of data-processing and shop-floor automation; to the second phase of MIS and DSS, to the third phase of using IT tools to change the very process of management. These tools comprise scalable platforms, 4GL languages, RDBMS, various gateways and the Internet. The new environment rewards motivation, adopts flexible organizational structure and focuses on both IT People with business skills and top executives with IT literacy.

The commitments for future management have to be for three S's: **Serendipity** to capitalize on individual to global interactions; **Synergy** for the organisations internal optimality; and **Symbiosis** for value-added inter-organisational networks. Strategic action plans must include electronic alliances, relationship marketing and creation as well as analysis on market intelligence on potential demands. Strategic networking of 'knowledge' covers an already-shrunk world under the WTO for developed and developing countries. In the new 'flat organisations', managers have to learn to coordinate activities of several "knowledge" workers, without multiple levels and work hierarchies. Information exchange will typically call for the Internet, Intranets and Extranets—applicable to the governments, corporations and basic public services. We provide a large number of examples for giant foreign banks, mammoth industries as well as huge hotel-chains, consumer-durable enterprises and even consumables for youngsters and children. Illustrations are also drawn from IT used as an 'engine' for growth.

IT Veneers

IT, over the years, has increasingly assumed a strategic role; which has gone far beyond what its progenitors would have conceived. This has been made possible for a number of veneers acquired by IT, out of which the following are easily discernible:

Openness

Information today is getting thoroughly commoditised—available, properly packaged and openly priced as a "product". Such products have the characteristics of being marketable through electronic media, regarding people, places, products, prices, programmes and projects. A classic example is the highly successful information 'product': *Encarta,* from the stables of Microsoft that provides the encyclopaedic information of a geographical kind, supplemented by rich historical, economic, political and natural information of the countries and regions of the world.

Accountability

Information 'products' have their own clear-cut accountability, with responsibility assigned for each task that brings overall clarity, identified interfaces, improved inter-person relations and assured organisational success.

Networking and Accessibility

Large databases are inter-connected today and freely accessed through telecommunication networks. Users can share and interconnect among themselves on a variety of information on: technology, trade, business, patents, travel, entertainment, finance, education and culture, to name a few. Specialists can get interconnected to enhance their 'knowledge' and share views. The best example is the Internet that operates as a *network of networks*, even if the latter arte dedicated or proprietorial in kind.

IT AS A STRATEGIC RESOURCE

IT is fast becoming an integral part of overall business planning: understood and managed at senior and top levels (including the board level) for management strategising and policy-making. The shift in emphasis has been from the earlier technical and efficiency issues to the more current business strategy and management issues. Right at the policy level, IT can operate as a catalyst: helping to achieve fundamental changes in managing business, which has to be *Right the first time, every time.* As we shall see, the end-users have become of paramount importance, who today is managers and 'knowledge' workers: dealing with information per se. For such usage, pure 'data' is replaced by "information", which is often further advanced to 'knowledge', not far short of 'wisdom'.

The above paradigm shift in user-environment has been in keeping with the general transition phenomena in society. As is well known, the pre-industrial society of farmers was based on land, labour and capital. The industrial society, comprising business and industry organisations, has the base of labour, capital and information—with the latter quietly replacing land. The post-industrial society, consisting of service organisations, is now witnessing an inexorable move to capital and information—with far less importance attached to labour. The now-unfolding further transition is towards information society, which covers leisure organisations, where the only dominant factor is information.

In the transition process of increasing importance for information, there have been a few phases. In the first phase, *data processing* handled routine accounting and accounting-oriented activities; *on-line query* dealt with capture, processing and transmission of fast and instantaneous data and information; and *process automaton* was concerned with large and repetitive jobs, like clerical record-keeping and shop floor systems. Payoffs in the first phase were simple and immediate. In the second phase, Management Information System (MIS) came into being with structured and periodic management reports, followed by Decision Support System (DSS) with semi/unstructured and occasional/ad-hoc management interaction. Both MIS and DSS relate to decision-making activities where 'screened/filtered' information is required: using selective, analytical, interlinked and integrative approach. *Automation* was attempted for random, discretionary and non-repetitive activities. Payoffs in the second phase have been subtle and complex. In the third phase, IT tools are getting used to change the very process of management itself: not only to automate the jobs, but also to perform jobs in essentially novel ways. Examples of these IT tools are: front-and back-office banking software; hardware systems with scalable platforms; relational databases and flat files; process control of any on-line transactions; and audit trails. Such tools are supported by interfaces and gateways, involving efficient message exchange system systems and the Internet.

Case 1: In a multi-national bank, IT tools were used for effective international settlement and global custody management service for cross-border investment strategies; especially formulated for the individual client. The bank also offered necessary functionality to closely monitor and manage domestic custody and sub-custody of money, material and documents. The important point to be noted is not the availability of the technology, but what the bank was doing with it. The lesson learnt was that technology has the capability to free us for: broadening the scope of the banking jobs; amplifying human potential and immensely enriching the workplace (by, say, projecting many possible future scenarios).

DEVELOPING COMMITMENT TO IT RESOURCES

The relevant development issues are the following:

Encouraging Environment

There should be an ambience for incentives and rewards as motivation, not threats and punishments.

Flexible Organisation Structure

The seniority-juniority hierarchy should not reflect a rigid architecture.

High Quality Skill and Interest

There should be a two-way traffic: IT people having strong business skills should intermingle with top/senior executives possessing IT literacy.

Human Factors

The major involvement should be with re-education, re-structuring jobs and necessary changes in job-roles. These factors are the most important IT resources.

Consistency in Management Style

Both the corporate leadership and the IT head should champion the cause of changeover and induct the 'culture'.

'3-S' OF FUTURE MANAGEMENT

For the emerging IT-executive interfaces, there is a need for *Synergy* of intra-organisation; *Symbiosis* of inter-organisation kind; and *Serendipity* of individual to universal kind. These are strong bonds and are exemplified below.

Case 2 (Synergy): Rollatainers, a leading laminating, tetra-packing and vacuum-packaging enterprise, developed a sophisticated vision of "one customer segment" across its many departments. Its Art department, for instance, performed the CAD function—tightly coupled with the marketing department, which brought in feedback and feed-forward of design information from customers. When the Indian skipper Kapil Dev made a world record of capturing Test wickets, there was a market-perception that his picture could appear on cigarette-packets, in order to capitalise on the public craze for him. This needed instant scrapping of a finalised design made earlier and overnight appearance of Kapil's photo on select packs! This was also tantamount to what is termed as the classic 'one-stop non-stop service'.

Case 3 (*Synergy*): Levi, the world-leader in Jeans manufacturing, transited into 'mass customisation' stage from the earlier two stages of 'individual production' (i.e., tailoring a garment, made-to-measure, as per individual's sizes) and 'mass production' (i.e., producing en-masse, using assembly lines). In 'mass customisation' stage, they could obtain an individual's sizes and special choices through IT in advance and still incorporate them in their assembly lines: to come out with products, which satisfy specific preferences.

Case 4 (*Symbiosis*): National Informatics Centre (NIC), the largest India-government IT network among the Centre and all the States, established 'close user-groups' on value-added network system (vans) on the ubiquitous NICNET, where the users could interact with each other and gain by sharing information, accessing consumers, etc.

Case 5 (*Symbiosis*): Reserve Bank of India's RBINET tied up with most BANKNETs (e.g., SBINET for the State Bank of India) and the proprietorial Society for Worldwide Interbank Financial Telecommunications (SWIFT), to exchange documents, share information and gain all-round knowledge.

Case 6 (*Symbiosis*): All the world's airlines used the proprietorial Society of International Traffic in Airways (SITA) to share and exchange, on a global basis, their passenger and cargo information for bookings, cancellations, transfers, etc.

Case 7 (*Serendipity*): The world's music industry, of late, downloaded music products—regardless of their points of origin—and sharing customers, clients, prices and quality on the Internet. While potential is endless and sky is the limit for individuals to go for global interaction, this has given rise to some thorny problems of 'intellectual property rights' (IPR), as discussed in Chapter 8-end.

STRATEGIC ACTION PLAN

IT tools are used nowadays to forge new kinds of strategic actions like 'electronic alliance' and 'relationship marketing'. Such actions involve databases and analysis on: clients, industry, products, prices, markets and quality. One of the prime constituents is *market intelligence* on potential demands that is dovetailed into future strategy.

Case 8: In Japan, Fujitsu established 'Laboratory of the Future' for research into emerging technologies and decide which ones out of them WERE worth pursuing for putting into production. The laboratory is located in Gotemba, in the picturesque foothills of Mount Fuji and is open to visitors.

In the strategic networking of 'knowledge', economics is also being networked and integrated, along with cultures and customs. On the whole, the world is shrinking—by obtaining a larger reach and evolving common standards—for trade and technology, for everyone to benefit from.

Case 9: The World Trade Organisation (WTO) provided an official forum for deciding on agricultural pricing strategy among developing countries like India and China, on one hand, and EU and the USA, on the other. The dialogues, yet to be concluded, included anti-dumping policy, tariff barriers, protection to farmers and pricing mechanisms, among others.

NEW MANAGEMENT STRUCTURE

The new managers have begun to coordinate activities of a large group of 'knowledge' workers, without necessarily multiple levels and work hierarchies. This is leading to increased productivity of new managers: bringing world resources at their information terminals, for improved costing, pricing, delivery and markets.

Such a structure is quite akin to an Orchestra, where the Conductor gives a sense of direction to coordinate and consolidate a large group of musicians. Neither does he spoon-feed them on their individual functions nor does he go through any intermediate levels.

NEW ORGANISATION STRUCTURE

In keeping with the new management philosophy, organisations have emerged which are neither hierarchical, nor do they need command and control from the top to reach people at the bottom performing routine jobs. Knowledge and information is available today at the bottom, hence the functions at the top can be limited only for communication and coordination.

These are essentially 'flat' organisations—with the broad working levels able to carry on with the IT-aided tools and the top confined to a limited coterie. This is in direct contrast to 'command-and-control' structures of the defence forces where several hierarchies exist to help percolate orders and interpret instructions.

Case 10: National Thermal Power Corporation (NTPC), a large *Nava Ratna* body, runs 11 coal-based and 7 gas-based thermal projects. NTPC had a strong hierarchy of functionaries: beginning with Assistant Engineers, Engineers and Senior Engineers at the lowest echelon; Assistant Managers, Managers and Senior Managers at the middle echelon; and General Managers and Directors at the top echelon. Now, with more IT-savvy Engineers making their entry at the junior levels, many of the erstwhile specialised jobs like Power Management System were virtually run at these levels, making much of the middle rungs superfluous. NTPC had to assess urgently how far, and how much, they would aim at a 'flat' organisation, in keeping with several such organisations emerging in the West.

On balance, new 'knowledge' organisations are run with 'knowledge' workers. The latter handle jobs with increasingly large information-content (in keeping with complexity) as critical components of their job. The large IT element, in turn, makes possible for new organisations to network their offices: locally through LAN and enterprise-wise through WAN. With the advent of the Internet, organisations are setting up 'Intranets' within the organisation (subsuming LAN and WAN), using TCP/IP protocols. The same philosophy is extended by two or more organisations for setting up 'Extranets' with identical protocols, but making them proprietorial and erecting 'firewalls' against information access by outside bodies. LANs, WANs, Intranets and Extranets favour today's organisations for the widest possible sharing of information—benefiting one and all.

ROLE OF GOVERNMENTS, CORPORATIONS AND PUBLIC SERVICES

Governments are shifting today from being 'controllers' into 'facilitators', because of the power of IT's geographical stretch. Fluidity of knowledge as well as fuzziness of geographical territories call for self-governance, skill building, self-discipline, mass education and institutional ethics. The wide prevalence of 'e-governance' in States like Andhra, Karnataka, Kerala, Tamil Nadu and elsewhere is a step in the right direction—calling for devolution of controls and delegation of power, with gradually reducing physical governance.

Corporations are set to redefine organisational processes so that they take advantage of 'consumers' whose preferences are defined as follows:

- Not by their orders within which they are identified and located;
- But by quality and quantity of information, regardless of their origin.

Public services are changing totally the concept of 'service': by utilising the power of IT and covering wide cross-sections of people served by the following:

- Learning about customer needs, based on available infrastructures; and
- Responding to customer-needs across the society: *better* and *faster* than before.

In providing such service, every customer is considered unique and treated as 'one customer segment'. At the mass consumption and mass supply levels, this implies: 'one-stop non-stop service'.

Case 11: In AP, 'information kiosks' were envisaged to provide 'one-stop service' to common citizens for birth/death/SC/ST/land records certificates. The idea, inherent in 'e-*seva*' project, was likely to be replicated in several other states.

IT APPLICATIONS FOR LONG-TERM RELATIONSHIP

The major point that needs reiteration is that IT has outlived its role for sales promotion, i.e., to work out temporary incentives to people for buying more of the same products. It is now envisaged, as illustrated below, for building long-term relationships, i.e., for 'brand building' to create new preferences and then satisfying them. Prime candidates for the latter two are new 'brands' of TV sets and cell-phone devices (equipped with digital camera) emerging everyday.

Case 12: Large foreign banks introduced, as a matter of routine, financial services for their clients that covered: investment portfolio; share baskets; and secured loans. These were typical 'one-stop non-stop service' using the concept of 'one customer segment'.

Case 13: Industry giants like Tata Steel that deal with consumer durables (medium and light structural, sections, joists, beams, angles, etc., in this case) introduced 'consume profile', 'product profile' and 'industry profile' databases for their steel products to keep track of the preferences and choices of the consumer industry and specific customers and anticipate their needs in a time-specific and location-specific manner.

Case 14: Large hotel chains like Taj Group and Oberoi Group initiated focussed cultivation for their frequently visiting guests in terms of their preferences for rooms, floors, food and beverage, so that their individual needs could be anticipated when they arrived. These were culled out from customer databases, as methodically built up from guest comment cards, feedback cards and hotel registers. Even smaller chains like Swoti hotel of Bhuvaneswar were using the idea.

Case 15: A medium-level software firm like Softek introduced the innovative idea of providing free updates for enhancements and new releases of their software packages, namely, Softword, Softbase, Softcalc, etc., which were much appreciated by their customers and helped them to procure new orders.

Case 16: Shyam Ahuja, the reputed designer and manufacturer of carpets, *durries* and related products, initiated databases of 4500 customers for making customer-friendly offers on their consumables for houses and apartments. These included first choice in 'Dewali' sales, free samples through core-decorator members and monthly check-up visits—backed up by 'Decorator Clubs' and special sales agents.

Case 17: Hawkins, the well-known pressure-cooker firm, built up a strong database on consumables for housewives, from their guarantee cards and sales records—to introduce Hawkins Miniatures, Hawkins Cookbooks and home-visit of cooks to demonstrate novel recipes to housewives.

Case 18: Parle Agro, the Frooti manufacturers, created a solid database for consumables sold to youngsters and brought in Frooti Function Clubs for: visits to plants; animation film shows; quiz contests; free take-away; and badges with wacky slogans.

Case 19: Barbie Dolls, the celebrated makers of consumables for children, especially young girls, went for 12000 active members in the 5-12 age group and had a 21000-strong database for both direct marketing and telemarketing. They encouraged children to write to 'Barbie' (a surrogate auntie) who would reply to every letter! This made for fun correspondence for children while learning, increased their involvement-levels and created attachment to new Barbie characters.

IT AS AN ENGINE FOR GROWTH

The role of IT as a 'growth-engine' can be clearly seen from its following detailed functions:

Tracking Facts on Transportation

- Giving up-to-date information on total 'route utilisation' for passengers and aircraft;
- Giving up to-date information on total 'route demand' for passengers and aircraft.

For the airlines, this is to improve equipment utilisation, crew scheduling and re-fuelling operation,-to manage price and reap in economic benefit.

Case 20: American Airways practised for long the last-minute passengers to be diverted to their next flight at a cheaper tariff, provided, she gave up her stake to board the designated flight.

Effective Personal Incentives

- Continually tallying sales commission and allowing salesmen to access their own records;
- Identifying and tracking contribution of a working unit (division, section);
- Fast information tracking to tell the enterprise whether work is on schedule, to meet the turnover quota,

This is applicable to any industry with a large sales force and distributed office networks.

Tracking Facts to Boost the Sale

- Spreading the sales network closer to customers at 'point of purchase' (POP), without relinquishing top-level control and automation:
    1. To use computers for executing transactions at remote sites,
    2. To instantly post them at centralised files, and
    3. To shift the POP nearer customer, while retaining timely records for access by top management,
- Giving opportunity to reshape customer relations through 'electronic ties'.

This is applicable to all large-scale process and engineering industry where centralised files are maintained, while keeping up the decentralised databases at the distributed sales force level.

Solutions for Production

- Installing sophisticated monitoring systems:
  1. To track errors per hour,
  2. To flag equipment downtime,
  3. To measure machine speeds,
  4. To assess workers' productivity;
- Allowing managers to remedy production problems in time before they become disastrous;
- Installing IT plus computer-aided design (CAD) and computer-aided manufacturing (CAM:):
  1. Specification of components to be expressed in standard geometric terms,
  2. Materials/parts specifications to be transmitted with total precision direct to suppliers/subcontractors for core equipment control programmes;
- Flexible automation:
  1. First-or second-generation robots,
  2. Computerised numerical control (CNC) machines.

These considerations are applicable for aerospace, automobile, electrical, electronic, heavy machinery, metal fabrication and casting industry.

Adaptation to Change

- Data architecture helping enterprise for:
  1. Adaptation to regulatory and other environmental changes,
  2. Fast response to any environmental change.

These factors are applicable to thermal/hydel power stations, gas generating units and agro-based industry.

Meaningful Budgeting

- Using IT to speed up budgeting process and examining:
  1. Wide range of 'what if' scenarios and comparison of outcomes.
  2. Immediate consolidation of unit budgets into overall 'corporate financial plan'.
  3. Ability to coordinate action under many alternatives, with small staff.

These points are useful for consumable and consumer durable industry.

Distributing Data

- Fast funnelling of new price information to sales office terminals,
- Monitoring customers' buying patterns at retail outlet terminals,
- Using IT for tightening corporate control:
    1. To lower inventory levels
    2. To boost sales turnover
    3. To match product-mix with customers' changing tastes
    4. To shift products from one or more stockyards
    5. To offer special sales promotion in a timely way.

These features are applicable for steel, cement, pharmaceutical, oil, fertiliser and coal industries.

Centralising Data

- Creating product centralised database at corporate office, containing details of every product sold and sending service representatives briefed in 'service history' for:
    1. To provide excellent service,
    2. To spot trouble before it gets out of hand.

These ideas are useful for truck, tractor, machine tool and heavy engineering enterprises.

Consolidating Data

- Suggesting customised profiles with customer details for:
    1. Better insurance coverage,
    2. Better bank service,
    3. Better investment suggestions.

Financial institutions can gain in long-term lending and medium-term lending services, while banks can benefit in commercial and competitive services.

CONCLUSION

As illustrated above with a number of cases, applications of IT as 'strategic resource' for business and industry are fast emerging as part of the new mindset of the computer-numerate managers and executives. For this generation of practitioners, IT is part of their upbringing and devising strategic applications with IT has to become a day-to-day policy and planning process.

Published in *CSI COMMUNICATIONS*, 2005

PART FIVE

WIDER MANAGEMENT ENVIRONMENT

12

EXECUTIVE COMPUTING: PLANNING AND SUPPORT

INTRODUCTION

The professional systems environment had always the ability to provide extensive support to the corporate enterprises and meet all their hardware and software requirements, But, since the advent of computer PCs in the 1980's, there arose computer-literate customers, as end-users, who were consumers at the less expensive end of the market, often new to computing and needed more hands—on help from the systems professionals belonging separately to what was 'EDP department', 'Systems department', 'Management Service' department' or 'MIS' department (referred as 'host computing' or simply 'host'). The new generation of end-users—mainly at the PC end—constituted what was termed the environment of *end-user computing* or *executive computing*. The latter was initially devoted to simple jobs involving word processing, spreadsheet, data base usage and PowerPoint presentation. As time advanced and customised software was increasingly available for this category of end-users, more complex jobs began being handled by them. In large organisations like NTPC, for instance, the entire Power Management System were handled and maintained by them, with marginal support—mostly in emergencies—by the Systems department. In the USA, *Information Centres* tended to appear when a 'critical mass' of 10 PCs was installed. Even small firms have gone far beyond that 'critical mass' today and the information centre is now a ubiquitous part of corporate life. In companies where information centres are not there, hosts serve the same purpose. The high point of executive computing is reached when it develops well-defined links and shares specific responsibility with host computing of Systems department in a client-server mode, as explained in Chapter 4.

LIFE CYCLE IN EXECUTIVE COMPUTING

Executive computing is still in an early stage of its life cycle. Just as people go through the life cycle of birth, childhood, adolescence, adulthood and death, most institutions also experience a series of changes over time. According to the life cycle concept, executive computing—as a specific product—undergoes five distinct life stages, as follows:

Introduction: with slow growth and uncertainty. Executive computing went through this stage in the early 1980's;

Growth: when the innovation is adopted, with the rate of growth increasing by the year;

Early Maturity: with growth continuing, but the rate falls every year;

Late Maturity: with negligible growth;

Decline: with executive computing decreasing with time and with marketing practices becoming more sophisticated to keep up with the end-user.

CRITICAL SUCCESS FACTORS

The information centre (or host, in its absence) is seen by the user-executive in terms of the services it offers. The *primary* services are consulting, training, trouble-shooting and research for selecting hardware and software. Of these first-tier services, only consulting and training are elaborate and time-consuming. The *secondary* services are application development, data extraction and news dissemination. Out of them, application development remains often a bone of contention between executive computing and host computing. In other words, getting users to develop their own systems have often faced difficulty at the user end.

The problems of executive computing fall into three major categories as follows:

Policy Problems: such as, lack of top management support, resistance by host computing and poor definition of responsibilities;

Service Problems: such as, high user expectation, inappropriate use of the centre and inability to reach top management;

Internal Problems: such as, high staff turnover and hardware incompatibility.

The following areas have been generally identified as of principal concern:

- User training — Security
- User awareness — Software limitations

- Management support
- Staffing
- Budget
- Hardware limitations
- Technical support
- User resistance.

The important factor, from the user point of view, is fast effective service. Users want help when they need it, look for knowledgeable assistance and wish to interact with people who speak their language and are service-oriented. The focus of this service is *trouble-shooting*, which reflects the major concern of users, having already selected their hardware and software, and having received their training. Today most end-users are in this position.

At the same time, information centre (or host) managers have such concerns as training, staffing, product selection, top management support and promotion of end-user computing. In general, host computing must neither be dominated by intense demands for help in specific applications, nor fail to give them support altogether.

BASIC STRATEGY OF EXECUTIVE COMPUTING

Cutting across many long-term and short-term decisions, a corporation needs to work out a strategy to establish the 'basic personality' of the end-user computing, based on the following questions:

- When should host computing be used, when should PCs be brought in and how should the two be viewed and interlinked in the long run?
- How does a corporation decide how much to spend for end-user computing in general and for specific applications?
- Should the management's role be to accelerate the end-user computing, control it, or do both?
- How should the end-user support be financed—directly by the organisation as an overhead expense, or, through charge-back?
- Should host computing develop applications for users? If so, under what procedures?

When, in the 1980s and 1990's, *host computing*—comprising data applications (retrieval and application development), analysis and graphics on mainframes—faced the explosion of PCs, relative advantages of hosts and PCs were often animatedly discussed and debated on the following lines:

Mainframes and Mega-minis

Processing Speeds: Mainframes or mega-minis with high MIPS could do processing jobs not feasible on PCs. Advanced PCs have eroded this advantage of mainframes.

Multi-user Operation: It meant that end-users, scattered across a corporation, could work together to develop a budget or keep track of product changes in a database. This multi-use capability continues to be needed.

Access to Corporate Data: This continues to be critical, as a large chunk of all applications need access to corporate data.

Simpler Technology: Hosts are not inherently simple, but the complexity of the hosts can be hidden from the end-users who do not have to configure the computer, add PCBs, or learn complexities of the operating system.

Operator Assistance: This is important in maintenance-type work, such as, taking backup, restoring archived files, loading printers and scanners, installing new software, adding updates to software and diagnosing problems. Few PC users are aware or even concerned about such activities.

PCs

Consistent Response Time: Action takes always the same amount of time, unlike the large hosts, whose processing power: being distributed over many end-users—changes from time to time. Also, executives do not suffer from low priority, lest the performance of the host system is degraded.

Simplified day-to-day Operations: End-users are not involved in the complex administrative work needed for the host computing. Also, application software is usually simpler than host software.

On balance, PCs are best for personal productivity applications that are often highly interactive and where efficient application software is available. In contrast, hosts should handle applications, needing vast amount of processing. Another consideration is of *data:* with PCs being preferred to purely personal or departmental data. Moreover, PCs—in the PS/2 Model 80 and Macintosh II range and beyond—now have so much computing power that the processing advantages of the host systems are being eroded rapidly. Again, hosts and PCs are getting converged towards a multi-user technical support structure, where PCs are standard workstations,—with hosts acting as file-servers for some applications, and the main processing source for others. Chapter 4 refers in this context.

COST-BENEFIT ANALYSIS FOR EXECUTIVE COMPUTING

According to some authorities, end-user computing should be restricted to applications in which quantifiable cost-benefit justification is available. According to others, line managers may be permitted to make decisions based on semi-quantified benefits. On the whole, there is a problem of end-user computing being unable to quantify most benefits, like a majority of business decisions.

Costs

Hardware costs: For PCs, base-prices are often quite low, compared to the add-ons that determine the ultimate price of the machines,—including peripherals and expansion boards.

Software Costs: These include all packaged software and costs to develop internal customised procedures for such packages.

Overhead Costs: These include maintenance charges, stationery costs and financial charges on capital purchases.

Human Costs: These are training costs, development of procedures and support costs.

Tangible Benefits

External time-sharing costs are often replaced by internal computing, with clear-cut savings. Though actual quantification may often be difficult, reasonable 'what if' analysis can throw up probable savings. *Time-savings* on host computing can also be justified even for 10 per cent of end-user time released by PC-based computing.

Intangible Benefits

Intangible savings are those like customer goodwill: gained by processing information rapidly and reducing backlogs. IBM lists three levels of justification: *Efficiency* at the lowest level for which costs and benefits are known; *Effectiveness* at the next level leading to better and faster decision-making; and *Innovation* at the highest level where the very way of individual or corporate working is restructured.

APPROVAL STRATEGY FOR EXECUTIVE COMPUTING

Experts have brought out the following steps in seeking approval:

- To identify primary applications;

- To name the end-user for each application;
- To tell how frequently each application will be used;
- To list software packages and see if they serve the need of end-user applications;
- To state how much time the PC will be used every month; and
- To note the benefits.

Quite often, the actual approval process consists in forcing end-users to think through their applications, rather than evaluate individual cases.

Support Tactics

These are recommended by experts as follows:

- To promote information flow with intense training and forming end-user groups;
- To subsidise costs to end-users, giving hardware and software free;
- To speed the acquisition process and provide a wide spectrum of good products; and
- To provide good user-support with consulting, training and finding solutions to users' problems.

Support services can be financed in two ways: (i) offering services through overhead, not charging the individual user-departments; and (ii) using charge-back, in which individual user-departments pay for the services received.

Control Tactics

These are advised by experts as follows for controlling acquisition:

- To restrict choice and use veto power;
- To forbid certain types of acquisition; and
- To offer a range of software support on the host, if mainframe is to be encouraged,

As has been observed, support and control are not mutually exclusive, but are independent dimensions of the corporate policy: with moderate degrees of both.

DEVELOPMENT STRATEGY FOR EXECUTIVE COMPUTING

Whether or not host should develop end-user applications and disseminate them is a hotly-debated issue. The main reasons for not developing end-user applications are the following:

1. It breeds dependency, instead of self-reliance among end-users.
2. There can be resource constraints, since all information centres (or host) operate on tight budgets.
3. There could be a legacy of backlog of data-processing applications with the hosts, and using time for developing end-user applications could be construed as counter-productive to host's original purpose.

In contrast, there are many strong reasons for offering application development as follows:

1. It can be more efficient for the host to do application development and delegated development would make sense for only if the end-user has one-time requirement.
2. If the information centre staff is active in application development, a great deal of money can be saved by producing mid-rang programs or by fine-tuning existing programs.
3. Application development can make the information centre more knowledgeable both of their products and of end-user needs.

On the whole, an important reason for doing development, especially for senior executives, is the impact of the development work on the image of the information centre. Quite often, it is a question for information centres (or host) of when, not whether, to do the application development. In any case, application development should not be considered the normal form of application support, but backed by a firm policy of requiring end-users to become more and more self-reliant. Eventually, the delegated development by end-users for other end-users can emerge to be a normal process and become the most prevalent development practice.

PLANNING STRATEGY OF EXECUTIVE COMPUTING

Priorities need to be set on the basis of user-needs, technical capabilities and broader organisational policy. This process can take the following shape:

Mission

It should first be decided broadly what is within the purview if planning and what is not.

Goals

Under broad directions of mission, measurable goals, such as, number of users to be trained, reduction of response time to a specific extent, etc., should be envisaged. Such goals are useful for evaluation purpose, later on.

User Needs

A fine degree of assessment of end-user needs is required even in a rapidly changing scenario: through periodic meetings.

Technology Forecast

This should cover price forecast for hardware and software, including prices for PCs, hourly charges for the host and networking charges. New products should also be factored in. A demand forecast is essential for corporate capital planning.

External Environment

Viewpoints of the information centre and/or the systems/MIS department and the corporate management are essential for good planning of end-user computing, since budgetting and planning are basically top-down activities in practice.

Actions

Based on goals and objectives on one hand, and user-needs, technologies and external environment on the other, programme of action should be laid down, including service offerings, training, purchase scheme of hardware and software, and numerous other details.

Resources

Staff, equipment and money should be budgeted. On the whole, needs, technology, environment, objectives, resources and actions are to be balanced by political negotiations within the enterprise. Higher management support is important for a compromise of different views and deciding on final resources.

IMPLEMENTATION STRATEGY FOR EXECUTIVE COMPUTING

Measurement, feedback and *accountability* are essential steps in an implementation process, keeping in view the following facts:

- Objectives can be scaled back;
- Resources can be increased; and
- New plans can be drawn to make more efficient usage of resources.

While planning refers to broad decision-making about what should be done, budgeting refers to establishing specified norms for funds and tracking actual spending. Planning is done first and detailed budgeting follows on the basis of planning decisions.

CONCLUSION

It should be noted that while most large organisations have information centres, smaller organisations may not have the same, but a communication process for the Systems/MIS department and corporate management must carry the entire organisation along, regarding the importance of end-user computing. Such a communication process should pay full heed to overall operations, namely, background and mission, terms and definitions, roles and responsibilities, strategic direction, education, data security and availability, package selection, package support, accountability for usage, and feedback considerations.

13

BUSINESS PROCESS OUTSOURCING (BPO): IT IN PUBLIC SERVICE

INTRODUCTION

Business Process Outsourcing (BPO) has had a burgeoning growth in recent years, especially for the English speaking market of India. This covers a whole range of client-assigned services, ranging from the low-end of medical transcriptions and call centres to the high-end of computer programming, architectural designs, life science databases, legal processing and animation arts. India's advantages are high density of English-language speakers; skilled workforce; cost-effective remunerations, emerging multimedia technology knowledge; open door policy and time-zone advantage.

BPO can be: internal outsourcing; partial outsourcing or complete outsourcing. Scores of companies have emerged at low, medium and high ends of BPO in India, where customer services offered by the **Service Provider** to the external **Client Organisations** follow clear-cut procedures. For third-party outsourcing and own back offices. Benefits of BPO can be both intensive and extensive: for both clients and service providers, with equally remarkable spin-off benefits to a wide range of ancillary services. There are problem-areas, too, for clients and service providers. Keeping in mind these considerations, specific S-W-O-T analysis can be done for both clients and service providers,—to see why BPO is gaining roots in the UK, USA and elsewhere with an astounding cost advantage to the clients: now allowed to concentrate on their core competence.

On the whole, BPO leads to a 'Win-Win' situation. where the client gains in substantial cost reduction and consequent profit enhancement; and the service provider gains in substantial foreign exchange earning and generation of employment. The recent spurt in BPO in the high-skill areas has only increased the all-round validity of the BPO philosophy.

THE CURRENT SCENARIO

Job Displacement

The current buzzword for jobs on the other side of the Atlantic is that it is 'destination India'. As the US companies are in a desperate rush to cut costs, one out of every 10 jobs in the US software industry has reportedly moved towards to such inexpensive emerging markets as India by 2004-end. Table 13.1 indicates the year-by-year status estimated for US jobs moving to India.

Advantages that India offers, as perceived in the West, are summarized as follows:

- 2nd largest English speaking country
- Availability of large skilled workforce
- Quality and cost effective workforce
- Growing need to be more cost-effective
- Emerging multimedia and e-commerce technologies further pushing call centres
- New technologies offering improved means of calls transportation
- Open-door policy
- Time-zone advantage and
- Higher company loyalty.

Table 13.1: US Jobs Moving Offshore to India

Job Category	*2000*	*2005*	*2010*	*2015*
Management	0	37,477	117,835	288,281
Business	10,787	61,252	161,722	348,028
Computer	27,171	108,991	276,954	472,632
Architecture	3,498	32,302	83,237	184,347
Life Sciences	0	3,677	14,478	36,770
Legal	1,793	14,220	34,673	74,642
Art, design	818	5,576	13,846	29,639
Sales	4,619	29,064	97,321	226,564
Office	53,987	295,034	791,034	1,659,310
Total	**102,674**	**587,592**	**1,591,101**	**3,320,213**

Source: US Department of Labour and Forrester Research Inc.

Recent Illustrations

According to the consultants Gemini Ernst and Young, 'Offshore outsourcing is no longer about call centres and low-content activities; it is about skilled work, too.' The following examples bring out this point for both USA and UK:

(i) *HSBC Bank, Lloyds TSB, Prudential, National Rail Enquiries and Bank of America, USA*: Joint plans to slash costs by moving thousands of jobs to India;
(ii) *Oracle, USA*: Plans to move 2,000 developed jobs from the US to India;
(iii) *Hewlett-Packard, USA*: Plans to close customer-service operations in Florida and send the operation's 1,200 jobs to India; and
(iv) *Forrester Research, UK*: 750,000 British jobs being moved offshore over the next decade.

Besides, 'International Partnership on Hydrogen Economy' (IPHE) has been launched recently, where the US is leading a group of 14 nations to a joint protocol, intended to make hydrogen a commercially viable future fuel. It is likely that India's cost-effective R&D facilities (in contrast to the costly intellectual capital standards of the US and European Union) will be made central in this partnership: slated to spend $1.7 billion by US alone in 2004-09 on production, storage and application to power its cars on hydrogen.

Outsourcing Today

Outsourcing is one operation that combines voice communication, data communication, data processing, video communication and allied technologies to enable a parent organization ('client') to entrust implementation of critical business strategies on development tactics to a partner organization ('service provider'): aimed at reducing costs and/or increasing revenues. The other two closely linked objectives are demanding and getting world-class service, and reducing the backlog of activities that the clients need not undertake and can offload with advantage to the service provider. This can be: *internally outsourced* (in the client's own back-office), or, *completely outsourced* (entirely a third-party service provider).

Instead of face-to-face real-time communication, outsourcing depends on a *broadband of connectivity*, such as. Telephone (as the "front door");

the Internet and WWW; multimedia call centres; VSNL-sold customer service centres and leased satellite links. Since the late-90's, these have worked through toll-free services; inbound call centres; outbound call centres; automatic call display (ACD); interactive voice response (IVR); fax on demand; computer-telephony integration (CTI), the Internet; e-mail; and voice over IP (VoIP). The number of people employed by 'service providers' has gained ground exponentially from the figure of 18 million in 2001. Web-enabled service providers are growing annually by 11.4 per cent by 2004 and the money spent has jumped from $23 billion in 1998 to $59 billion in 2003.

Table 13.2 indicates an illustrative list of service provider companies, functioning currently in India. Many like GE Capital, DELL, American Express (AMEX), Citibank, World Bank and Hughes run their own back-offices in India. Others like Teletech and Convergys are third-party operations. There are a few companies with mixed operations, too.

Table 13.2: Service Provider Companies

Bechtel	Ansals
HCL	Infosys
Data Tree	IBM
Healthscribe	Talisma
Selectronic	Daksh.com
CBay India	I & A International
British Airways	Care Technologies
IDLX	Sankhya Infotech
Cybiz Tech	Infotech Enterprises
Saffron	Stracon
Minerva	Hero Corporate Services
Flex	HDFC
WIPRO Spectramind	GE Capital
DELL	Tata Elxi
Accenture	Progeon
eFunds	Exel
Citibank	World Bank
Amex	Teletech
Hughes	Sutherland Chem.
Convergys	TCS

Source: Fonet Consultants & others

NATURE OF BPO

1. Customer Service

This is a primary and indispensable caring need, ranging over help-desk, *IVR* (computerized voice-recording which guides the caller through a series of menu-options); *ACD* (receiving calls from telephone or the Internet and distributing them among the available agents of service provider); CTI (integrating computer technology and telephone systems), page pushing (also done by CTI); text-based chat; call-me buttons; and e-mail and multimedia calls). Medical transcription is an important instance in point.

2. Transaction Processing

This can be applicable for purchase-orders; reservations; loan-applications; mortgages'; insurance-premium payment; insurance monitoring; banking activities; etc.

3. Tele-marketing

This can involve credit-card marketing; home-shopping (or catalogue-shopping); Cell phone scheme marketing; and a variety of promotional activities.

4. Payroll Processing

This is an outsourced feature of GE Capital.

5. Database and Knowledge Management

This is a recent development offering and selling databases relating to a lot of fields, areas and domains. Microsoft has been a principal player.

6. Password Resetting

This is also from the stables of Microsoft for integrating customers better into business.

Apart from the above, many medium-skill to high-skill areas are emerging everyday like draftsman's work (following engineering blue-prints); animation-frames; pre-printing in mass media; program coding on code-sheets (following specifications defined by system analysis); and a host of others.

STRATEGY FOR BPO

The Modus operandi for BPO can be divided into two parts: for third-party outsourcing and for internal back-office.

Third Party Outsourcing

A. The client, while looking at increasing profit and outsourcing whatever does not need one-to-one interaction, needs to identify a suitable service provider with reference to the following:
   - Credibility
   - Careful-handling capacity
   - Infrastructure (location, building, technology) and
   - Manpower

B. The 'requirement for information' (RFI) is to be followed up by 'requirement for proposal' (RFP), by its *business development team*: by assessing service provider's service and price-quotation.

C. Agreement is needed on pricing, after discounting the usual flowery reports by service provider, preferably by spot-checking.

D. 'Due diligence process' is undertaken with regard to papers and documents for service provider.

E. *Transition team* is floated by service provider to go to client-site and check out the processes.

F. *Operation team* then learns the processes, to ensure that there is no difference in customer perception and the processes are transparent.

G. Manpower forecasting is finally done, depending on how much business the client has to outsource.

Internal Back-office

A. The client, wishing to cut costs, looks at own outsourcing possibilities.

B. It examines various locations for opening back-offices, with reference to manpower (*primary* need) and legalities (*secondary* need), followed by infrastructure and capital available.

C. While no business development team is needed, strong transition and operation teams are called for, to assess manpower with reference to the business volume.

D. Budget forecasting is finalized, based on estimated revenue earning after meeting the costs of manpower, infrastructure and technology.

Benefits of BPO

In the process of osmosis of benefits, both the client and service provider stand to gain:

Client

- Cost saved in terms of reduced delay, overhead expenses, technology, and infrastructure; and
- Productivity and efficiency achieved in the mainstream activities.

Service Provider

- Revenue earned in dollars;
- Jobs and employment generated;
- International practices inducted;
- Earnings even at the lower end of education (12th standard) surpassing those of the engineers in any other industry, provided, English language proficiency exists;
- Client technology having multiplier effect;
- Ancillary development happening (e.g. 10,000 employees using 3,000 computers per shift, thus boosting PC manufactures);
- Boosting of the hardware industry as a whole, by upkeep of technology, cascading to other IT companies; and
- Using the cutting-edge hubs, routers, satellites, under-ocean cables and underground optical fibres: in order to bring voice, images and data from the oversea clients to service providers.

Problem-areas for BPO

Client

- Agitation against outsourcing in the parent country for loss of jobs;
- Sharing a lot of confidential material with service provider, which could break the back of the business, if leaked;
- Potential loss of customers permanently, if outsourcing not competently handled; and
- Legislative impact of the country of service provider (e.g.,

women working in night-shifts, which can be legislated against by the government), resulting in breakdown of business.

Service Provider

- Insecurity in business, as the client can shift shop any day, if not feeling comfortable;
- High employee turnover, since most people work on stop-gap basis, but do not make their career;
- Perception as glorified 'Operation' jobs or a clerical nature;
- Compromise on traditional values (such as, spending evening time with family) and occurrence of night shifts;
- Frequent price negotiations, since prices widely fluctuate due to competition from the Far East (China, Philippines and Malaysia) who offers cheaper prices. Last four years, however, have seen stable prices. GE China has often been more competitive than GE India;
- Legal and socio-economic issues like night shifts for, seven-day working per week, etc.

S-W-O-T ANALYSIS FOR BPO

Client

1. **Strength (S)**
   - Availability of functions that do not require face-to-face interaction;
   - Ability to concentrate on core competencies.
2. **Weakness (W)**
   - Perception of loss of job opportunity;
   - High turnover of service providers affecting profit.
3. **Opportunities (O)**
   - Getting jobs at a cheaper rate, thereby increasing profit.
4. **Threat (T)**
   - Sharing of confidential material with service provider, that could be used by the latter against the client;
   - Legislative blockade in service provider's country;
   - Losing customers due to poor handling by service provider.

Service Provider

1. **Strength (S)**
   - High English-speaking population;

- Cheap labour;
- Government support, by way of many exceptions, since the industry is generating so much employment; and
- Availability of good analytical skills and technical capabilities.

2. **Weakness (W)**
   - Poor infrastructure;
   - Lack of updated technology and exposure;
   - Unfavourable socio-economic environment; and
   - Quality of spoken English not upto the mark, hence lot of feeort needed to teach good spoken English (with training time and money).

3. **Opportunities (O)**
   - Revenue earning in international countries (dollars, pound-starlings);
   - National employment generation at low level;
   - Desire of client companies to outsource;
   - Variety of business spectrum available; and
   - Elevating clerical services into knowledge service (e.g., Ingenuity, Inc., USA, dealing in molecular biology,—creating database on mammalian organs, etc., leading to new demands).

4. **Threat (T)**
   - High employment turnover;
   - Sudden loss of business; and
   - Competition from east and south-east Asian countries;
   - Change in legislation.

IT TOOLS FOR BPO

CTI is an important feature where guidance is provided on telephone in response to a customer query. Service provider finds the solution process by pushing the customer screen on-line ('page pushing'). Another feature is *Computer-enabled dialling*, where dialler software dials telephone lines to get contacts @ 50-100 calls per minute, Out of, say, 50,000 accounts, the agent is able to talk to specific customers, with reference to data seen on the screen. Available options are:

- *Answering Machine*: Call going to the agent and message left on the machine;

- *Disconnected Tone*: Call going as "disconnected" and put on "keep trace" mode;
- *Privacy Message*: Unable to go to human voice, screen is used;
- *No dial-tone*: Re-dialing at specified frequency; and
- '*Fast Busy*' and '*busy*': according to the strategy of the facilitator.

In effect, different dispositions and strategies are devised to maximise the possibility of contact at any given point of time—within the local standard time between, say, 8 am to 9 pm.

CONCERNS OVER PRIVACY AND CONFIDENTIALITY IN BPO

As observed by NASSCOM, "Earlier, information, even if sensitive, was travelling only between two companies, or, only between two nations. But today, business transfers confidential banking, medical and other data via the Internet. Information sometimes passes through several countries, so it is important to have privacy laws that can stand this test of time and technology." Medical transcription is a typical example of sending sensitive hospital data over the Internet.

Since poor privacy protection laws could affect India's prospects in this business, the Ministry of Information Technology and Communication are taking a second look at the older data protection law, based on the European Union (EU) norms defined in the pre-Internet days. They are working on another proposal to protect interests of Indian service providers and the oversea clients that send their data here.

The problem has been aggravated by the report of a Pakistani call centre operator that one Ms. Lubna, who recently extorted sizeable (but undisclosed) amount of money from her agents by threatening to post confidential patient records on the Internet. Ms. Lubna, who allegedly made herself rich through threatening e-mails, might have jeopardized an otherwise successful Asian business model and cost India alone a sizeable share of the $4 billion medical transcription market!

PRESENT STATUS AND FUTURE FORECAST IN BPO

Over the years, the IT-enabled Services (ITES) have demonstrated superiority, sustained cost-advantage and fundamentally powered value propositions The shift from low-end business processes to higher-value, knowledge-based processes is having a positive impact on industry growth. Indian companies are enhancing their global-service delivery-reach through a combination of Greenfield initiatives, besides partnerships and alliances with local players. This apart, global software-product giants, such as, Mcrosoft, Oracle, SAP, etc., have established their captive development centres in India. The US financial giant Citigroup has its affiliate e-Serve International in Mumbai, with 2000 people engaged for value-added document-processing of 'Letters of Credits' (LC) and other services, while *Amazon.com* has retrenched customer-service jobs in Seattle by adding positions in Gurgaon through *Daksh.com*. On the whole, ITES industry has grown by 28 per cent by providing 3 million job-opportunities and its contribution to the national GDP has risen from 1.2 per cent in 1999-2000 to around 4.8 per cent during 2005-2006.

Michael Dertouzos, the Director of MIT Laboratory of Computer Science, USA, has recently anticipated that if some 50 million English-speaking Indians out their mind into 'office work proffered across space and time', they could each earn $20,000 per year. This would make a total of $1.0 trillion, i.e., *twice* the current Indian GDP!

CONCLUSION

Business process outsourcing, on the whole, remains a *Win-Win* situation: where the 'client' gains in substantial cost reduction on activities not needing one-to-one interaction and much enhanced attention to its core competencies; and the 'service provider' gains in substantial foreign exchange earning and generation of employment at a low level of education. The recent spectrum of outsourcing even in the high-skill areas has only increased its dollar-earning potential. The substantial time-cycle advantage between India and the USA only adds to the piquancy of the business operations. According to reports quoted above, the potential of Indian outsourcing revenue—already very substantial—could reach the staggering proportions comparable to the magnitude of her GDP, at the current rate of business growth. On all counts, it is an optimistic scenario, once the petty irritants like privacy, confidentially and legislative impediments are taken care of.

14

E-GOVERNANCE: FROM POLICY TO PRACTICE

INTRODUCTION

E-governance is the basic transformation of the governance processes resulting from the continually increasing applications in society of advanced digital technologies along with the Internet and mobile computing. It focuses on how these technologies can strengthen public voice as a force to re-shape democratic processes and re-focus the management structure of government to better serve the public interest. E-governance fosters partnerships and collaborations of those who have a stake in enriching the processes, after an essential re-engineering and re-deployment of governance processes and evolving their structural and functional adaptation. In the Indian context, the important question is whether the e-governance policy framework facilitates IT being an effective instrument for enhancing economic growth, improving quality of governance and affecting the core of all planning; reduction of poverty.

COMPONENTS OF E-GOVERNANCE

Government-to-government (G2G): This sector represents the backbone of e-governance, as it is expected that governments must at all levels enhance and update their own internal systems and procedures, before e-transactions with citizens and business can be successful. This involves electronic exchanges at the national, state and local levels, to improve programme delivery with availability of timely and accurate data.

Government-to-Business (G2B): Receiving high degree of attention, this sector includes both sale of surplus government goods to the public and official procurement of goods and services, by better leveraging e-business technologies for communication.

Government-to-Citizen (G2C): This sector is designed to facilitate citizen interaction with government,—as often viewed as the primary goal of e-governance, so that citizens can avail of one-stop point-of-contact and access high-quality government service.

STRATEGY OF E-GOVERNANCE

The goal of e-governance is to provide the following:

- *Better Service*: Convenient, reliable, high-quality, with low-compliance cost;
- *Cost Effectiveness and Efficiency*: Cheaper information and services for customers and better value for taxpayers;
- *Participation by People in Government*: Making it easy for people to contribute to governance;
- *Leadership*: Support to the 'knowledge society' through innovation; and
- *National reputation*: Giving India a better image as a modern nation and an attractive global location for people and business.

In order that government can serve the digital society toward the above goal, the strategy should include:

1. A vision that succinctly expresses the concept and plan for e-governance;
2. Specific objectives that can be monitored and measured;
3. Identified tasks that are necessary to support e-governance;
4. A methodology for determining organisational readiness;
5. A process for prioritising e-governance initiatives; and
6. A business model to sustain e-governance initiatives.

Following UNPAN Global Survey, stages of e-governance in India can be as visualised below:

Stage 1: *Emerging Web Presence*: Permitting formal and wider presence of independent government Websites to serve as public information kiosks/sources;

Stage 2: *Enhanced Web Presence*: With expanded and regularly updated Web presence for user-access to dynamic and specialised information;

Stage 3: *Interactive Web Presence*: With government Websites acting not only as portals by directly linking users to ministries, departments and agencies, but also allowing formal interaction between citizens and service providers, based on specific user-interests and user-needs;

Stage 4: *Transactional Web Presence*: With user-ability to conduct complete and secure transactions on-line like obtaining visas, passports, birth and death certificates, licenses, permits and other specialised government services; and

Stage 5: *Seamless, Fully Integrated Web Presence*: With clustered government services for common needs through one-stop-shop portal, where all transactional services offered by government will be available on-line.

POLICY OF E-GOVERNANCE: BARRIERS AND REMEDY

A recent report by McKinsey pointed out that there are some constraints on economic growth in India and the growth-rate of the national GDP can be increased by 4 per cent if the following barriers were removed:

- *Product-market Barriers*: These exist in sectors like power, telephone, banking, hotel, retailing, manufacturing, mining, sugar and textile,—accounting for about 2.3 per cent of the obstacles to growth.
- *Land-Market Barriers*: They relate to construction, hotel resorts and retail trade industries,—accounting for 1.3 per cent of the barrier factor; and
- *Government Ownership*: The balance 0.4 per cent comprises barriers in government-controlled owned like airlines, banking, insurance, manufacturing, mining, power and telecom.

By systematic removal of these barriers, it is possible to raise growth-rate from 6 per cent to 10 per cent in Indian economy.

In order to remove some of the barriers as above, the following policy statement was made at Secretary level in 2001-end:

- '*IT-for-masses*' *Policy*: Indian population has three layers: a *top* layer of 30 per cent, affluent and educated, who can handle advanced computing systems; a *middle* layer of 30 per cent, educated but not affluent often yet to cross the poverty line), who can be taught the elementary computing skills; and a *bottom* layer of 40 per cent, uneducated, poor and living in scattered locations, whose barriers of illiteracy, poverty and energy make their needs most difficult to solve. The Media Lab Asia, an ambitious, 10-year, Rs. 5000-crore plan, has a distributed paradigm: driven from the bottom up projects and operating from the village level.

- *User-friendly Computing-device Policy*: In order that the illiterate can make use of computers, one should have menu-driven, icon-oriented, touch-activated and voice-supportive devices, such as, those used by the XEROX Media Lab in their 'Banaras' project (using bicycle-handle, among others, to have 'walk-through' among the landmarks).
- *E-infrastructure policy*: There is a vision for providing information-infrastructure, capable of delivering bandwidths to every school, college and university, besides engineering and medical colleges, and suitably networking them.
- *E-security Policy*: Relevant laws on foolproof security for transactions, occurring in e-payment systems and 'smart cards' are now in position under the IT Act of 2000, where the 'digital signature' regime—along with 'public key infrastructure' (PKI)—has been envisaged.
- *Telemetry Policy*: This would involve electronic metering applications for automated billing and terrestrial broadcasting for reducing manpower in public broadcasting.
- *Geographical Information System (GIS) Policy*: This is of great relevance for watershed development and management, soil conservation programmes, frontline management, disaster management and traffic control, besides urban public utility planning as worked out by NIC for metropolitan Delhi using GIS.

PRACTICE OF E-GOVERNANCE: GENERAL

Early Attempts

A laudable part of e-governance in India has been put in place by the NIC over the last 25 years, involving the Central government, state governments and their various ministries, departments and agencies,— for information exchange, creation of databanks and planning models. Some pioneering efforts by the NIC and separately by the states are as follows:

1. *National*: Supreme Court records onwards to prominent High Courts; 27 parameters of 'district information system' (DISNIC); 'Information Superhighway'; 'medical information system' (MEDLARS);

on-line posting of examination results by CBSE and other boards; 'information kiosks' for publicising of prices for 800 *Mandis* (agro-markets, out of 7,000 *Mandis* in India) for agricultural commodities and other perishable products: are some of the NIC handiworks, involving networks for regional, state, district and even block-level databases.

2. *Andhra Pradesh* (*AP*): Establishing 'information kiosks' under the *E-Seva* project, where the government has completely re-engineered the *G2C* by bringing 30 different services under a single window,—foe electricity bill, water bill, house tax, driving licence, land records, etc.

3. *Karnataka*: Introducing programmes like *Bhoomi* for land-records; *Sarikaj* for RTO offices; and *Reshmi* for the silk sector; beside the concept of 'anchor company' like Infosys, Microland, TCS, Motorola, Cisco, Texas Instruments, etc., to play a lead role in Karnataka's going forward.

4. *Kerala*: Intending to make it an 'intelligent state' with 100 per cent literacy, reducing unemployment and making administration citizen-friendly.

5. *Madhya Pradesh* (*MP*): Attempting social engineering through *Soochanalaya* (information centre) on low-cost commercial lines and an Intranet for *Gyandoot* (purveyor of knowledge) to cover government departments and Panchayats.

6. *Maharashtra*: Introducing mandatory use of e-mail between state and districts, and also between Taluk headquarters and districts.

7. *Orissa*: With the Orissa Computer Application Centre (OCAC) working for land-records computerisation, using the *Sarala* Oriya script.

8. *Uttar Pradesh* (*UP*): Computerising the treasury system, besides working on crime records database and fingerprint database.

Current Decade

All the 487 blocks of the northeast India—covering its eight states of Arunachal, Assam, Manipur, Meghalaya, Mizoram, Nagaland, Sikkim and Tripura—have now installed 'information kiosks' with PCs, TVs and satellite connectivity, to provide IT-based services like *Mandi* (agro-market) prices; certificates on birth, death, caste, income, etc.; and land records. Considering the time of 10 days taken by a letter to reach northeast, a time of 10 minutes of e-mail is no mean achievement, to reduce drastically the distance-perception!

Using the Internet and powerful computer facilities, 'citizen service gateways' under the NIC have widely opened their portals. Whether it is a ration card in HP or on-line property valuation in Tamil Nadu or computerised MTNL telephone directories for Mumbai and Delhi or downloading of pension forms in MP, the list is already endless. A small illustrative list is provided below:

- On-line Ticket Booking at Indian Airlines
- Indian Airlines Flight Schedule
- Travel Offers from Indian Airlines
- Auction of Air Tickets by Indian Airlines
- Online Reservation Status at Indian Airlines
- 'Desh Videsh' Offer from Indian Airlines Write to the Prime Minister
- List of Identified Jobs for Persons with disabilities
- Daily Orders and Cause List from District Courts of Delhi
- On-line Status of Supreme Court Cases
- Cause Lists of Indian Courts
- Daily Orders of the Indian Courts
- Daily Cause list of the Supreme Court
- On-line Employment Status from Government of Tamil Nadu
- On-line Licenses and Permits in Andhra Pradesh
- Book a Railway Ticket through the Net
- Application Forms for getting a driving license in Delhi
- Acts and Rules from Ministry of Information and Broadcasting
- Latest News from CBSE
- Schemes from Sports Authority of India
- On-line Citizen Services from Himachal Pradesh Government
- Citizen Charter of the Ministry of Coal
- Model Code of Conduct from Election Commission of India
- Missing Persons List from CBI
- On-line Application forms from Rural Electrification Corporation Ltd.
- Citizen Charter of the Ministry of Shipping
- 'Get your new vehicle registered on-line' Scheme in Delhi
- Modernisation Scheme for Central Government Organisations
- Tourist Information from Jammu & Kashmir
- Schemes for Rural Development in Assam
- On-line Market Prices from National Horticulture Board
- Recruitment Information from UPSC
- Central Govt. Health schemes
- On-line Application Forms from Department of Atomic Energy
- Programmes & Schemes for Rural Development
- Get a Digital Certificate from DGFT
- Lodge your grievance with Delhi Minorities Commission

- Strategies and Policies on AIDS Prevention
- Research Activities of National Tuberculosis Institute
- Database of *Haj* Pilgrims
- Research in Homeopathy
- 'Wanted Persons' List from CBI
- *Pradhan Mantri Gram Sadak Yojona* (PMGSY)
- Investor Information from Coal India Ltd.
- Quality Certification for Agricultural Products (APEDA)
- On-line Publications from Department of Education
- Personal Banking from State Bank of India
- Welfare Schemes for tribals
- On-line Prices for Agricultural Commodities
- Citizen Services of Gujarat Government
- Check your IEC Status Online (Customs)
- Civil Services (Main) Exam Result
- National Vocational Training Information Service
- Holiday Ideas for Tourists
- Parliament Bills and Acts Online
- Weekly Wholesale Prices of Food Commodities (NAFED)
- Download Application forms of IDBI Services
- Exam-Related Announcements by CBSE
- Empanelment of Chartered Accountants at CAG
- Database Services of Ministry of Labour
- Passport Forms
- Online Status Enquiry for Passport
- Formats of Affidavits required for a Passport
- PNR Status Online (Indian Railways)
- Foreign Currency Reference Rates (RBI)
- Agricultural Marketing Information Network
- NCERT Books On-line
- On-line Announcements from UPSC
- Route Map for Delhi Metro Rail Service
- *Pravasi Bharatiya Divas* for Indian Diaspora
- Electronic Data Info Filing and Retrieval System (SEBI)
- Training Courses at Indian Space Research Organisation (ISRO)
- Travel Offers from Air India
- 'Companion Travel Offer' from Air India
- Know About Service Taxes in India
- Child Development Schemes
- On-line Application Forms from Directorate General Civil Aviation I

- Forms for export of Horticulture Products from North-East
- Obtain a certificate for physical handicap in Delhi
- Multimedia Interactive Study Material for schools
- Income Tax Forms
- Online Registration of Public Grievances
- ICAR Guidelines for filing Patent Applications
- Online Judgement Information System
- Govt. Accommodation allotment waiting list
- Online Forms from Employees Provident Fund Organisation
- E-mail Facility for Indian Learning and teaching Community
- Guidelines for Foreigners Visiting Delhi
- UPSC Examination Information
- Daily Orders of Supreme Court and High Courts
- New Schemes of Central Government Housing Welfare Organisation
- Agricultural Marketing Information Network
- Indian Railway's Train Schedule
- Multimedia
- Interactive Study Material (School)
- Access Parliament Codes Online
- Forms for filing Income Tax Returns Online Judgement Information System
- Agricultural Marketing Information Network
- Get a Voter's Card in Delhi
- Lodge a complaint with Central Vigilance Commission
- Case Status of Supreme Court of India
- Central Govt. Employees Housing Schemes
- Income Tax Forms
- Submit Opinion on Review of Working of Constitution
- Stolen Vehicles Status (Delhi Police)
- File a Human Rights Violation Complaint
- Access Parliament Acts Online
- Check out your IEC Status sat Custom (BIN)
- Form for opening a Post Office Savings Bank Account
- Obtain an Excise Control Code (ECC) Number
- Announcements for Overseas Study Scholarships
- Application Forms for PM's *Rozgar Yojana.*

PRACTICE FOR E-GOVERNANCE: STATE EFFORTS

First Case—AP

The AP government's initiative is based on an overarching role: *Simple,*

Moral, Accountable, Responsible and *Transparent,* called SMART government. In order to achieve this vision, the strategies selected are: *technology framework* at the base; *resource framework* for garnering resources in a short time-span; *prioritisation framework* for different areas of governance; and *implementation framework* upto the *tahsildar* and agricultural officer levels. For ICT architecture, for instance, comprehensive technology architecture has been built based on 300 technology components recommended. In order to create a suitable framework with a multiplier effect, Public Private Partnership (PPP) model has been worked out. They have also adopted a *6C model,* covering the operational details of what a project needs to achieve success.

Implementation since 2001 began on a large scale,—with five core applications based on a survey of 50 major departments. The *core initiatives* contain three elements: *core policies, core infrastructure* and *core applications.* On the 'core policy' side, they include IT architecture, Internet information security policy and change management policy. On the 'core infrastructure' side, they have 'AP State-wide Network' (APSWAN), which would eventually consolidate and spread 'AP Secretariat Campus Area Network' (APSCAN) and other existing networks. Among five 'core applications' is 'Data Centre' to create servers for each of the departments, security and so on,—for the entire government. 'Single window' for citizens on the lines of *E-Seva,* mentioned earlier, would be positioned. 'Single window' for business would imply that either for *starting* or *running* or *closing* any business, one has to visit only one site.

By the 'social benefit' system, *digital divide* issues would be addressed, by taking some core initiatives along those lines, with integrated LAN databases, GIS and related issues. E-procurement would be next in line, for which RFPs would be rolled out. HR management system for 500,000 employees would lead to a generic office management system. The other application is an Integrated Financial Information System. Most of the new initiatives would cut across departments. Among 50 departments mentioned, there is, incidentally, a logical and functional grouping, such as, Municipal Group, Transport Group, and so on, apart from the individual applications.

Second Case—MP

As the second largest state in the Indian Union, it receives funding from Central Government for 10 mission-mode projects to cover such state departments as land records, transport, police, treasuries, property registration, employment exchange, agriculture, municipalities, Gram Panchayat and commercial taxes. The state's e-governance initiatives are as follows:

1. *Integrated Treasury Computerisation Project (ITCP)*: This is a major e-governance initiative covering the entire state and 8,000 drawing and disbursement officers. *Cyber treasuries* permit the registered payees the facility to make payments on-line to the state through the Internet., using the net banking facilities.

2. *Transport*: The *E-Seva* portal of the transport department provides all the forms needed by the citizens for various applications, issue of certificates and dynamic information on the Internet. The department has recently started SMS-based and Internet-based *E-Seva,* free of cost, related to any vehicle registered in MP. Therefore, the traffic enforcement wing of the police department can collect any information from the net.

3. *Land Records*: This has alleviated needless running around by the landowner from pillar to post for retrieval of a particular record. *Bhu-Abhilekh* has the largest database of land records in the country with all textual records in digital form: storing plot-wise and owner-wise detail of land, crops, revenue, irrigation, demand, collection, land-type, tenancy and so on. *Bhu-Abhilekh* is a *G2C* and *G2G* application software used by the commissioner, land records and settlement department of revenue in the government.

4. *Urban Administration and Development*: This relates to computerisation of municipal activities—resulting in improved public service delivery and increase of revenue—in cities like Khandwa, Indore, Bhopal, Ujjain, Gwalior, Jabalpur, Raisen, Dabra and Balaghat,

5. *Electronic Tendering*: As a *G2B* activity, e-tender aims at streamlining all procurement activities, including contract for supply, customisation, installation, operation, maintenance and training with Web-enabled procurement system,—handled by a consortium of Wipro and Nex Tenders (I). This enables a degree of transparency in the tendering process, besides time and cost advantages.

6. *Social E-governance*: The state initiative in education sector are: *Fund-a School*, with a Website seeking contributions from society for promoting school activities and development of school infrastructure; *GIAN* (General Information Access Network) for citizens with computers and Internet facilities at each district headquarter; and *EDUSAT* connecting 730 isolated schools with the ISRO satellite and IGNOU to support elementary education programmes.

7. *MP On-line*: In another *G2C* activity, the government has launched on-line citizen service-portal, in a joint venture with TCS,—for providing service in the areas of health, education, agriculture, government services and business to the masses in the rural MP.

8. *Police*: With assistance from Central government, 'automated fingerprint identification system' (AFIS) has been established to digitize all fingerprints in all districts, apart from the fulfillment of the RTI Act, lodging complaint and status feedback, vehicle enquiry, and tenders status.

Third Case—Tamil Nadu

The e-government initiatives in Tamil Nadu are characterised by their accent on local languages. These are illustrated below:

1. *Land Records*: In keeping with the land records systems implemented by the NIC in various states under names like *Bhoolekh, Bhoomi, Dharni, Himbhoomi,* etc., effective at the field level in local language contents, *Tamil Nilam* (Tamil Nadu Infosystem on Land Administration and Management) is in use in all 201 taluk offices in Tamil Nadu. It maintains, in Tamil, data for more than 250 million land records and 350 million land subdivisions,—mostly with 'touch-screen' kiosks and with voice messages in Tamil to guide the users into the 'touch-screen's system.

2. *Birth Certificate*: The same taluk system maintains birth certificate particulars of the villages and is a widely-used facility over last few years.

3. *Registration*: Project STAR (simplified and transparent administration of registration) relates to property transactions with data over 13 years, captured in 300 computerised offices in Tamil. The project makes the certificate generated user-friendly for the rural public and has been enhanced with REGiNET facility for providing 'encumbrance certificate' over the Web.

4. *Electoral Roll*: In keeping with electoral roll database maintained in local languages in most of the states, data for all 234 'assembly constituencies' of Tamil Nadu, captured in Tamil, are hosted in the Website of the election department.

5. *Civil Supplies*: Civil supplies and consumer protection department maintains a huge database, in Tamil, on citizens with family details of more than 180 million families in Tamil Nadu.

6. *Judiciary*: More than 100 computers are in use in city civil court, metropolitan courts and court of small causes in Chennai city,—primarily to speed up judgment entry and issue of certified copies. Judgments, pronounced in Tamil, are stored in Tamil using these systems.

7. *Tamil Virtual University*: It is a unique organisation for promoting Tamil language, using the power and reach of the Web. It stores almost all the ancient and contemporary Tamil literature in digital form

Fourth Case—Inter-state Comparison

In a national conference held in February 2007, it was unanimously felt that, in the present-day scenario, e-governance was no more restricted to G2G or G2B, but its true essence lay in delivering to the citizens as G2C. Therefore, it must transform itself to e-democracy! In order to promote excellence and encourage the best e-governance applications, an inter-state comparison was made and Table 14.1 enumerates the applications selected for awards

Table 14.1: Inter-state Awards of Excellence for E-governance

Categories		*Awardees*
■ Excellence in Governance Process Re-engineering	I.	On-line Student Management System: Education Department, NCT of Delhi
	II.	*E-Khazana*: Science & Technology Department, Finance Department and NIC, Bihar, Bihar
	III.	Double-Entry Accounting Information System, HP Institute of Publishing Administration
■ Exemplary Horizontal Transfer of ICT-based Best Practice	I.	Integrated Treasuries Computerisation System, Directorate of Treasuries, MP
	II.	*Sarthi* and *Vahan*, Department of IT and Dept. of Transport, Punjab
■ Outstanding Performance in Citizen-centric Service Delivery	I.	Citizen Felicitation Centres: Kalyan, Dombivili Munic. Corporation, Mumbai
	II.	Computer-Aided Learning: Education. Department, NCT of Delhi
	III.	*E-Mitra*: Department of IT and Communication, Rajasthan
■ Exemplary Usage of ICT by PSUs	I.	File Monitoring: System (FMS) and Barcode Solutions: Delhi State Industrial and Infrastructure Development Corporation Ltd.

(Contd.)

(*Contd.*)

Categories	*Awardees*
	II. Medical Cards and Referral Slips Software: NTPC III. Delivery of Customer Services through single window system FALCOM: BSNL, Karnataka
■ Best Government Website	I. 'Incredible India': Ministry of Tourism, Government of India II. National Portal of India: NIC, Headquarters III. Orissa Priority Education Programme Authority
■ Special Mention Innovative Technology Usage in e-governance	Litigation Information Tracking and Evaluation System (LITES): Justice Department and The Department of IT and Communication, Rajasthan

NATIONAL POLICY FOR E-GOVERNANCE

While islands of e-governance appeared in the past decade or so, 'National e-Governance Plan' (NeGP) has been formulated in 2007 to make *G2C* a reality: by improving the quality of basic governance, on a massive scale: in area of concern to the common man and make government services accessible to his locality. The plan is especially relevant for *aam admi,* particularly for those living in rural areas.

The Goal

NeGP has the potential aim to transform the socio-economic landscape of rural India.

The Tasks

- To set up 100,000 'common service centres' (CSCs) in 6 lakh villages;
- To make CSCs Internet-linked and broadband—enabled;
- To provide a host of government and private-sector services;
- To provide a platform for the RTI Act,—to meet the need for an accountable government; and
- To use technology as a trigger for reform and as a strategy to increase the transparency in the functioning of government body.

CONCLUSION

In fine, IT can be viewed now as having reached the common man. The benefits of *computerised voice-mails* for telephones, railways and airlines are at one's doorsteps and taken almost for granted. Media Lab Asia, launched at the government initiative, plans to work with industry, NGOs, government departments and public systems to bring IT innovations to every village. A few core projects are:

- *Chikan-CAD Project:* for developing technologies and products, at IIT, Kanpur, for handicrafts, especially in *Chikan* embroidery;
- *Rapid Terra Cotta Project:* for using Gangetic clay in a rapid-modelling process, in order to create computer-controlled 3-D structures for patterns and tools;
- *Khasra-Pengachu Project*: for developing a low-cost computer (on *Pengachu* platform), dedicated to have secure and incorruptible land-records, with sound and picture interfaces; and
- *Swara-Gandhar Project*: for developing a low-cost music synthesizer, helpful in retaining the rich heritage of classical music, providing training and composing raga-based music.

The latest announcements are about the 'Challan Identification Number' (CIN), made mandatory like PAN,—with every payment uniquely identified and resulting in correct credit for payment by every taxpayer. In the outer space, the European Union, 'Galileo' satellite programme—with 30 satellites encircling at medium-level orbits at altitudes around, 24,000 km to cover the entire earth surface by 2008—can generate for India enormous benefits in sectors like transport, telecommunications, agriculture and fisheries. Sky is literally the limit for IT for common man!

15

ISO-9000, SIX SIGMA, CAPABILITY MATURITY MODEL: INDIAN EXPERIENCE

INTRODUCTION

The world over, IT is making phenomenal progress with both hardware and software crossing new frontiers of excellence every other day. Since software remains the driving force, it has been increasingly felt that there should be universally recognised, three-fold criteria for *quality, standards* and *maturity* that would ensure its acceptance by professionals on one side and managers, executives and administrators on the other side: without any qualms of conscience. As we shall see, International Standards Organisation (ISO)—9000, *Six Sigma* and *Capability Maturity Model* (CMM) are steps in that direction, all of which have become trademarks of excellence in a quality-conscious society: with universal recognition and acceptance.

CERTIFICATION OF QUALITY AND STANDARDS

Quality

Quality may be defined as the 'ability of a set of inherent characteristics of a *product, system* or *process*, to fulfill requirements of customers and other interested parties'. It should be noted that 'product' remains the end-point output of any 'system', which follows pre-designated steps or 'processes' in order to reach the end-point starting from specified inputs. 'Quality' is a measure of excellence for all these elements,—every time, everywhere.

It has been always felt that their should a global scheme of certification of quality by an international body, which would not merely complement the national bodies responsible for such certification, but will also supervise them. Accordingly, International Standards Organisation (ISO), was established in 1947 in the non-government sector as a worldwide federation of national standard organisations of the member-countries. ISO was created to achieve the following aims:

- To promote the development of standardisation and related activities; and
- To develop cooperation in the spheres of intellectual, scientific, technological and economic activities.

Standards

Standards, in this context, are viewed as 'documented agreements that contain technical and other precise criteria, and are used as rules, guidelines and characteristics, and ensure that material, products and services are fit for their purpose'. Several formats that have become commonplace are derived from an ISO international standard. An example is the format of *credit (or debit) cards* that must follow unambiguous practices of entering name, photo, serial number, signature and other details of the card-holder,—which are adopted in a uniform pattern across the globe. Another example is the *phone card* that again has similar pre-designed formats obeyed all over the world.

Membership of ISO is divided among the following three categories:

1. *A member body*: national agency that represents standardisation in its country.
2. *A correspondent member:* organisation in a country that does not have a fully developed national activity, but are still participative in a collective effort towards standardisation.
3. *Subscriber membership:* for countries with very small economy.

ISO has a highly decentralised method for executing its technical work, The technical activities of the ISO are carried out in a hierarchy of some 2,850 technical committees, sub-committees and working groups. Some 30,000 experts participate in meetings every year. The ISO hierarchy is indicated in Figure 15.1.

SCOPE OF ISO 9000

ISO 9000 is the international standard for quality management system. Some equivalent national standards are as follows:

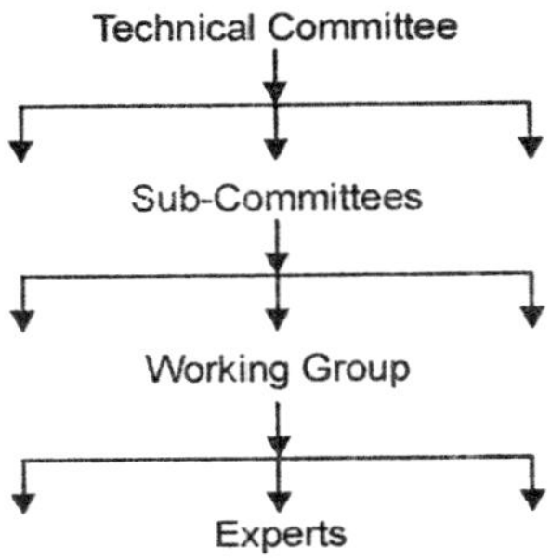

Fig. 15.1: ISO Hierarchy

- IS/ISO 9000
- BS/EN/ISO 9000
- EN/ISO 9000
- ANSI/QC 90

ISO 9000:2000 standards are quality management and quality assurance standards for implementation, assessment and verification of quality systems. These are *process* standards, not *product* standards and provide only basic requirements. The companies have to establish their own rules, procedures and policies to comply with these standards. Good business practices are defined in a set of sections under **ISO 9000.**

ISO 9001 specifies requirements for quality management system where an organisation needs to demonstrate the following:

- Ability to provide, consistently, products that meet customer and regulatory requirements; and
- Customer satisfaction through the effective application of the system, including processes for continual improvement and prevention of non-conformity.

ISO 9001 can be illustrated through a process of specifications, designs or tests for implementation that lead to the ultimate product for the customer.

STRATEGY OF QUALITY MANAGEMENT

The quality management principles involve the following:

- Customer-focussed organisation
- Leadership

- Involvement of people
- Process approach
- System approach to management
- Continual improvement
- Factual approach to decision-making
- Mutually beneficial supplier relationships.

The quality management implementation needs the following guidance factors:

- Supportive management decision and continued commitment
- Study and understanding of ISO 9001:2000
- Analysis of current practices
    - Are required processes established?
    - Are established processes in compliance with ISO 9001: 2000?
    - Are procedures defined?

The quality management documentation has to ensure the following:

- Are the documents available?
- Are they compliant with ISO 9001: 2000?
- Are there adequate records to demonstrate compliance?

Besides the above, internal audits and self-assessment procedures should be in place.

SIX SIGMA

Six Sigma, as a spin-off from Total Quality Management (TQM), was first introduced in the 1980's by Bill Smith of Motorola,—as a statistical approach aimed at increasing profitability by reducing defects. **Six Sigma** is a scientific, innovative and strategic way of doing business by taking decisions based on *systems data* and by taking into account the variations in the system through the statistical measure of *standard deviation.* Initially, a single metric, *total defects per unit* (DPU), was established and the corporate goal of Motorola was re-stated in *parts per million* (PMU). By 1988, Motorola received a national quality award for excellence and 'Six Sigma Research Institute' was established in 1990.

Critical to satisfaction (CTS) of customer is first determined for aspects that give him sufficient confidence in the company, such as,

- Critical bugs fixed within a stipulated time,
- Productivity in terms of *number of transactions per unit time,*
- Other measures like *critical to cost* (CTC) and *critical to delivery* (CTD)

The above measures are then utilised to derive *critical to quality* (CTQ), with 'CTQ Tree' as a tool to aid in translating customer expectations into quantified requirements for product/service. *Quality,* following the ISO definition, is the totality of features and characteristics of product/service, which satisfies the customer's stated and implied needs. In Six Sigma, *quality* is a state in which value-entitlement is realised for the customer, and the company, in every aspect of the business-relationship covering the entire supply chain,—in a 'Win-Win' approach for all.

Cost of Quality

Obviously, 'cost of quality' includes: *appraisal cost*, like inspection, training, etc.; and *prevention cost,* like planning, training, etc. But the hidden cost of failing to meet customer requirement constitutes 'cost of poor quality'. The latter comprises *external failures* like delayed submission of developed modules, customer dissatisfaction, etc., and *internal failures* like redesign of modules, reworking on effort estimation, loss inn productivity, etc.

While *process* is the sequence of activities resulting in product/service, *value-addition* is the activity for which the customer is willing to pay. While *key process input variables* (KPIV) influence the output of a process (e.g., time and temperature as key inputs for heat treatment process), *key process output variables* (KPIV) influence the performance of CTQ. *Metric* represents the quantified performance of a process, product or service. *Defect* is a feature in process/product that causes dissatisfaction to a customer. *Defect per unit* (DPU) of product/process/software/transaction) is total number of defects in a sample,—divided by total number of units in that sample.

A unit may have more than one type of defect, where defects are different opportunities. *Defects per million opportunities* (DPMO), *throughputs yield* and *aggregated (rolled) throughput yield* are some measures of poor quality.

Strategy of Six Sigma

For **Six Sigma,** the five-phase approach, called DMAIC, can be enumerated as follows:

- **'D'** for 'define' for the project based on customer, business, operational needs,—having a link to the company's strategy;
- **'M'** for 'measure' as tools applied to the baseline of current performance;
- **'A'** for 'analyse' for identifying the sources of variations and developing relationship between process variables and the product;
- **'I'** for 'improve' and 'innovation' to get desired benefits; and
- **'C'** for 'control' of the results and improvements achieved and consolidated through a system approach.

Define Phase

1. 'Recognise' the customer voice
2. 'Translate' it into company language
3. 'Establish' customer CTQs
4. 'Identify' the processes.

Measure Phase

1. 'Select' performance variables
2. 'Create' process map
3. 'Establish' measurement system
4. 'Measure' performance variables
5. 'Establish' performance capability.

Analysis Phase

1. 'Select' performance variables
2. 'Benchmark' performance 'metric'
3. 'Conduct' gap and root cause analysis
4. 'Identify' success factors
5. 'Define' performance goals.

Improve Phase

1. 'Select' and 'diagnose' performance variables
2. 'Develop' strategy for improvement
3. 'Propose' and 'confirm' variables
4. 'Establish' operating limits
5. 'Verify' performance improvement.

Contral Phase

1. 'Select' variable for control
2. 'Define' control system
3. 'Validate' control system
4. 'Audit' control system
5. 'Monitor' performance metric

Quality improvement can be achieved through project approach adopting DMAIC methodology. **Six Sigma** implementation is not complete unless the cultural change is effected in the company. *Quality function deployment* (QFD) is a tool that links a company with its customers and assists the organisation in its planning process. 'QFD matrix' uses a matrix format to capture *horizontally* information relating to customer, and *vertically* technical information that responds to customer inputs. The voice of the customer is the basic input of QFD process and the gap between the company effort estimation and customer effort estimation should not be more than 2 per cent.

The next step is to understand the customer's competitive evaluation of the company's deliverables,—in order to understand the customer's rating in 1-to-5 scales. Once the customer portion in QFD is understood, the next step is to develop steps to respond to each voice of the customer. The centre of the QFD matrix is to show the customer and technical requirements intersection,—to provide an opportunity to record the presence and strength of relationships. QFD's targets represent how much is required to meet, or excel, the customer's wants and needs.

Process Maps help in improvements in process by: removal of decision blocks; parallel processing of activities; and re-engineering to reduce time/cost. *Cause and effect diagramme* is a diagnostic tool—in order to link the various causes with the effect in a diagrammatic form. Failure Mode Effect Analysis (FMEA) is another diagnostic technique,—to identify and eliminate known and/or potential failures, problems and errors. While failure is viewed as 'any non-compliance in meeting the customers requirements'—stated or unstated—the analysis can be used everywhere and it does not necessarily need data. FMEA has three parts: occurrence, severity and ability to detect, using an ordinal scale of 1-10 to denote each part. On the whole, authentic data removes the opinionated version and ensures objectivity in decision-making process. The latter is as valid for software projects as for any other business activity.

CAPABILITY MATURITY MODEL

The Capability Maturity Model for Software (**CMM** or **SW-CMM**) can be understood as an amalgam of quality, standards and maturity, as shown in Figure 15.2.

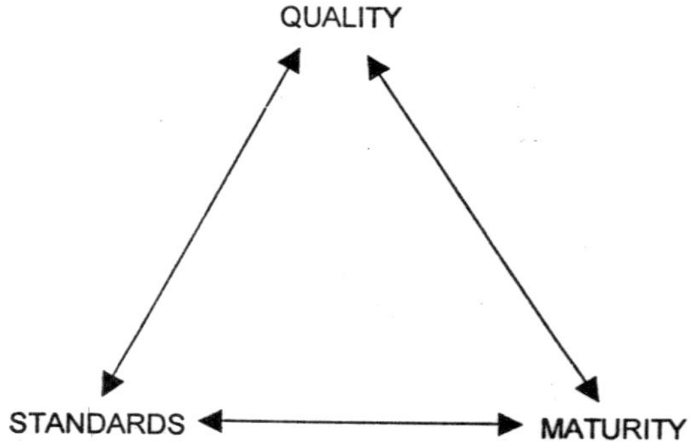

Fig. 15.2: Capability Maturity Model

CMM, created in the late 1980's, is a model for judging the maturity of the software processes of an organisation and for certifying the key practices that are required to increase the maturity of these processes. It was developed by the software community with stewardship of the Software Engineering Institute (SEI), Carnegie-Melon University, Pittsburgh, USA, under Dr. Watts Humphrey. Incidentally, SEI was established by the US Department of Defense to improve state of software practice and once CMM was developed, US Government began using it as a basis for selecting software developers with the lowest overall risk for the following purposes:

I. Planning and tracking work;
II. Using best technical methods; and
III. Measuring and managing quality.

The software **CMM** has become a de-facto standard for assessing and improving software processes. Through **CMM,** the SEI and software community have put in place an effective means for modelling, defining and measuring the maturity of the processes used by software professionals. It guides organisations in adopting sound management practice, but not directly affecting products. Incidentally, **CMM** is one of the three models that provide the initial **CMM Integration** (CMMI) product suite.

Software process model, like any other process, can be illustrated as in Figure 15.3.

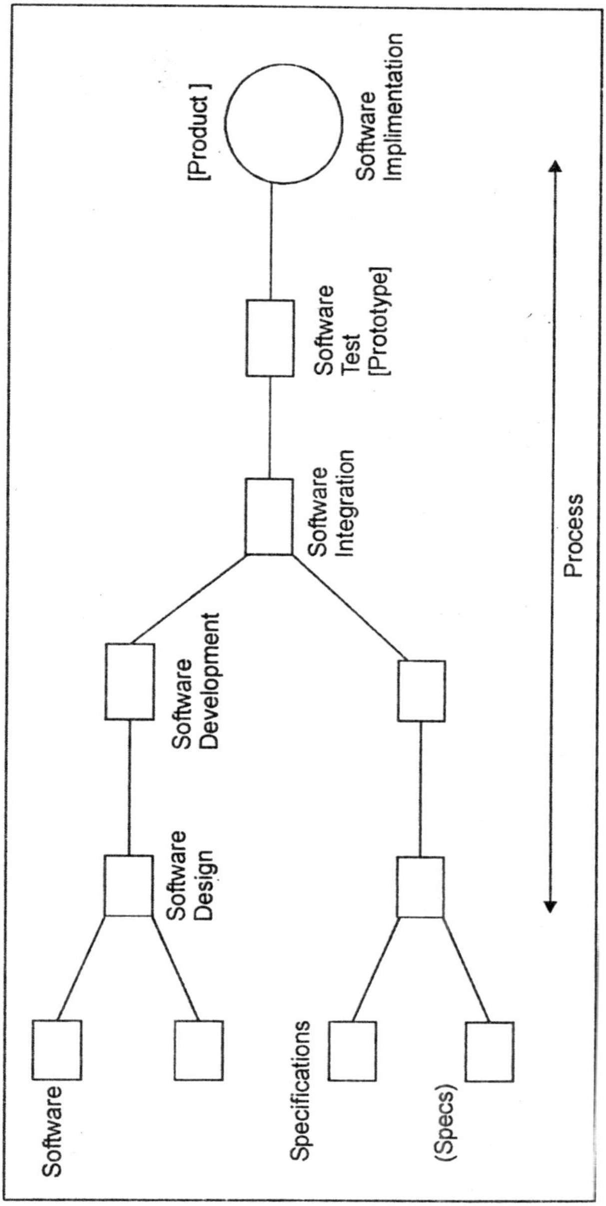

Fig. 15.3: Softwarte Process Model

The unique way of looking at software through the prism of **CMM** is to consider it as a bundle of *process, procedure, performance* and *practice* that all lead on to software *products.* While Phil Crosby was responsible for creating **quality management maturity grid,** Watts Humphrey took it forward as **CMM** for **continuous process improvement**. This software discipline has now been well inculcated and put into the educational curriculum of such premiere institutions of the USA as Carnegie-Melon University, Purdue University, Pittsburgh University and Stanford University.

CMM—as developed at SEI—is now a globally acknowledged mechanism for evaluating **software process capability**. The model has been developed for **software process, software professionals** and **software acquisition,** and assisted in the evaluation of **systems engineering** and **integrated product development. CMM** also assists organisations in maturing their *people*, *process* and *technology* assets and improves their *long-term business performance*. It is eminently useful for understanding the following:

- Where we are today—*Baseline;*
- Where do we want to be—*Target;*
- How do we get there—*Plan;*
- Have we reached there—*Measurement.*

The above can be viewed as in Figure 15.4.

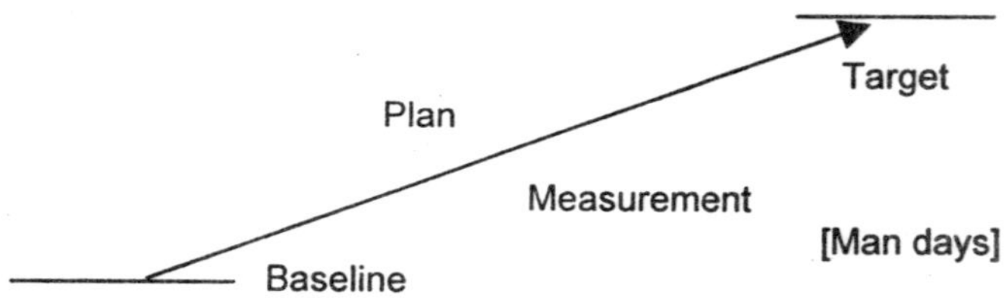

Fig. 15.4: CMM Model

LEVELS OF CMM

The all-important levels of **CMM**, which software companies strive to achieve across the globe—with or without uniform success—are enumerated below for careful observation:

Level One

This is the **initial** level, which is considered status-wise as **average.**

Indeed, almost all software companies start at this level. The process is considered **normal** and the process characteristics are **ad-hoc**, i.e., one-time, available as needed.

Level Two

This is the **repeatable** level, which is status-wise considered as **good.** Quite a large number of the world's software agencies have achieved this level already by graduating from the **initial** stage. Project management practices at this level are institutionalised and the process characteristics are **disciplined,** i.e., clearly understood.

Level Three

This is the **defined** level, which is status-wise considered **very good.** Many countries, including China, have their software houses clustered around this level and are quite praiseworthy. Technical practices at this level are well integrated with management practices and the process characteristics are **consistent,** i.e., uniform, homogenous.

Level Four

This is the **managed** level, which is status-wise considered **excellent.** Relatively few countries have achieved this coveted status and India has creditably a couple of hundred such software companies. Products and processes are quantitatively controlled and the process characteristics are **predictable**, i.e., can be anticipated.

Level Five

This is the highest, **optimising** level, which is status-wise considered **outstanding.** This is a highly coveted status and India, among the comity of nations, has attained the enviable feat of crossing 50 early this millennium. Process involvement is institutionalised and the process characteristics are **continuously improving,** i.e., dynamically achieving progress.

STRATEGY REQUIRED FOR CMM

The following steps required to achieve **CMM** may be noted:

1. There are no standard methods laid down, barring observing other organisations' experience, in order to prepare for **CMM**;

2. The organisation that wishes to go for **CMM** has to choose a SEI authorised lead-assessor (or consultant) who guides the organisation towards achieving **CMM**;
3. It depends on this lead assessor (or consultant) to decide on a path for the aspiring organisation.

India's software prowess, which is deservedly world-recognised, has let the country achieve the 'Level Five (Optimising)' status for a very substantial number of companies. An illustrative list of Indian 'Level Five' companies, as published early this decade, is indicated below:

- Applitech Solutions Limited., Ahmedabad
- CBS India
- CG-Smith Software Limited, PB No.5814, Bangalore, India
- Citicorp Overseas Software Limited, Mumbai, India
- DCM Technologies
- Eastern Software Systems Ltd.
- Engineering Analysis Center of Excellence Pvt. Ltd. (EACoE), India
- Future Software Ltd., Chennai, India
- HCL Perot Systems, Noida, India
- HCL Technologies Ltd., India
- Hewlett Packard India Software Operations Ltd, Bangalore, India
- Hexaware Technologies Ltd., Chennai and Mumbai Centres
- IBM Global Services, India
- I-flex Solutions Limited, IT Services Divisions at Mumbai and Bangalore, India
- Information Technologies (India) Ltd.
- Infosy Technologies Ltd. Bangalore, India
- Intelligroup Asia Pvt. Ltd., Advanced Development Centre, Hyderabad, India
- International Computers (India) Ltd.
- IT Solutions (India) Pvt. Ltd, Bangalore & Chennai Development Centres
- ITC Infotech India Ltd., India
- Larsen & Tubro Infotech Ltd., EUROPE SBU at Mumbai and Navi Mumbai
- L G Soft India Pvt. Ltd., Bangalore, India
- Mastek Ltd., Mumbai
- Motorola India Electronic Ltd.(MIEL), India
- Network Systems & Technologies (P) Ltd., Trivandrum

- NIIT, Knowledge Solutions Business (KSB)
- NIIT Software Solutions
- Patni Computer Systems Ltd.
- Sasken Communication Technologies Ltd., Bangalore, India
- Satyam Computer Services Ltd.
- Siemens Information Systems Ltd., Telecom & Major Projects Division
- Signal Tree Solutions (India) Ltd.
- Sky TECH Solutions Pvt. Ltd.
- Sonata Software Ltd, Bangalore, India
- SSI Technologies, Chennai, India
- Syntel, Inc. (India)
- Tata Consultancy Services (23 centres in India)
- Tata Elxsi Ltd.
- Tata Interactive Systems
- TCG Software Services Pvt, Ltd., Kolkata, India
- US Technology Resources India, US Software Pvt. Ltd., Trivundrum, India
- Wipro Infotech, enterprise Solutions Division.

As mentioned earlier, the list has already crossed 50 and the respective software company Websites display their latest status. Normally, the process of certification is a fairly uphill task,—marked by more than one failure. Figure 15.5 displays the typical certificate awarded to a 'Level Five' company like XANSA (India) limited, which was quite exemplary to obtain the certification at a *single* attempt!

COMPARISON BETWEEN ISO AND CMM

ISO sets forth the minimum requirements for a quality system, whereas CMM underlies the need for *continuous process improvement.* Many large companies require their subcontractors to be **ISO 9000** compliant, while other companies use **CMM** to define their process strengths and weaknesses, and to improve on their existing process.

Currently, there are no mandatory *quality* standards that govern software development. Both **ISO 9000** certification and **CMM** audits are optional, although companies find them useful for finding problem areas with development processes. The existing **ISO** and **CMM** standards provide a starting point for developing these standards. Since both **ISO** and **CMM** share common concerns for **quality** and **process** management, they are highly correlated.

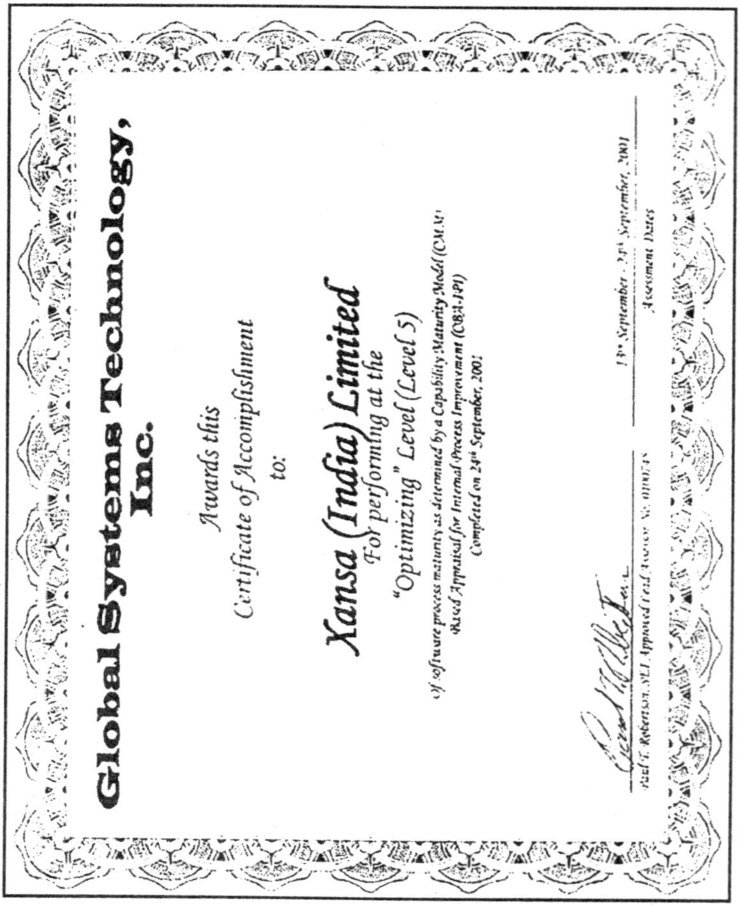

Fig. 15.5: Certificate for a 'Level 5' Company

Some distinguishing features between **CMM** and **ISO 9000** are mentioned below:

- All practices in **CMM** are not addressable in **ISO 9000;**
- It is possible for a 'Level One' organisation to receive **ISO 9000** registration;
- There are areas not addressed in **CMM,** but addressed in **ISO,** like documentation rigours;
- Although an **ISO 9001**-compliant organisation would not necessarily all of the 'Level Two' key process areas;, it would satisfy most of the 'Level Two' goals and many 'Level Three' goals. This is because there are practices in **CMM** that are not addressed in **ISO 9000**; and
- A 'Level Three' organisation would have little difficulty in obtaining **ISO 9001** certification and a 'Level Two' organisation would have significant advantages in obtaining certification of **ISO 9001.**

CONCLUSION

The above is an integrated view of methodology—adopted internationally—on categorisation methods like **ISO 9000,** statistical decision-making tools like **Six Sigma** and continuous process-improvement criteria like **CMM.** As stated above, they are all voluntarily adoptive and have no mandatory dictum, although their usage makes life easier, especially with internationally-accepted norms: on quality, standard and maturity.

PART SIX

SEVEN CASE STUDIES

(Conducted at International Management Institute)

16

CRITICAL OVERVIEW

Having slowly and fairly efficiently entered the new millennium, it is natural to look beyond the horizons. Men and women are typified by their curiosity and their hunger for knowledge. It is natural, therefore, that one tries to peep into tomorrow–that is an inherent embodiment of human nature – and into the effect that Information Technology or IT will have on tomorrow's management frontiers. The recent growth and subsequent attention that IT has garnered makes it a natural subject for a futurist's curiosity.

Among the *case studies*, there are *two* on EDI – *Electronic Data Interchange*. The first report deals with the application of EDI in the automotive industry. The second deals with EDI and the Digital Equipment Corporation and their efforts to provide software support for this emerging area of development. In either case, the subject of Electronic Data Interchange is discussed from the basics. While the report on the use of EDI in the automotive industry concentrates more on the efforts utilized by various organisations and individuals to popularize this concept and to try to sell this idea to others, the report on EDI and Digital Equipment Corporation concentrates more on the technical aspects of certain issues that are relatively important in the implementation of EDI. Therefore, while the report about the use of EDI in the automotive industry is more "commercially oriented"– to the extent that the commercial and organizational aspects of the implementation of EDI are very thoroughly discussed – the report on EDI and Digital Equipment Corporation concentrates more on "how" of the issues involved. The fact that EDI has immense potential and provides valuable advantages has been pointed out in both papers, but while the efforts of the report on the application of EDI in the automotive industry have been identify the important

---

Contributed by Prof. Dinkar, Asia-Pacific Institute of Management, covering seven case studies, together with 'Comments' for each.

components of EDI and identify–and recommend–the steps that must be taken to implement the system fully, the effort in the report on EDI and Digital Equipment Corporation is to understand the technological issues fully and to explain how Digital Equipment Corporation provides the necessary software to implement EDI efficiently. *It, however, does deal with and explain the technological issues involved.* In the report on the *use of EDI in the Automotive Industry,* the issue of standardisation and the use of appropriate protocols has been thoroughly dealt with, the use of appropriate networking standards has been extensively dealt with in the report on EDI and Digital Equipment Corporation. At the same time, in the report on EDI in the automotive industry, the important position of Value-Added Networks (VANs) is dealt with elaborately. Therefore, it would be appropriate to say that the two reports are not conflicting, indeed ideally they are complimentary to each other and they deal with the different aspects of EDI and the importance of different aspects of EDI in the overall scheme of implementation of EDI, particularly in India.

Internet Banking is the subject of *the third report.* In recent years, the terms Internet Banking, On-line Banking, Electronic Fund Transfer, etc., have not only gained currency but also respectability in terms of likely banking development. However, most of the report is about the development of Internet Banking in the United States. But then if we follow the popular adage "What the US does today, India does tomorrow", the paper appears quite contemporary. Since most of the data on these topics available today is about the US, as development in this field have largely taken place in the United States, it is natural that the events there should form the centerpiece for discussions on this topic. While it is exciting to talk about the "Build Your Own Bank (BYOB)" service provided by Bank of America, it is important to concentrate on the security features available, not only in Internet Banking but in all the related issues that have been mentioned above. It must be remembered that in India, unlike in the United States, it is unlikely that the bankers and particularly the central bank–Reserve Bank of India–will buy any modernity only if security is absolutely guaranteed. Therefore, a degree of analysis of the security requirement needed to enable the regulators to allow what many old fashioned politicians are liable to call "new-fangled ideas." While security issues have been discussed, they have been discussed at a theoretical level. Likely cases and their solutions are probably required to be demonstrated to convince the regulators to encourage more banks to take up Internet Banking. On the other hand, ICICI Bank has started Internet Banking service in India and is using a software package called INFINITY. This is a package that provides convenient remote banking service and the facilities that it offers has been discussed in considerable detail. Here again, a while the security provided by INFINITY is discussed, as peripheral. For example, detailed capabilities of the firewall and the filtering routers as well as the specifications need to be given, if interest is to be created amongst conservative bankers. Another issue that needs to be dealt with, in case of on-line banking or Internet Banking is that of digital signature. As an appetiser, the report on Internet Banking is very good. Therefore, as a report that will arouse interest in using Internet Banking in India, the report is excellent and will form a format or guideline of how to go about implementing Internet Banking in India.

The Fourth Case Study on Electronic Modes of Payment and Related Issues is closely related to both Electronic Data Interchange and Internet Banking because when Electronic Data Interchange is used to transmit financial information and payments in electronic form it is usually referred to as *Electronic Funds Transfer* (EFT) and one of the precepts aimed at in Internet Banking is to transfer funds safely and securely. Therefore, all the papers that have been discussed so far are pushing the frontiers of application of Information Technology in Management. From that point of view, Electronic Data Interchange should be looked at more closely. In fact the situation is analogous to the time during the early development of computers when the terms real time, multi tasking, multi programming and on-line, although each is distinct in meaning, were often used interchangeably–without any loss in understanding. If one looks at the development of EDI, Internet Banking and Electronic Modes of Payment, it can be seen easily that the basic aim in the developments taking place in these are the areas all relate to cutting down paperwork and the idle time and effort lost in transit, not to mention the additional chances of error that creep in because of repeated handling. To that extent, all these reports have highlighted these points admirably. The only criticism that can be leveled against them is probably that there are a few areas that needed to be researched and commented upon while discussing the many relevant related issues. The issue of security has already been brought up and is relevant in all the papers that have been discussed as indeed for some of the reports that have not been considered so far. The second issue is that of digital signatures. This topic is again relevant to all the four papers that have been discussed so far. In the paper on *Electronic Modes of Payment and Related Issues* a great deal effort has been made to explain the theoretical issues involved. At the same time considerable effort has been taken to identify certain products and to present their characteristics in a very farourable light (these products are probably the most marketable in the present state of developments for example Mondex.)

The fifth case study is on *Security on the Internet* that deals with the theoretical issues involved in creating and maintaining security on the Internet. While sources of security involved have been looked at quite thoroughly, system vulnerabilities have been explained very well and effective security tools suggested, the report seems to lack drive because the firewalls to be used have not been explained in detail. Since adequate cases have not been taken up as examples, the theoretical report seems to lack authenticity and the flavour of solid research. However, such a theoretical report has its place. And in fact, in the plan proposed below, this theoretical report will also have its special relevance.

It would be interesting to consider as exercise, a composite plan–to be made out after incorporating the issues discussed in all the five papers discussed above–and to include in this composite proposal the small factors that have been missed in the various papers as mentioned above. Such a comprehensive plan could be presented to the IT ministry and the IT ministry could be persuaded to sell such a plan to the Finance Ministry, the Commerce ministry, the Industries ministry and other agencies relevant to the comprehensive plan. It has been seen that the Commerce ministry and ICICI Bank have already attempted to implement some of the ideas that have been discussed so far. But a comprehensive plan that could be implemented in phases would be ideal for the IT ministry to formulate and implement.

Consider the advantages of implementing such a plan. It is obvious that the cost of implementation will be substantially lower than if each of the modules is implemented separately. Secondly, all the principals involved will be doing it together. Therefore, there will be less confusion and quicker implementation. Since all the principals will be involved, decisions can be immediate. Therefore, implementation of the various modules will be quick. However, all this may be merely a 'pipe dream' its the government does not act quickly, even though they may agree with the concept.

There are reports that concern other issues such as the *sixth case study* on *Planning and Implement of ERP*. The report has included the standard features required in such reports: the history of ERP, the penetration of ERP in India, the benefits of using ERP, pre and post implementation steps for ERP, ERP evaluation and mention of the important vendors of ERP. All these topics may be considered standard for such a paper. However, there are two interesting features in this report. These are "the role of BPR in ERP" and "the future of MRP III." These two topics are worthy of some discussion. Business Process Re-engineering (BPR) has, in many ways, been a precursor to the use of ERP and today it is generally believed that before implementing an ERP in an organisation, it is worthwhile to go through the steps of BPR. Whether this is true or not is a debatable point and this debate has been carried on in quite a lively fashion in this paper. This is a very interesting part of the paper. The second interesting issue is the hypothetical discussion about the likely development of ERP. A hypothetical MRP III has been imagined as the likely next generation addition to ERP. MRP III is an acronym for the Money Resource Planer. There is a lively and interesting discussion about MRP III. These two special additional topics make this paper on planning and implementation of ERP a particularly interesting and useful study.

The *seventh case study* on *Supply Chain Management* deals with a refreshingly different subject. Supply chain analysis has become a recent analytical tool that has received a great deal of attention and indeed acclaim and it is logical to believe that it will be one of the managing frontiers of Information Technology in the future. As before (the report on ERP and its implementation), the report is comprehensive in as much as all the necessary issues have been dealt with and is unlikely to raise any ones hackles. However, as a document that can persuade people to adopt the methods of supply chain analysis, it is unlikely to succeed. Two of the suggestions – to develop a *Vendor Supply Chain Strategic Matrix* and build a win-win vendor compliance program do have considerable merit for consideration as subjects for lively discussions, but are unlikely to move people that matter to adopt supply chain analysis in spite of the unquestioned technical appeal of the process.

17

EDI and Its Implementation in the Indian Automotive Industry

COMMENTS

This is a report on the use of Electronic Data Interchange with a special reference to the Automobile industry. The article begins with an explanation of Electronic Data and it identifies a related concept Electronics Funds Transfer (EFT) and points out the different aspects of the two-Electronic Data Interchange (EDI) and Electronics Funds Transfer (EFT). The origins, importance and benefits of EDI are discussed at length. The elements involved are explained. The focus shifts to its applications in industry-particularly in the automobile industry. While the entire discussion is India-specific, the developments in EDI worldwide have been considered although fleetingly.

The report highlights the various requirements to suitably implement EDI in India. The prime requirement is the presence of a Value-Added Network. The obvious considerations here are the facts that the network should be easily accessible so that all the parties involved—the manufacturer, the raw material supplier, the customer and all the intermediate links—can access the network easily. At the same time, there must be adequate precautions so that the safety, secrecy and reliability of the documents involved can be maintained. These aspects, although discussed have not been fully clarified. However, adequate explanations have been given of the EDI software required, the status of EDI in India-particularly in the automotive industry and the results of a pilot study.

A great deal of importance has been placed on the suggestions for the growth of EDI in India. The authors appear to be committed to the continued use of EDI in India and these have been illustrated by discussions of the pilot study undertaken at the behest of a team consisting

---

Contributed by Avijit Basu, K.P. Ramakrishnan and Nalin Tripathi.

of President AIAM (Association of Indian Automobile Manufacturers), Joint Secretary Commerce Ministry and Managing Director, Maruti Udyog Ltd., among others. The results—the identification of the problems in implementation and serious suggestions made for the development and use of EDI in India are discussed. The committed suggestions are the best part of this report.

ELECTRONIC DATA INTERCHANGE (EDI): INTRODUCTION

Electronic Data Interchange (EDI) deals with the exchange of standard business documents electronically, instead of paper-forms between business partners, that is, customers and suppliers. It allows data produced by one computer to be transmitted to another system and processed there without any prior manual intervention. EDI can be used to electronically transmit documents such as purchase order, invoices, shipping notices, receiving advice and other standard business correspondence between trading partners. EDI can also be used to transmit financial information and payments in electronic form, when used in this application; EDI is usually referred to as Electronic Funds Transfer (EFT).

EDI is often viewed as simply a way of replacing paper documents with electronic documents and replacing traditional methods of transmission such as mail, phone, or in-person delivery with electronic transmission. However, EDI should not be seen as an "end" but as a "means" to streamline procedures and improve efficiency and productivity. EDI allows a new look at the "processes" within an organisation, with a view to re-engineer them in what has come to be known as Business Process Re-engineering (BPR).

Once data are entered into the buyer's computer system, the same data are electronically entered into the seller's computer, without the need for rekeying or re-entry. This is normally referred to as application-to-application EDI. When EDI is fully integrated with application programmes, not only does data flow electronically between trading partners without the need for re-keying, data also flows electronically between internal applications of each of the trading partners

The repeated keying of identical information in the traditional paper-based method of business communication creates a number of problems that can be eliminated or significantly reduced through the usage of EDI. These problems include:

- Increased Time
- Low Accuracy

- High Labour Charges
- Increased uncertainty.

EDI has become a major means of business communication among large companies. EDI can do for communications between unrelated companies what Electronic Funds Transfer (EFT) has done for transactions between large financial organizations—substitute electronic transactions for paper ones.

EDI consists of standardised electronic message formats for common business documents such as Request for Quotation, Purchase Order, Purchase Change Order, Bill of Lading, Receiving advice, Invoice and similar documents. These electronic transactions set enables the computer in an organisation to communicate with the computer in another organization without actually producing paper documents. The human effort required to read, sort and physically transport such documents is eliminated. The documents just mentioned, for which standard EDI formats are either in existence or under development, constitute about 85 per cent of the official communications associated with commercial transactions between business, government, educational institutions and non-profit establishments in the US and most of the industrialised world. It is estimated that in the developing countries also, the preponderance of these documents is in similar position.

To take full advantage of EDI's benefits, a company must computerise its basic accounting records. Trading partners are individual organisations that agree to exchange EDI transactions. EDI cannot be undertaken unilaterally but requires the co-operation and active participation of trading partners. 'Trading Partners' normally consist of an organisation's principal supplier's are among the early supporters of and participants in EDI. In the manufacturing Sector, EDI has enabled the concept of Just-In-Inventory to be implemented. JIT reduces inventory and operating capital requirements.

Origins of EDI

It should be realised that EDI is not a revolutionary concept. As the powers of computing and telecommunications have grown, EDI technologies have evolved as a natural data carrier replacing the paper document. Nor is EDI a new concept or a new practice. It has existed for over two decades in Europe and North America in industry sectors with products or services having a short shelf life but a high unit price. The airline industry is a classic example of this type of industry-empty seats mean lost revenue opportunities.

More recently, however, the profile of EDI has been increasing and it would appear at the moment to be very much 'flavour of the month'. In fact the situation amounts to much more than this. A number of factors, including drastically reduced costs of computing hardware, software and telecommunications combined with the lifting of trade barriers across Europe mean that EDI is moving from an embryonic, innovative phase into a phase of exponential global growth, a classic market life cycle. Another major factor is the increasing realisation of the role of EDI as a business enabler in increasingly competitive and dynamic markets. Increasingly EDI is seen as much more than a way of automating tedious, manual, 'paper-aden' processes. EDI is a tool which allows business processes to be executed using more effective but totally different approaches.

Importance of EDI

There is no doubt that EDI can bring significant benefits to organisations. These can generally be classified into strategic, operational and opportunity benefits and will vary in emphasis across different organisations, depending on why and how EDI has been implemented. Initial EDI applications have concentrated on corporate efficiency by improving data flow and error reduction. In these instances the business case for EDI was based primarily on direct cost savings.

With EDI, businesses can eliminate the need to re-enter data from paper documents and thus prevent clerical errors. Estimates suggest that 70 per cent of all computer input has previously been output from another computer. Each re-entry of data is a potential source of error. It has also been estimated that the cost of processing an electronic requisition can be one tenth the cost of handling its paper equivalent. In addition, EDI can reduce the need for personnel involved in orders and accounts processing.

EDI systems can shorten the lead time between receipt and fulfilment of orders. When scheduling information is transmitted with ordering data, companies can plan production more accurately and thus reduce stock inventories. Reduction in inventory can result in major savings. Use of EDI to transmit invoice data and payments can improve a company's cash flow and may increase the amount of working capital as accounts can be dealt with more efficiently.

Trading information obtained from historical data built up from EDI transactions is an invaluable source of market research and strategic planning information. The process of working with trading partners to implement EDI can also result in the benefit of closer working relationships with trading partners. It has now become apparent that the greatest value of EDI will emerge in strategic areas such as the provision of better levels of customer service and improved marketing competitiveness.

EDI Benefits

Strategic Benefits

- Faster trading cycle
- Ability to adopt new business processes such as Just-in-Time manufacturing techniques
- Ability to win new business or retain existing customers leading to improvements in business efficiency
- Ability to respond to highly competitive new market entrants.

Operational Benefits

- Reduced costs of paper and postage bills, cut-reduction in money tied up in stock-manual processing costs (e.g., associated with verification, keying and rekeying of documents and the cost of manual filing systems).
- Improved cash flow
- Security and error reduction
- Acknowledgment of receipt.

Opportunity Benefits

- Enhanced image
- Competitive edge
- Improved corporate trading relationships.

The other benefits can be summarised as follows:

- Convenient exchange of business documents during business or non-business hours
- Reduced business transaction costs
- Reduced information float
- Improved customer service
- Faster inventory replenishment and better overall inventory control
- Reduced inventory-carrying costs
- Reduced safety stock

- Advanced notification of shortages, cuts and substitutions
- More reliable forecasting at both supplier and vendor ends
- Improved shipping, receiving, and cargo tracking
- Quick, accurate and automatic reconciliation of documents
- Opportunity to negotiate better discounts on payment terms
- Improved cash flow management
- More efficient data flow at both intra-and inter-company levels
- More productive trading partner relationships
- Stronger bottom lines.

....For instance, Hewlett Packard achieved 40 per cent reduction in returned merchandise within Just twelve months of EDI Implementation.

....Levi's, the Jeans manufacturer, reduced inventory levels by 27 per cent, while simultaneously increasing sales by 27 per cent after implementing EDI.

Elements of an EDI System

- Computer System
- Telecommunication Access
- Value-Added Network Services
- Translation Software
- EDI Standards
- Article Numbering
- Trading Partners.

In the EDI environment, the basic steps are:

- Preparation of the document by in-house application system. This work is the same as in conventional system, except that paper need not be printed.
- The electronic document is translated to a standard pre-defined format and mailed electronically to a value-added network, which acknowledges the document immediately.
- The electronic document is then delivered to Addressee Company by the Value Added Network (VAN), using its own communication network.
- The receiving company will have translation support to convert this standard format to suit the internal format and will update the application system with the information; the document can also be printed, if needed.

- The major benefit is that the information availability is almost instantaneous depending on the network coverage in the Value Added Network. Also at the receiving company, the manual efforts are eliminated.

Applications in Industry

In short, the field of application of EDI extends to all trade and trade related activities. It is relevant to everything from printing to shipbuilding. Activities can be categorised into two main areas:

Commerce

- trade and industry
- manufacturing
- finance/banking
- tourism/travel

Transport

- road, rail, air, sea
- forwarding/dispatching
- warehousing Governmental
- customs and excise
- national/international trade
- statistics.

As far as particular industries are concerned, EDI is being viewed increasingly as a business-enabling technology facilitating any of the above activities. In the automotive industry the take-up of EDI has been primarily because of *Just-in-Time* (JIT) With JIT, a manufacturer orders and then receives parts from suppliers just before it needs those parts on the assembly line, EDI provides the transaction link that JIT requires. Where a transaction involves import or export, customs declaration documents can now be submitted using EDI messages which greatly expedite the whole process. For example, Singapore harbour implemented EDI techniques in the 80's. Now renowned for its competitiveness, customs transactions which previously took a day to complete can now be done in minutes.

Banks and other financial institutions are supporting usage of EDI in the area of Electronic Funds Transfer (EFT). EDI enables these organisations to operate much more efficiently by eliminating significant amounts of paperwork. This of course results in a beneficial knock-on effect on their Customers. Governments view EDI as an enabler to help manage such areas as procurement, taxation, logistics and so on. Government departments deal with vast quantities of information and EDI can facilitate the tracking and, therefore, management of this information. Governments intend using EFT to process payments such as social insurance and unemployment benefit to individuals and other organisations.

Implementing EDI: Major Considerations

EDI involves five main processes:

- **extracting** data from a computer system
- **translating** the data into a transmittable format
- **transmitting** the message
- **translating/interpreting** the message at the receiving end
- **downloading** the data in the receiving computer application.

Technically there are three main components which must be addressed before electronic trading can commence. As these components are the subject of separate articles in EDI Aware, they will not be examined in detail at this point in time:

1. *Standards*: EDI standards define the techniques for structuring data into the electronic message equivalents of paper-based documents. In other words, there are standardised methods for describing the components that make up a trading document, e.g., product code, price, name, address, etc.

2. *Software*: EDI management or translation software is designed to translate messages as they move from standard to internal formats and vice versa. Trading partners' EDI messages can move directly to the application software.

3. *Communications*: The communications infrastructure required for electronic trading is analogous to the pipeline of the process, that enables messages to flow between trading partners. Different options are available, such as exchange of discs or tapes, private networks etc, although most UK organisations now engage the services of third party network operators known as VANs.

What is just as important as these technical considerations is the fact that EDI implementation is primarily a managerial activity and therefore requires buy in at management level. The technical elements of EDI are relatively straightforward and easy to implement. Realisation of strategic gains through EDI will not be achieved from the elimination of paper flow within individual processes. The real opportunity depends upon the original analysis of the business and how EDI has been implemented so as to change the way the organisation does business. Once management commitment to move forward with EDI has been gained, the next step is to establish a phased plan for EDI implementation. The following gives an idea of how to approach this planning activity:

- Nominate EDI project champion
- Form EDI project team
- Analyse and prioritise internal and external information flows
- Analyse manual/automated generation of information
- Identify key trading partners in terms of order quantity, value, etc.
- Identify and exploit EDI information sources
  - awareness centres
  - trade associations
  - EDI software vendors and network operators
  - EDI consultants
  - EDI users
- Assess costs of implementation
- Plan pilot implementation based on all of above
- Agree pilot with selected trading partner(s)
- Assess and agree requirements for software, communications and standards
- Run pilot and subsequently evaluate outcome
- Progress implementation to new business partners and new business functions.

EDI is implemented in the same way as any other major business strategy, that is to say, in a piecemeal fashion proving and evaluating each stage before moving to the next. To prevent any disruption to key processes, it is advised not to attempt to switch to full EDI operation overnight. It takes time for people, systems and processes to adapt to the new methodology. Once the business is ready for full implementation of EDI there are further considerations in relation to interchange agreements with trading partners, specific responsibilities of each trading partner, etc.

EDI Standards

EDI is an inter-company and inter-industry project and, therefore, requires a common approach. Otherwise, there could be unmanageable volume of non-uniform electronic messages, which would become impossible for deciphering and use between applications. The exchange of business documents in a commonly-agreed structured format—necessitated the development of EDI standards. EDI standards are basically data standards in that they lay down the syntax and semantics of the data being exchanged.

The first thing to know about EDI standards is that there is more than one or, more accurately, more than one syntax upon which the EDI messages are built. The syntax comprises the rules that define how a message is assembled for exchange. Three syntax standards dominate in the world of EDI: ANSI ASC X.121 (often called ANSI X.12), UNTDI2 and EDIFACT3. ANSI X.12 is the dominant standard in North America and is also widely used in Australia and New Zealand. UNTDI used to dominate in Western Europe, and messages using this syntax are still widely used in the UK as a part of the TRADACOMS message set. However, the only international syntax standard is EDIFACT.

EDIFACT was born in 1985 as a merge having the best features of UNTDI and ANSI X.12 and out of recognition that in the world of commerce, transportation and administration there could no longer be national or regional syntax standards. In fact, no new messages were developed using ANSI X.12 after 1995 and in many industry sectors, such as insurance, travel and leisure, statistics, health and social administration, the only syntax standard used is EDIFACT. CEN4 will start to adopt chosen EDIFACT messages as the European standard from this year. An EDI standard comprises the syntax, the message design rules (i.e. the technical rules which must be followed when designing a message) and the directories (i.e., the messages themselves and the building blocks of the messages: segments, data elements and codes).

Messages and Directories

Messages are designed by technical experts in conjunction with the users. This is done in national or international committees set-up for this purpose by the relevant maintenance authority for the directories. New business functions will require changes to the directories: e.g., new messages; changes to segments; new data elements; new codes. These changes must be agreed by the maintenance organisation for the directories. Depending on the type of maintenance organisation this can take a long time (two or three years in the case of international EDIFACT directories because the directory changes must be agreed at the international level).

When the new directory has been formally approved, then the new message or new version of an existing message is available for use. The international body responsible for the maintenance of the EDIFACT directories is the UN/ECE trade procedures committee-Working Party 4 (WP.4). Although many user communities often prefer to wait for formal approval of a message before it is used, this is not always the case. For instance, the message used for the INTRASTAT declaration for trading within the E.U. was not formally approved until many months after the live date of the INTRASTAT regulation (1st January 1993) and the live use of the message.

Industry Standards

In the US, the transportation industry was one of the forerunners in development of EDI standards. Certain large segments of the retail industry also saw the advantage of EDI and proceeded to develop unique standards. This was not purposefully done to be different, but the earlier standards efforts of the transportation industry were not adequate to accommodate some of the retailer's requirements. The Uniform Communications Standard (UCS) was devised by the grocery segment and adopted by them and several other retail sectors. Meanwhile in Europe, other set of standards was developing. These were Trade Data Interchange (TDI) for warehousing, Organisation for Data Exchange by Tele Transmission in Europe (ODETTE) for the automobile industry and Data Interchange for shipping (DISH).

- ANSI X.12
- UN/EDIFACT (EDI for Administration, Commerce and Transport)
- EANCOM (National & International Trade & Distribution)
- EDIBUILD (Construction Sector)
- CEFIC (Chemical Industry)
- EDIFICE (Computing and Electronic Sector)
- EMEDI (Health Sector).

The basic unit of communication among EDI trading partners defined by EDIFACT is an **Interchange**.

An Interchange consists of functional groups of messages. Each functional group may contain many messages of the same type. Every message consists of a collection of segments with each segment comprising data elements-both composite and otherwise. Special delimiters are specified in the service segment to be used as separators for segments, components data elements and (composite) data elements.

Structure of EDIFACT

EDIFACT Structure	*Text Structure*
Data Elements	Words
Segments	Sentences
Syntax	Grammar
Message	Text
Envelope	Envelope

It is easy to understand EDIFACT structure, if we use a simple analogy between EDIFACT basics and text basics.

Data elements are the lowest entity in EDI, equivalent to words in a text. Group of similar data elements form a segment, which is equal to a sentence in text, and syntax of EDIFACT are Grammar equivalents of text. Group of segments form a message. For example, invoice is a message in EDI, which is a full-text message in text structure. Messages are enveloped electronically with information like from whom and to whom and message, to compete the business transaction. Envelopes are nothing but header and trailer information.

VALUE-ADDED NETWORK SERVICES

One of the key components of EDI is the communications medium used to enable the electronic transmission of business documents, between a large number of different organisations, throughout the country or indeed the world. In the early days of EDI, organisations like the Article Number Association evaluated the most appropriate means to support electronic communications between companies. As a result of their deliberations, it was determined that the use of magnetic tapes or disks for the exchange of data was limited in potential and would inevitably cause problems when the volumes of data exchange grew. They also decided that direct communications links between organisations would be difficult to manage in large numbers. They therefore determined that the way forward was to use a VAN to support the communications requirements of industry and commerce.

Documents exchanged between various different partners need a common postal service. This is true in the case of electronic data

Interchange also, the business partners need common electronic postal addresses, inter-connectivity between trading partners to transact documents electronically and adequate software to handle UN/EDIFACT standards. All these call for a common agency and Value Added Network Service to fulfill this role.

The Maior VANS

The result of this initial EDI activity, driven by the ANA, was that ICL was asked to develop an EDI service. This service known as TRADANET has become the most widely used VAN in the UK. Subsequently, ICL set up a joint venture company with {HYPERLINK "http://www.ge.com/geis/index.htmr'} (General Electric Information Services) to focus on the electronic messaging services business. This company, INS, is now completely owned by GEIS; between both companies they claim to support the largest community of EDI users in the world. Another VAN which was originally UK-based was Istel. Istel developed a substantial user community initially in the automotive and engineering sectors. It was then acquired by global communications company {HYPERLINK "http://www.att.com"} Thus both of the early market leaders in EDI are now part of international organisations offering global messaging services.

The other main contenders in the UK market are BT and IBM. {HYPERLINK "http://www.ibm.com/globalnetwork/"} has had a presence in the market for some time and in particular has established a niche for itself in the insurance business. More recently the company's presence has developed in other industry sectors and in particular, it has developed its European base of users. The last major player to enter the market was BT. Its presence within the messaging business was also enhanced by it's acquisition of Tymnet. Tymnet was one of the leaders in the US marketplace and gave BT a source of EDI products and services from which to evolve and expand into other markets including the UK. All of the above companies are expanding their market presence in an increasingly wide variety of industry sectors. In addition, there are some smaller organisations which offer niche services in particular industries, e.g., AutoLink and DIALnet. These are often targeted at the smaller companies.

Benefits of a VAN

It was clear from the outset that any organisation, committed to EDI, would have a requirement to communicate with a large number of different organisations, whether these be suppliers, distributors, banks etc. The function of the VAN, therefore, was to provide a single channel to facilitate this type of communication. The VANs support links to their networks for all of the main computer hardware and software operating environments. When you join a VAN, you need only be concerned about your link to the network, whether it be from a PC, mini or mainframe computer. What your customers or suppliers use to connect to the network need not concern you at all, as the VAN will take care of all these individual connections to their services.

Until recently, there was a limiting factor, in that there were no connections between networks. As a result, the early development of user communities tended to focus around a particular VAN. If you are in the retail sector the most widely used VAN is the INS TRADANET service, the automotive industry, on the other hand, developed around the AT&T Easylink network. If your organisation had trading partners on different networks, you had no option other then to join more than one. Now, all of the networks support interconnections with each other. It is possible, therefore, to join one network and communicate with trading partners on any of the networks via these interconnections. It should be noted, however, that the current nature of these interconnections means that the full end-to-end Audit Trail capabilities of VANs outlined below do not exist when you send data across more than one network.

In addition to providing the benefits of a single communications link to multiple trading partners, VANs fulfil the following functions:

Mailboxing

At its core, the VAN is essentially an electronic post office. It receives electronic messages which may be orders, invoices, etc., reads the addressing information contained in the EDI envelope surrounding these messages and posts them into the mailbox of the recipient.

All of this can be accomplished within a matter of seconds ensuring that critical business documents can be received by trading partners within minutes.

Security

Because of the commercial sensitivity to some of the documents transmitted, e.g., invoices, the VANs suppliers have ensured that high levels of security are maintained. Access to VANs services are password controlled and most VANs offer Trading Relationship validation as a means of ensuring that only authorised messages or documents may be transmitted. Therefore, if you wish to send an order electronically to a supplier, you will have to specify this in detail and set it up on the VAN. Equally your supplier will have to set up his side of the relationship and thus accept receipt of that order document. Only once matching relationships have been set up will the network allow transmission and receipt of the document by the consenting parties.

In addition to password and trading relationship control, some networks check the integrity of each transmission into their service. This level of checking ensures that the transmission conforms to the mandatory data elements of the EDI standard being used (see the article on EDI Standards). Also, if the transmission is incomplete, perhaps due to failure of the communications link, it is usually possible either to re-send the missing data or to re-send the whole transmission. The networks do not allow duplicate transmissions and will, therefore, automatically discard any of the data which has previously been delivered to the recipients mailbox.

Audit Control

It will be obvious from the above that it is essential for the VANs to provide the end-user with a full audit trail, so that users have information at their disposal with which to manage their use of such services. These facilities are now generally available as a function both of the networks themselves and the software packages that are available for connection to the networks. As previously stated, however, detailed end-to-end audit is not available via most of the available network interconnections.

In general, there can be no doubt that the overall levels of security maintained by the VAN operators is very high and that sending important business documents electronically is much more secure than using paper-based systems.

How do You Choose a VANs ?

In reality, most first-time EDI users will be influenced by factors other than the variations between VANs services. One of the main factors which influences most companies is an analysis of which VANs are used by their main trading partners. As stated earlier, different industries tend to have developed their use of EDI around a particular VAN, therefore, a useful starting point is to evaluate which of your key trading partners are on which network. It is also often the case that if your interest in EDI is being driven by a request to trade electronically with a particular customer that this organisation will have developed a start up package for you with a particular network. This again will influence your decision.

If you are not influenced by either of the above factors, you can evaluate each of the VANs in turn and make up your own mind. There are differences in approach and the costs associated with joining each network and you should, therefore, determine which VAN best supports the needs of your organisation. Your local Awareness Centre will help you if you need further information about VANs. VANs provide three distinct classes of services:

- Communication Network
- EDI Engine
- End-User Support

Communication Network

Private Data Communication Network for connectivity, as public lines will be expensive for intercity traffic. The situation perhaps could be different in future, when the Internet could become a way of normal communication channel, but this is not still an answer for immediate short term.

EDI Engine

This is hardware and software set to handle collection and distribution of documents electronically. The role of the EDI engine is to provide post box and mailboxes; authentication of trade relationship between partners; checking data integrity, e.g.: mandatory fields; end-to-end tracking for guaranteed delivery; audit trails and archival system.

End-user Support

This includes software at the front end to user for translation, mapping and communication. the communication portion handles the dial-up, transfer and confirmation.

VANs provide through their support engine, electronic post office service similar to conventional post office service for receiving and transmitting mails/documents.

EDI SOFTWARE

Of the three components that make up the EDI triangle, standards, software and communications, it is the software that will present the greatest number of variables and must be carefully selected to accommodate the way EDI systems are used. In terms of basic functionality, for messages to be transmitted, an EDI software package must be able to perform two fundamental tasks, it must incorporate a facility to allow data to be entered and encoded into an EDI standard format. Conversely, for incoming messages, the software must allow data to be decoded and entered into an in-house format. These two features define EDI software. The choice of software is of crucial importance to the effectiveness of an EDI system. The process of software selection cannot, however, be made in isolation from the rest of the EDI system and should be viewed in parallel with intrinsically related business and technical issues.

Which Software for the SME's ?

Many SMEs currently using EDI will have been advised by a major EDI user, commonly known as a 'hub', to purchase a preferred pre-packaged solution. While the larger company in this scenario may enjoy the multiple benefits of EDI, the smaller partner's sole motivation for implementation is usually to fulfil it as a condition of trading. This arranged solution usually specifies the hub's preferred software supplier, a VAN, and can sometimes impose an uncomfortably short timescale for implementation. The software supplier will already be familiar with the hub's messaging requirements and will, therefore, personalise the software to the specific nuances of the established user. Examples of this are the EDI rollout programme of Marks & Spencer which specifies a pre-configured version of {HYPERLINK "http://www.cityscape.co.uk/users/ew48/index.htm"} EDI software for use by SMEs or Safeway's policy of recommending Meadowhouse Bar Laser as its preferred software supplier.

The advantage of this type of system are that the software development costs are shared by the hub's supplier network. Conversely, the disadvantage of this type of software can be network connectivity restrictions that may become a limiting factor, if further expansion is required. As a result of the time restraint, in this situation, the SME often has very little time to investigate alternative options and will usually adopt the suggested solution while keeping the system as open as possible in order to accommodate further trading partners or networks. Some large hubs are not so aggressive in their EDI rollout programmes and will offer SMEs some general advice and a more reasonable timescale in which suppliers can decide how they may most effectively incorporate EDI into their IT systems. In this case, the SME will have some time to analyse the

EDI software market and select the system that is not only suitable for present but also future business needs.

The process of singling out one piece of software is a time-consuming task and can be a major exercise in decision making. The main reason for this is that there are now over 40 EDI software packages available in the UK market alone. To most companies, in particular the relatively computer illiterate SME, this often presents the single most difficult decision to be made as part of the EDI implementation process. For the typical SME one question for software selection, 'Make or Buy?', will already be answered by the limitations of in-house resources and so the question will become, 'Which off-the-shelf package to buy?' Since it is never the case that any two EDI systems' configuration will be identical, there are no hard-and-fast rules for software selection. However, by considering a number of criteria, a shortlist of prospective suppliers can be drawn up. In some ways, the process of EDI software selection is similar to choosing from a range of software for any other application.

EDI Software Selection Factors

EDI software consists of computer instructions that translate the information from unstructured, company specific format to the structured EDI format, and then communicate the EDI message. EDI software also receives the message and translates from standard format to company specific format. Thus the major functions of the EDI software are data conversion, data formatting and message communication. The EDI software is available for Mainframes, Mini computers and microcomputers.

Though technically any file transfer protocol can be used to transport standard EDI messages. The main considerations are:

Ease of Upgrade

As an EDI trading relationship matures, there is likely to be a requirement for enhancement of the existing system. This may arise as a result of additional messages, changes to existing messages or standards, addition of new trading partners and their individual messaging requirements, inclusion of additional network connections, or an increase in the number of business applications to be integrated.

Network Connectivity

Some EDI software packages are restrictive since they do not allow users to be connected to all of the major EDI networks. This is the exception to the rule, since most of the most popular packages have multi-network connectivity.

Multi-standards Capability

In cases where a supplier is faced with a situation in which two or more trading partners require EDI messages to be transmitted using different standards, the software must be able to accommodate this need.

Print/Report Generator

Many SMEs will not be integrating their EDI application with their in-house systems and will require a hard copy print of any incoming EDI messages. This facility must be available on the software under consideration.

INDIAN PERSPECTIVE

The Indian government liberalised its trade policies in the early 90's, which has given boost to trade its international trade. Indian exports to Europe, U.S.A and Asia have increased considerably. Since many of these countries are using ED, it has already exposed the Indian exporters to the benefits of EDI. The coordinated EDI implementation in most of the government regulatory agencies has given fillip to EDI, as exporters have stared utilising electronic submission of documents using EDI services provided by departments like Customs in ports at some cities in India.

EDI in India

EDI, or the inter-company computer-to-computer communication of standard business transaction in a standard format, is gaining impetus not only regards domestic business transactions, but also for overseas transactions. Videsh Sanchar Nigam Ltd.(VSNL) was the first to offer EDI Infrastructure in India. VSNL installed the EDI system in Mumbai, with access nodes at New Delhi, Calcutta and Chennai in 1993 and offered this service to potential users.

According to software industry forum, NASSCOM, EDI services are initially integrated into industry operations through purchaser function. EDI is the preferred mode of business in the automotive, railroad, chemical and pharmaceutical industries. There is a push to integrate EDI in paper products, metals, oils and gas, chemicals and pharmaceuticals industries and medical supplies industries.

The Ministry of Commerce stared EDI implementation in 1995-96, in some of the customs houses and the Directorate General of Foreign Trade. Kochi is the first Indian port to implement EDI transactions. The systems in Kochi is based on the UN/EDIFACT (EDI for Administration, Commerce and transport) standard, accepted worldwide. Kochi had entered into an agreement with Satyam Infoway for trial runs of the EDI system. With major international ports such as Singapore, Colombo, Hong Kong and Dubai announcing that all their trading partners should switch over to EDI before the year-end, or risk losing business, the Ministry of Commerce has directed that all ports should be brought within the EDI network before the year-end.

EDI Market Size and Growth

While the global market is lucrative, domestic potential is equally exciting. Conservative estimates indicate that the Indian Market for EDI solution and Services would be more than Rs.500 Crore very shortly. Export revenues would be extra. The implementation of EDI at customs and initiatives in this direction by various port trusts, Directorate General of Foreign Trade, Airport Authority of India, Container Corporation of India Limited, Banks, Shipping Agents, manufacturers, educational institutions have created a stimulating environment and has attracted a number of users towards adapting EDI. As a special initiative, Government of India under the perview of the 9th Five-year Plan, has decide to give strong impetus to infrastructure and core sector in order to improve and consolidate the competitiveness of the country. In this effort. EDI would act as a facilitator and value enhancer. Special focus would be given to transport sector and ports in use of EDI services for sharper international perspective to trade and commerce.

VAN Operators in India

Major VAN operators within the country are:

- National Informatics Centre (NIC)
- Videsh Sanchar Nigam Limited (VSNL)

Private VAN operators include:

- Global Telecom Services Ltd.

- Mahindra Network Services Pvt. Ltd.
- Satyam Infoway Pvt Ltd.
- Wipro Systems who has started in collaboration with—General Electric Information Systems (GEIS)
- Singapore Network Services Pvt. Ltd.
- Sterling Commerce USA,
- Netscape

Solution Providers

Sensing the market waiting to be tapped and opportunity soon to be unleashed, many IT majors have decided to jump in the fray and gain an early foothold in the Electronic commerce marketplace. The offerings would range from providing equipment to systems integration, from constancy to services, from translators to editors. There are more than 75 companies actively involved in offering products/services relating to electronic commerce. Some of the players are:

- CMC Ltd.
- A.B. Infosys
- Global Telecom
- Mahindra Network Services
- Satyam Infoway
- Verifone
- Wipro Systems Ltd.
- Tata-IBM
- Microsoft
- Manipal Control Data Systems Digital Equipment(I) Ltd.
- National Informatics Centre
- Oracle

UN/EDIFACT Security

Security issues have also been addressed under UN/EDIFACT by the UN Security Joint working Group (UN-SJWG). It proposes to provide an open standard to cover end-to-end solutions integrating security with EDI. Not only are these solutions network protocol independent, they are also independent of the security algorithm being used. Two approaches have been taken to provide integrity, authentication and non-repudiation of origin.

Integrated Approach

In the integrated approach, special security headers and trailers are included as a part of the UN/EDIFACT message.

Separated Approach

In the separated approach, an UN/EDIFACT message has been developed to provide these services.

UN/EDIFACT messages are also being developed to provide for non-repudiation of receipt, confidentiality and key management for security. services.

ELECTRONIC DATA INTERCHANGE IN THE INDIAN AUTOMOTIVE INDUSTRY

In India of early, '97, in a few top automobile manufacturing companies initiated some laudable efforts to introduce EDI among their suppliers on a pilot scale. While the attempt was to conduct transactions electronically, the methodology did not possess all the ingredients or infrastructure to ensure complete success, hence, while in a limited way they could make some progress, there were impediments when they attempted to include some partners. Some of the issues that threatened progress were the following:

- Lack of standard messaging structures
- Lack of connectivity among the service providers which can give flexibility to the users
- Non-uniform method of operation among the service providers
- A tariff structure which was observed to be not entirely rational
- Lack of legal sanction for the for documentless transactions
- Absence of other critical agencies such as banks, and excise who are an integral part of the transaction
- Absence of an agency to co-ordinate the movement
- Issues of security.

Role of the Commerce Ministry

Realising the importance of the whole exercise, AIAM and ACMA decided to jointly approach the Commerce Ministry for assistance. In April 1997, the Commerce Ministry nominated a group of users from the industry to form a team, which was given an exposure to EDI under the auspices of

UN-ESCAP at Bangkok which is co-ordinating this activity in the Asia Pacific region. It was decided to conduct a pilot trial on a National scale so that the experience can be incorporated.

Work of the Auto Team

Since then, the Auto Team has met several times and discussed various issues and has made progress in several areas. Prominent among them are the following:

- Arrived at standards for two critical documents, delivery schedules and Invoices for the Auto Industry as a whole
- Organised connectivity on a pilot scale with the help of VSNL and the Commerce Industry
- Arrived at common approaches to evaluate service providers
- Conducted a pilot trial invoicing several leading automotive organisations
- Given feedback to service providers to enable them to improve their services
- Launched a second trial to confirm that service providers are able to meet the requirements.

The Task Ahead

Auto Team's prime task is to check and confirm that the system can operate with a high degree of reliability. Parallelly, the other issues such as bar coding and induction of banks are also being discussed, where necessary, as in the case of legality, the assistance of the Commerce Ministry is being enlisted to achieve progress. It is expected that with this task force and win-win approach, the problems would get sorted out in the coming period and that the Auto industry would be in a position to demonstrate success through a co-operative endeavour.

Implementation Issues in the Indian Context

The following issues were seen as having a bearing on the implementation of EDI in India.

Country-specific Issues

- The position of EAN, it's role and structure
- Development of norms for the working of Value. Added Network service providers

- The availability of interconnectivity between the service providers
- Development of messaging standards; this may have to be evolved for each industry
- Development of rational tariff structure among the service providers
- Development of a mechanism for dealing with disputes not only among partners in EDI implementation, but also among other agencies which will interface with the system.

Some of the issues are elaborated below:

Normally, business partners are spread all over the country and also extend overseas. Electronic transfers cannot be cost-effective if we have to use public phone lines to individually interact with all the partners, national and international. So, VAN service providers need to offer this service cost effectively. They lease the DOT lines to connect between cities, hoping to pool the transactions across many companies for cost sharing.

When private operators are allowed, we normally get more than one choice. For example, in Japan, it is said there are over 1300 such VAN Service Providers. In such cases, there is a need to interconnect these services in order to facilitate that a company be with only one VAN service provider, irrespective of the choice with its business partners. Standards and finalisation of common electronic document forms are major steps which require a common forum. International connectivity and legal issues require coordination with Government agencies. e.g.: Acceptance of electronic documents without personal signature.

Company-Specific Issues

Format Translation

- Each company has its own internal format.
- EDI standards have a different format.
- Data for company specific mapping to service provider.

Skill Development to Manage Internal/External Issues

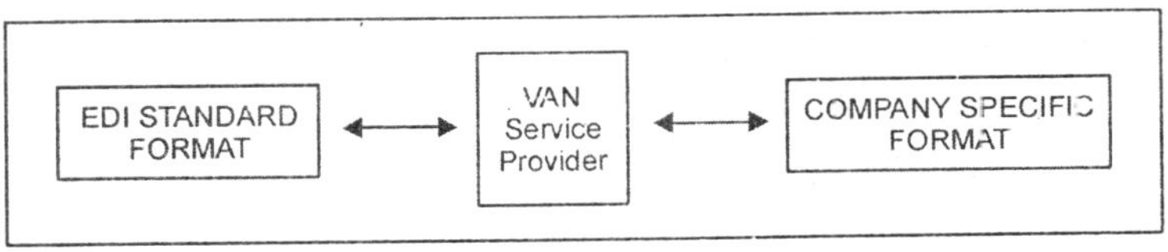

Internally, each company will have its own application systems and document formats to handle business transactions, based on culture and comfort. EDI standards would be different in format and therefore an internal mapping and translation is needed at the front end as interface, in order not to disturb the internal application systems. Normally, the VAN service provider should support this aspect.

EDI Initiatives in India

Early Initiatives

Ministry of Commerce(MOC) became the nodal agency and thus the focal point for all activities in the area of EDI/EC in India. Various institutional arrangements like EDI Council of India, etc., has been created by the MOC for the development of this sector. India is a member of the Asia EDIFACT Board.

As early as the beginning of 1997, some of the Original Equipment Manufacturers (OEMs) in the vehicle manufacturing industry such as TELCO, Bajaj Auto, Mahindra and Mahindra and Maruti Udyog took the initiative to experiment with EDI and implement on a pilot scale among their local suppliers. The idea was to send delivery schedules to suppliers electronically thereby cutting the response time for the system. In turn, suppliers were requested to send a copy of their invoice through EDI so that pipeline stocks could be immediately reckoned into calculation. In these efforts, they were assisted by service providers, such as, Global Telecom Services Ltd. Through their perseverance, they were able to commence operations on a small scale.

Problems Faced

However, when the need arose to expand this operation, some of the implementation issues gained prominence. These were issues, such as, the following:

- Need for common formats across the industry. In the absence of this commonality, component manufacturer's would have to deal with many formats for the same transaction.
- Need for connectivity among the service providers so that the participating companies need not be compelled to select any particular service provider.
- Need to develop some criteria to assist in the selection of a service provider.

- Some gaps in the knowledge of EDIFACT standards and in the implementation strategy.
- Fixation of tariffs.

Role of the Commerce Ministry

Some of the ancillary organisations, notably MICO and Sundaram Fasteners Ltd., brought these to the attention of the Automotive Component Manufacturers Association (ACMA) and the Association of Indian Automobile Manufacturers (AIAM) who decided to make a joint representation to the Commerce Ministry to assist in the matter. Recognising the need to have an organised approach, the Commerce Ministry initiated the following:

A formal launch of EDI implementation in India was organised by the Commerce Ministry in July'97, attended by President AIAM, Joint Secretary, Commerce Ministry and President EAN India, Managing Director, Maruti Udyog Ltd. and others. At the conclusion of this session, it was decided to form appropriate committees to handle the task of implementation through a pilot project for the auto industry. The important ingredients of these efforts were as under:

Objectives

- There will be a participation at 3 levels namely, the OEM, the component supplier to the OEM, and the component or the raw material supplier to the component manufacturer
- The idea will be to have a Win-Win approach whereby all the parties in the effort will stand to gain
- International practices will be adopted
- There would be one set of standards for all
- The system would be cost-effective
- Choice of service providers would be ensured
- All the connected agencies will be brought into the fold E.g Bankers, Transport Organisations, Ports and Government departments which interface with the industry such as Customs and Central excise.

Formation of Teams

The following teams were formed:

A Steering Committee comprising of two members each from AIAM and ACMA with the Commerce Ministry in the chair.

MESSAGE DESIGN FOR THE AUTO INDUSTRY

Even though there are a lot of messages that can be transacted between the trading partners, they decide to restrict the interchange to a few messages. These messages were selected based on the ease of understanding and the volume of transactions between the trading partners. Since the external agencies such as banks, tax authorities etc. are not yet EDI-enabled, messages to third parties were not considered. The recommended structure is only a guideline so that a uniform mapping can be made and the mapping costs can be eliminated. This will also help in ensuring that the EDI implementation can be fast and effective. However, if the sizes of the data elements is greater than the recommended size or if you have some other data which is not in the auto industry standard, you have to have a separate mapping made for your document and you should also ensure that your trading partners are able to understand the enlarged size and the new fields. If your size is greater than the maximum size, you must modify your application so that you at least adhere to the maximum size. The following documents were considered for the first phase of the autopilot:

1. Delivery Schedule (DELFOR)
2. Despatch Advice (DESADV)
3. Receipt Advice (RECADV)
4. Invoice (INVOIC)
5. Multiple Payment Advice (PAYMUL)
6. Multiple Credit Advice (CREMUL).

Delivery Schedule (DELFOR)

This can be used to specify delivery schedules requested by the buyer. In addition, the message may also be used for transferring:

- The total acceptance or rejection of a previous delivery schedule message
- The total rejection of information at for individual product(s)
- Amend the data of a previous delivery schedule message.

There are two methods in which a delivery schedule message can be used:

1. *Location-Driven*: a delivery location (or locations) may be provided with multiple products and delivery quantities for the locations identified.

2. *Product-driven*: a product (or products) and quantities may be identified with multiple delivery locations for the product(s).

It is not possible to club the location-driven approach and the product-driven approach in the same interchange. The delivery schedule structure will depend upon the approach you decide to take.

Invoice (INVOIC)

The invoice message can be used to claim the payment for goods or services supplied under consideration under conditions agreed between the buyer and the vendor. Even though the same document can be used to pass a credit note/debit note between the trading partners. Also export considerations were not considered

Despatch Advice (DESADV)

The despatch advice specifies details for goods despatched or ready for Despatch under agreed conditions. The message may be used either to indicate the Despatch of goods being delivered or to indicate the Despatch of goods being returned.

The despatch advice message should always be sent before goods are physically delivered or returned. This makes it possible for the receiving party to use the data to prepare efficiently for the reception of the goods. The despatch advice can be complex and indicate the hierarchies in the shipment and give details until the lowest level. However, a simple despatch advice should be adopted which consists in specifying the items to be despatched and the relevant information per item.

Receipt Advice Message (RECADV)

The receiving advice message addresses the business needs related to the receipt of goods. It is used either to confirm reception of goods or in conjunction with the despatch advice messages to confirm receipt or to advice discrepancies following the reception of goods. It allows the buyer or recipient of goods to provide the supplier or respective agent with:

- A confirmation of the receipt of goods
- A notification on discrepancies between the number of items received and accepted and the number of items despatched.

The buyer should always send the receiving advice to the supplier or their respective agents after the goods are physically received and inspected.

The message should be sent within a commercially agreeable time period, e.g., 24 hours after the receipt of goods. This makes it possible for the supplier to check the data of the despatch, adjust the invoice or issue a credit note, control internal procedures, count shock, etc. Even though the receiving advice message can be used to address complex scenarios they considered only the simple scenarios. Within this scenario, the receipt advice message is only used to confirm or advice discrepancies related to the despatch advice or Note. When only confirming the reception and acceptance of goods, only the header section of the receiving advice message needs to be transmitted.

USE OF EAN SYSTEM IN INDIAN AUTOMOTIVE INDUSTRY

EAN-India, a number of EAN International, Brussels, has been very actively associated with the Auto pilot project since the beginning. Participating in the EAN Mission in Bangkok.

The EAN Toolbox, broadly, comprises the following:

1. Identification of products—EAN-13 and EAN-8 numbering systems with corresponding barcode standards.
2. Identification of Despatch Units—EAN-13, DUN-14 numbering systems with corresponding EAN-ITF-14 barcode standards. For despatch units as well as logistics unit, EAN/UCC 128 standards with corresponding barcode standard are also recommended for a variety of information with regard to weights, quantities, prices, dates, locations, etc., to be incorporated as barcodes for greater flexibility.
3. Identification of locations—EAN-13 system of identifying locations uniquely is recommended for use in EDI as well as use in distribution through EAN/UCC Application Identifier system.
4. EDI Standards—EANCOM standard (the latest version released in July 1997 contains 42 messages) developed by EAN-International within the framework of UN/EDIFACT Standard adopted by India could form the basis for development of standard messages for the use of communities, for instance, auto industry, deciding to use EDI.

Specific Recommendation for Auto Pilot

1. *Use of EAN Location System*

A location number is simply a numeric code that indentifies any legal, function of physical location within a business or organisational entities, for instance, companies, subsidiaries, divisions, customers, departments, warehouse, delivery points, transmission points, etc. Each location is allocated a unique identification number. Such a number is reference key for retrieving information from databases such as postal address, type of Location, region, telephone, fax numbers, contact person, etc, EAN location numbers are governed by strict rule to guarantee that each one is unique world-wide.

(a) EDI

Location numbers are key concept in EDI. They provide a unique, unambiguous and efficient identification of all locations relevant to the EDI transaction. This is a pre-requisite to electronic trading as it is obviously preferable if all trading companies use the same location coding standard. Networks are then able to route EDI messages accurately to the designated mail box, work station or application. Names, addresses and information about particular locations do not need to be communicated for every transaction, the necessary information is communicated for every transaction, the necessary information is communicated only once, entered on to computer files and subsequently retrieved by referring to a unique, standard location number.

(b) Goods Physical Flow Process and Transport

EAN location number can be presented in barcode format and physically marked on to:

(i) Trade units to identify the parties involved in the transaction (buyer, supplier);
(ii) Transport units (Consignor, Consignee);
(iii) Physical locations (place of delivery, place of departure)

The symbology for this is the UCC/EAN-128 which can be scanned to automatically capture the details referred to the location: EAN Location Number vis-a-vis Internal System:

It is possible for any company to design its own internal system and code structure to identify all the locations covering it operating requirements. Although an internal solution might seem to be the easiest and fastest way for word, this solution is beset with several problems when information is exchanged between computers of distinct companies, viz., duplication, complexity and significance. The problems may render the internal solution ineffective and chaotic.

The use of EAN location numbers, on the other hand, provides companies with a method of identifying locations, within and outside their company, that are unique, multisectoral and international.

2. *Results of A Pilot Study*

In March 1997, trials took place among the various partners and documents were exchanged, It was observed that interconnection was possible through the X.400 gateway, though all the combinations could not be successfully tested due to other software issues.

While selected documents were exchanged amongst almost all the partners, the following deficiencies were noticed:

Technical Traceability

- Facilities for acknowledgement and notification required improvement
- Logs of VANs needed fine tuning so that traceability would be easy.

Speed

Speed of transaction must be enhanced.

Errors

- The following types of errors were noticed
- Documents missing during transmission
- Difficulty in deciphering contents at time
- Duplication of documents
- Appearance of unauthorised transaction.

Deficiency at the Presentation Level

In some cases, the ability of VAN software to access the database and convert it to EDIFACT format advice was not wel! developed. From the Managerial perspective, the following needed attention: Some enhancement in the competence of the VANs. Experience had to be brought into the effort and this can be bridged, since the VANs have access to a Global knowledge base.

Inter-VAN cooperation protocol needs to be tied up. The mode and manner of operation of each VAN differed and there was a need to have a common understanding on the manner of operation between them. This included aspects, such as, Network Management with VSNL as a player, hand shake procedures, and other assorted software issue.

ACTIONS TAKEN AND TASKS AHEAD

Short Term

- Complete a second round of trials and rectify any deficiencies that may be notices.
- Generate a report based on the experiences gained and publish guidelines for implementation.
- Finalise a procedure for introduction of the new standards in companies which are already on EDI.
- Organise unique location numbers for each Organisation based on the EAN numbering system. Communicate the numbering pattern to the EDIFACT board.
- Finalise the draft interchange agreement.
- Organise permanent connectivity with cooperation from VSNL.
- Conclude the pilot stage.

Long Term

- Select messages for the next phase of activity. This will relate to the financial part of EDI transactions such as payments and bank advices. Develop standard for messages.
- Widen the scope and operation of EAN numbering systems.
- Get legal clearance to operate exclusively through EDI without the need for a hard copy back up. This will have to be done with the assistance of the Commerce Ministry who need to get it cleared with the Ministry of Law, Justice and Company Affairs.
- Study the Tariff angle to check if there is some uniformity and consistency in the tariff structures of the VANs.
- Build a mechanism so that, on an ongoing basis, there is a method to keep track of and further develop EDI in the auto industry. It is proposed to do this by having representation of the auto industry in the EDI council.
- Bring banks and the other agencies interfacing with the auto industry into the EDI circle so that there is a benefit over a wider perspective. e.g.; Customers, Central Excise and the Port Trust.

In particular, the induction of banks will help in expediting payments of component manufacturers.

- Develop a mechanism so that when other agencies are brought into the movement, there is a participate and win-win approach.
- Develop a framework so that all the VANs would be able to sort out issues pertaining to connectivity among themselves in an amicable manner.
- Bar coding issues to be taken up on lines similar to EDI so that there is some anoint of uniformity in the application of bar coded systems in the country and also there is scope to integrate with the rest of the world in this critical area

ACTIONS ORIGINALLY PROPOSED

Realising the enormous implication of the introduction of EDI systems in the country, the Government and the auto industry decided to form a working group to study and recommend the desirable course of action. In addition to looking into all aspects of the issue, it was decided to conduct pilot trials to the proposed system to satisfy all the connected agencies. The crucial area requiring clarity were identified as under:

- Uniform messaging standards
- Connectivity among VANs
- Assessment of VAN capability
- Legal cover of EDI
- Tariff structure of VANs .
- Participation of other player
- Development of EAN numbering and bar coding systems.

Actions Completed

Towards the above, sub-committees were formed and the following were achieved:

- Uniform message formats for delivery schedules and Invoices
- Checklists for the following
- To assess VAN capabilities, termed as guidelines for VANs
- For conducting auto pilot trials and arriving at meaningful conclusions
- Temporary connectivity with the assistance of VSNL
- Pilot trials were conducted and the results have been shared with all the partners for ensuring corrective action

Action to be Taken

- Permanent connectivity
- Legal cover of EDI
- Study of tariffs
- Extension to other messages and to other partners
- EAN numbering and Bar Coding System.

PROBLEMS IN IMPLEMENTATION

1. Lack of required level of coordination among the competitors and participants
2. The EDI system used abroad and the system used in India modelled on it.
3. A lot of errors detected:
   - Documents missing during transactions
   - Content deciphering difficult at times
   - Duplication of documents
   - Unauthorised transactions appeared
   - Failure at times of the VAN software to access the database
4. Inability to develop and finalise a set of new standards in existing companies already on the EDI
5. Need to organise a permanent connectivity with cooperation from VSNL
6. Lack of uniformity and consistency in the tariff structures of VAN's.
7. How to connect the rest-manufacturers and suppliers (VAN's, connections compatibility, VSNL, etc.).
8. Small manufacturers found non-EDI compliant. This means that large investments in the big companies results in a lop-sided framework. Small Companies lack the resources to be EDI compliant.

SUGGESTIONS

1. The primary suggestion would be to institute this system among all large players. This requires aggressive selling of the benefits and advantages of the system.
2. Nurturing the small companies in a gradual transition towards the EDI for all. Small companies need to be adjusted for.

3. Speeding up the trouble-shooting process to ensure a bug-free system.
4. Gaining government approval and concessions for using VSNL for Data transfer.

INTERCONNECTIVITY CHART FOR THE AUTO PILOT

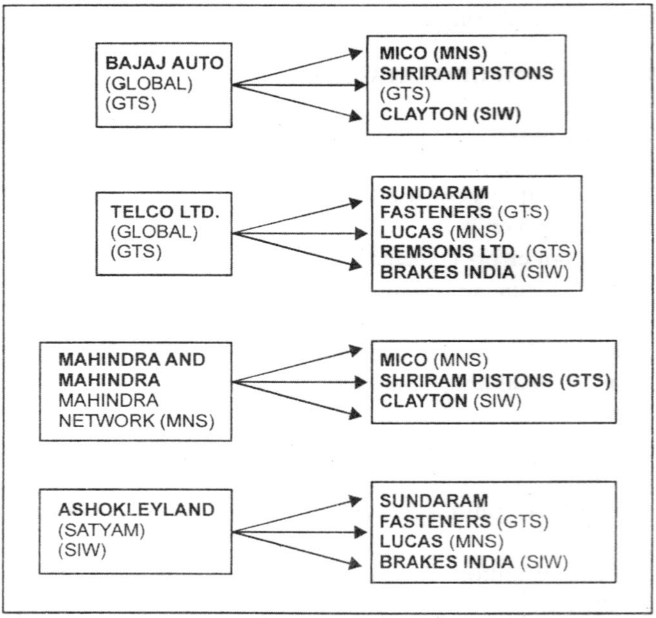

18

EDI and Digital Equipment Corporation

COMMENTS

This paper is an approach to Electronic Data Interchange (EDI) as propagated by Digital Equipment Corporation. Electronic Interchanges are explained and the important issues involved are highlighted, particularly major legal concerns. Also, the protocols and syntaxes involved (X12 and EDIFACT) are discussed and explained during the course of the paper.

The discussion begins with an explanation of EDI—a computer-to-computer exchange of business-related documents—and the understanding of EDI in terms of a trading partner. The trading partner may be a supplier, a customer, subsidiary or any other organisation with which the enterprise does business. The history of EDI explains many of the issues and their importance today—early electronic interchanges were based on proprietory formats agreed between two trading partners. Finally in the 70s, users and vendors agreed to create EDI standards. The focus shifts to EDI in India and it is pointed out that there are more than seventy-five companies offering products and services suitable for EDI in India. Of these, the prominent six have been identified and one—Digital Equipment India Ltd. and their products and services concerning EDI has been taken up for detailed discussion. The EDI market consists of two types of buyers: entry-level EDI users and existing EDI users whose requirements are more sophisticated. At present, most EDI users are in the "basic EDI" stage of implementation. New value chain programmes are being implemented by senior managements in many organisations, such as, Just-in-Time manufacturing and retailing, Total Quality Management, Quick Response Systems, Enterprise Resource Planning Systems, and this requires EDI integration. Therefore, EDI buyers' needs have to be carefully looked into.

Contributed by Pooja Murti, Ruby Khanduza, Gunjan Vacchar and Olina Sen.

Then there are technical issues, starting with standardisation of EDI. There are two prevailing standards: X12 from ABC, which is popular in North America and EDIFACT from UN, which is popular in Europe and most of Asia. The status of acceptance of each is discussed. This is followed by a discussion of Basic EDI Software functions, syntax and semantic conversion, grouping and splitting and application integration. Other issues discussed are data mapping and different communications options, including X25 Public Packet Switched Networks X400. A discussion of legal issues and security concerns regarding EDI conclude this general discussion on EDI.

This generic discussion of EDI is followed by a discussion on Digital Equipment Corporation and its Indian subsidiary, Digital Equipment (India) Ltd. and their EDI solution and its key features and benefits. Digital's EDI Services are highlighted. The report finally concludes with Digital's EDI Integration Services.

INTRODUCTION

EDI is the computer-to-computer exchange of business related documents such as Purchase orders, invoices, payment remittances and shipping notices between an enterprise and its trading partners. Early electronic interchanges were based on proprietory formats agreed between two trading partners; therefore, with the implementation of EDI, two translation standards were developed: (SASC) XI2 and UN/EDIFACT. EDI implementation process entails following steps: customer engagement process understanding, new process design and process change. The market is made up of two types of buyers: Entry level EDI users and Existing EDI users who need more sophisticated EDI capabilities to meet changing business needs. Buyers typically choose among three approaches to address their EDI needs: Basic EDI, Integrated EDI and Electronic Commerce. For systems development options, firms would have to consider the choice of EDI software and approaches to introduce application software. EDI software refers to the translation and communication software required to generate and interpret EDI transactions and prepare for communication to the trading partner. The basic aspects that need to be considered when looking at EDI software are: data extraction, data encoding, data transmission, data receipt, data decoding and data incertion.

The major legal concerns of EDI are regarding the authenticity of information, the question of when an acceptance of an offer becomes effective, the permanent storage of information, satisfaction of laws which requires certain legal information to be written and signed, risk of errors amongst trading partners and intermediary networks. Protection of the information within the trading partners organizations—is considered to be—a separate local issue. EDI security in this context focuses on the vulnerability of the messages during transmission. Loss of confidentiality, integrity, availability, threat from unknown users, service disruption due to viruses are the main EDI security problems. To counter the above-mentioned threats, some EDI security requirements have been identified.

Digital Equipment Corporation, founded in 1957 in USA, gives its Customer a winning internet advantage through a comprehensive portfolio off Internet solutions based on award winning systems, advanced networking infrastructure, innovative software and industry applications. Incorporated in 1988, Digital India is a subsidiary of the Digital USA. It is one of top five multinationals IT vendors in the country. It has become most sought after total solution provider in the field of manufacturing, finance, telecommunications and Internet. It introduced its first EDI software, DEC/EDI in 1990. It offers business a complete portfolio of EDI and electronic commerce services. Electronic Commerce from Digital integrates X.435-based mail, fax routing. X.500—based directory services, Internet access via SMTP and more. It offers EDI services to organisations in various forms viz. EDI orientation services, packaged EDI services, customised EDI services & DEC/EDI integration services.

WHAT IS EDI?

EDI is the computer-to-computer exchange of business-related documents such as purchase orders, invoices, payment remittances and shipping notices between an enterprise and its 'trading partners.' A trading partner, in EDI parlance, is a supplier, customer, subsidiary, or any other organization with which an enterprise does business.

HISTORY OF EDI

Although the business computer enabled companies to store and process data electronically, companies needed an expedient method to communicate the data. This method was realised by the widespread use of computer telecommunications. Using telecommunications, companies could transmit data electronically over telephone lines, and have the data input directly into a trading partner's business application. These electronic interchanges improved response time, reduced paperwork, and eliminated the potential for transcription errors. Computer telecommunications, however, only solved part of the problem. Early electronic interchanges were based on proprietary formats agreed between two trading partners. Due to differing document formats, it was difficult for a company to exchange data electronically with many trading partners. What was needed was a standard format for the data being exchanged.

In the 1960's, a cooperative effort between industry groups produced a first attempt at these common data formats. The formats, however, were only for purchasing, transportation, and finance data, and were used primarily for intra-industry transactions. It was not until the late 1970's that work began for national Electronic Data Interchange (EDI) standards. Both users and vendors input their requirements to create a set of standard data formats that:

- were hardware independent
- were unambiguous, such that they could be used by all trading partners
- reduced the labor-intensive tasks of exchanging data (e.g., data re-entry)
- allowed the sender of the data to control the exchange, including knowing if and when the recipient received the transaction.

Although today there are many syntaxes for EDI, only **two are widely recognised:** XI2 and the EDIFACT (Electronic Data Interchange For Administration, Commerce, and Transport).

EDI IN THE BUSINESS PLAN

It is possible that a firm is being pressurised by a customer to transmit electronic documents "now". The best route is to anticipate and plan for a timely and efficient EDI implementation. If introducing EDI is a business issue, then having a formal EDI strategic plan helps. It ensures that EDI is incorporated into the firms overall business plan and that top management is actively involved. The EDI strategic plan forces management to take a hard look at the firms operational efficiencies as they relate to fulfilment of the business plan. As a result EDI becomes a management tool.

EDI IN INDIA

With the Electronic Commerce market burgeoning its way in India, the IT majors in India have jumped into the EDI bandwagon to gain an early foothold in the marketplace. The multiple offerings in the marketplace range from providing equipment for systems integration from consultancy to services, from translators to editors. There are presently more than 75 companies actively involved in offering products and services. Some of these are:

- Global Telecom
- Satyam Infoway
- Verifone
- Wipro Systems Ltd.
- Tata-IBM
- Digital Equipment India Ltd.

EDI MARKET OVERVIEW

EDI allows entire business documents to be translated into internationally defined standard electronic formats that are readable by different computer systems.

The market is made up of two types of buyers:

- Entry-level EDI users
- Existing EDI users who need more sophisticated EDI capabilities to meet changing business needs.

Frost and Sullivan in a recent study released in March 1995, projects the worldwide EDI market will more than quadruple from $700 million in 1994 to $3.2 billion by the year 2001. Analysts at Gartner Group in Stamford, Conn., estimate that 75 per cent of the market is currently in the U.S., but that U.S. sales will represent only half of total sales by the end of the decade. Much of the growth predicted will come as data from trading partners is woven more tightly into the business fabric of the enterprise.

Market Dynamics —Where is EDI Today and What Lies Ahead?

Today, many organisations are in the "basic EDI" stage of implementation, and face replacement of their existing system for one of two reasons. For some, the addition of more trading partners and new transaction sets is straining the capacity of existing EDI systems. For these organisations, the need is about mores horsepower"—increased EDI processing automation efficiencies to handle growing volumes without making corresponding changes in business processes. In other organisations, senior management is instituting new value chain programs, such as:

- Just-in-Time (JIT) manufacturing and retailing
- Total Quality Management (TQM)
- Efficient consumer response systems
- Quick response systems
- Enterprise Resource Planning (ERP) systems
- Evaluated receipts settlements

For these organisations, just adding "more horsepower" is not the answer. The move to integrated EDI requires more than a faster translator running on a bigger box. A new software architecture utilising client/server technology to support multi-platformed EDI-enabled business applications and communications flexibility is required.

EDI BUYER NEEDS

Buyers typically choose among three approaches to address their EDI needs. In order of increasing levels of business integration and expected payoff, they are Basic EDI, Integrated EDI, and Electronic Commerce.

Basic EDI

Many firms initially implement EDI in response to requests from major customers. The implementation is usually limited to the requirements of a few trading partners and involves basic EDI transaction sets such as purchase orders and invoices. Technology costs for basic EDI implementations are relatively low. They require EDI "Translation" software that converts data to and from standard EDI formats and a connection to a Value-Added Network (VAN), which allows for the data transfer to the trading partner.

Simple file transfer tools are sometimes used to move EDI data between business applications and the EDI translator. Otherwise, a manual re-keying process, which is often error-prone and people-intensive, is used to transfer the data. The result is a paper replacement system with little strategic impact on the overall operations of the organization.

Integrated EDI

As relationships develop between strategic trading partners, business processes are re-examined according to the value to both trading partners. Processes which do not add real value are eliminated or streamlined. In this environment, EDI linkages become the de facto method for all trading partner interactions and are used to "complete the loop" for the entire business cycle.

As an example of EDI business integration in the consumer-packaged goods industry, product activity data is taken directly from retail stores, replacing traditional purchase orders. Advance Ship Notices to customers are produced—including shipping container barcode information—eliminating the need for invoicing. "Quick response" systems are created by integrating EDI product activity data into replenishment planning systems. Key technology requirements to support integrated EDI include:

- Faster, more reliable communication connections to support Just-in-Time (JIT) requirements of multiple trading partners in a comprehensive Value Chain Environment.
- Core business applications, such as, order management, that are "EDI-enabled" to support electronic trading processes.
- Sophisticated applications integration tools to move and manage EDI data from "EDI-enabled" business applications which may reside on multiple computer platforms.
- Linking together of "EDI-enabled" business applications running on multiple platforms to form a single EDI gateway providing common system management complete data integrity and powerful transaction auditing across the entire network.

Electronic Commerce

Electronic Commerce requires tight linkages between trading partners plus integration among various internal functions. Here, EDI initiates the internal information linkage across the organization. EDI data is linked at the system level with all business applications, networks and electronic messaging systems. Routing and electronic mail are integrated to support the streamlined electronic trading environment. Key requirements include:

- EDI as an integral component of an enterprise-wide messaging backbone.
- Support for multi-media business documents.
- Integration of workflow, information sharing, electronic mail and EDI.
- Security, reliability and audit trails for supporting automated "mission critical" business operations.

To support these requirements, utilisation of the new X.435 EDI communications standard will become a business necessity.

EDI TECHNICAL ISSUES

Status of EDI Standardisation

There are two prevailing EDI translation standards: the Accredited Standards Committee (ASC) X12 is the standard in North America and UN/EDEFACT is the standard in Europe and most of Asia. The format and the data that can be exchanged for any particular EDI business transaction are defined in the X12 EDI standards and EDIFACT EDI standards. The EDI standards development process has evolved continuously since the first standards were conceived by X12 in 1979. By 1987 there were only 16 (transaction sets) X12 standards. However, since 1987 tremendous growth in EDI has taken place in standards committee members, users, transaction standards and the complexity of EDI standards. As of the end of 1991, there were 600 committee members, 12 sub-committees, an estimated 20,000 users, 106 approved (transaction sets) standards and 150 new (transaction sets) development projects. The EDIFACT standards are following the path of X12, but the EDIFACT progress is behind X12.

Although XI2 is the prevailing standard in North America, EDIFACT is becoming more widely 'accepted internationally, especially in Europe and in Asia. The Department of Commerce has published Federal Information Processing Standard (FIPS) 161 on EDI [26]. This FIPS states that federal agencies that are implementing EDI shall use ANSI XI2 and/or UN EDIFACT in the cases where these standards provide EDI form definitions that meet the agencies EDI requirements. Very recently, the XI2 committee voted (as a whole) to adopt the EDIFACT syntax by 1997.

Thus, the original XI2 syntax will be phased out for many, or most, or possibly all applications. Exactly how this will occur is not known now, since implementation plans have not yet been put into place. (FIPS 161 is currently being revised to reflect this situation.) X.435 is the only existing International Standard for communicating EDI transactions. X.435 defines and standardises an EDI message which can be communicated over X.400 based electronic mail systems.

X12 and EDIFACT

In 1979, the American National Standards Institute (ANSI) chartered the Accredited Standards Committee (ASC) XI2 to create a set of standards to facilitate the electronic exchange of business information. These standards define the data formats and encoding rules required for a multitude of business transactions, including order placement and processing, shipping and receiving, invoicing, payment, and many others. The X12 and EDIFACT standards perform equivalent functions (e.g., both define a purchase order). Although X12 is more mature and provides functions not present in EDIFACT (e.g., acknowledgements, security), many of these functions are under development in the EDIFACT organisation.

Issues and Options in EDI Systems Development

Under the systems development options, firms would have to consider the choice of EDI software and approaches to introduce application software. EDI software refers to the translation and communication software required to generate and interpret EDI transactions and prepare for communication to the trading partner. The basic aspects that need to be considered when looking at EDI software are: data extraction, data encoding, data transmission, data receipt, data decoding, data-insertion.

Options to Acquiring EDI software

There are basically three approaches a user to acquiring EDI enabling software:

1. *Write the software yourself*

This approach will require sufficient in-house resources and will be time consuming and expensive. It is, therefore, only justifiable if the next two options would require too much customization to meet the application requirements in a commercial way.

2. *Purchase an off-the-shelf EDI enabling software package*

The option has the advantage of acquiring a relatively low-cost, high performance specialized software but has the disadvantage that it will leave the more complex task of integrating the outgoing and incoming EDI messages with the business applications up to the user, either through in-house resources or through a specialized EDI consultancy.

3. *Purchase an off-the-shelf business application package with integrated EDI capability*

Unfortunately, there are not many packages of this type available on the market today. Application software suppliers claim that the cost of developing such modules could be too high pricing their products out of the market, especially for smaller businesses, many of which are still unaware of EDI and its benefits.

Basic EDI Software Functions

Software functionality can be broken down into six different categories summarized as follows:

Conversion: The primary function of EDI enabling software is to convert firm A's data files structure according to internal formats to an EDI standard message, a process commonly refer as construction of EDI messages, and to convert EDI messages into data files structured according into firm B's Internal format, a process commonly known as *EDI message translation*. The greater the flexibility offered by the converter on Internal file format, the more likely that modifications to the application will not be required.

Syntax Conversion: Internal format messages are constructed towards statically correct message standard according standards syntax rules (control of service segment, segment tags, separator character, character sets repetition and mandatory statuses).

- It is important to consider the number of message standards to be supported by the software (e.g., EDIFACT, ANSI X12, TRADACOMS etc.)
- Whether the package is delivered with the standard' message definitions and updated with new message versions.
- The package should be able to handle specific trading partner conversion and multiple versions for one message to suit the requirement of individual trading partners.
- The converter should also be able to automatically include default and standard data element values such as control counts, hash totals, and reference numbers by means of tables and parameter files, a useful functionality, as this type is usually not contained within the application data base.

Semantic Conversion: Semantic conversion relates mainly to representation and meaning of data to and from a standard message. EDI message standard use standard codes which will not correspond to the codes used in the in-house application. Code conversion is, therefore, an important semantic function. This is especially true for qualifiers. Code values which give a specific meaning to generic data element in an EDI message standard. As in-house application do not use qualifiers, semantic conversion converting application data to appropriate qualifier and vice versa.

Semantic conversion will also include the conversion of units, lower case capital conversion and adjustment of length and precision of data element (variable length fields common in EDI message standards to fixed length fields of an application).

Grouping and Splitting: This includes the grouping of outgoing messages into one interchange/per recipient for sending in one communication session and the splitting of incoming interchanges into messages, which may be sent as one file to the relevant application for processing.

Application Integration: There are generally three different approaches to application integration. It is the most common approach is to submit the application data through an intermediary program, sometimes refered to as an application service program, which will prepare and in-house file or flat file which can be processed by software converter. Going through an application program interface (API), available in some common software packages. Through an API, the application directly communicates with the EDI software through a standard interface. The principal difference between service program and an API is that the first is driven by the application data while API is driven by application processes and is therefore more interactive.

The final approach is to key-enter data manually through an input screen in the EDI software. Although this is no longer EDI as it involves human intervention and not a direct application-to-application exchange of data, it may be practical solution until full application integration is finalised or to provide data needed in EDI transaction, but which can not be generated or is not contained within the Business application. Some EDI software packages may contain Data entry modules for this function.

Data Mapping

Data mapping is required to get data sent in the form of an EDI message standard into the database of a business application software. Although message standard implementation guidelines such as EANCOM have eased the problem of data mapping by specifying the type of data which might be sent and where in the message it should be sent, trading partners still have to know and agree on data the customer is going to send and the data the supplier will need to receive. It is this aspect of EDI which helps to build trading partner relationship and enhances collaboration and cooperation in the business relationship. Companies need to discuss their business at the data level which leads to a much better understanding of their business processes and relationships and provides an opportunity to understand and resolve underlying problems. As a result, data mapping cannot be left solely to the technical people. It is a business problem that business functional managers who understands the business, supported by data-processing staff, must resolve.

Management-Reporting and Maintenance

If EDI involve business transaction and the objective is to have an unattended, automated process, it is essential form the software to be able to control the message flow. The following functionality should be included:

(a) *Logging of files* as an overview of messages sent and received. The information should include the type of data and, time sent and received, the employees and applications concerned, length and status of messages etc.
(b) *Audit Trails*: a complete of each separates message.
(c) *Recovery files*: Contain the actual contents of message and can be used to reprocess the message sin case of errors.
(d) *Error reports*: Usually produced by the software indicating errors that occurred during the translation process.
(e) *Status report*: Reflecting the status of messages, i.e., the message received, translated etc.
(f) *Purging and Archiving*: Storing and deleting completely processed messages. Maintenance of EDI software is also of importance and the user should ensure proper maintainance of translation tables for new messages or new message version, etc. either through an editor, a module in the software listed or by software vendor.

COMMUNICATIONS

The software may include a communication module which provides an interface with the network/s use to send messages. Most EDI users subscribe to services of Value-Added Network (VAN). EDI software should enable the connection to one another VANs, may support various different communication protocols used by different trading partner and also support direct point to point communications with specific trading partners. The software will control communications sessions, including modems and logon's and logoff's from the network service.

Communication Option Analysis

Along with the message standards and EDI enabling software, there is the requirement to transmit the information contained in a message from sender to receiver. Telecommunications, being part of the EDI concept, the focus is on five basic options within this domain as follows:

Point-to-Point Private Leased Lines: This option requires the greatest implementation effort due to the need to establish for each individual link common speeds of synchronous or asynchronous transmissions, data encoding formats and communication protocols. It is used for very high volumes of data exchange between stable long term communicating partners, e.g., intra-organisational communications.

Public Telephone Networks: The public telephone networks have the advantage of being widely available with no interconnection constraints. Although most telephone networks are slowly upgrading to full digital networks, the large amount of analogue equipment still limits the speed and volume of communications and is more prone to errors due to noise distortions on communication lines. Parameters like common transmission speeds, types and protocols will still have to be agreed upon between the trading partners. Both private leased lines or public telephone networks will usually require the use of modem to convert the digital data from a computer to the analogue form before transmission. The introduction of ISDN will form the basis of the next generation of public telephone networks offering full voice and data communications across high-density, low-distortion telecommunication lines.

Public Packet Switched Networks, X.25: X.25 has become the standard interface between a data communications terminal and most public packet switching networks. The international packet switched networks now provide coverage in most countries and since they are based on the X.25 recommendations, an international, most are interconnected. X.25 hardware and software interfaces are also available as standard products. Despite the advantages of X.25, it does not address any functions concerned with the identification and movement of information, so there's still a requirement to implement a common file transfer protocol on top of the communications functionality provided by X.25.

X.400: is the international standard for electronic messaging. The services provided by X.400 involve the enveloping of messages across one or more administrative or private domains and the receipt of the message at the destinations. X.400 provides a reliable, non-real time service with multiple levels of priority, X.400 users being able to specify the urgency and security of their messages. X.400 offers many user benefits:

- Through X.400 message tracing, a sender knows the transmission status of the transaction through delivery notification, even if the recipient is on a different X.400 network.
- A single message can be sent to multiple trading partners, providing not only convenience, but also saving transaction costs.
- X.400 offers a more comprehensive level of data security through secondary log-ons and password sequences and through data encryption.
- It provides global accessibility with most major VANs supporting X.400 standards and supporting interconnections with other X.400 VANs.
- Value-Added Networks—VANs: No communication protocols are needed in this case, which is a major benefit compared to other communication protocols. Also, it reduces the problems in terms of communications co-ordination and availability of trading partners.

Data Entry and Output Functions

Data entry will be usually driven by input screens whose format should be able to be defined by the user. Features which should be included in the data entry module are the holding of standard and default values in tables, menu-driven code selection, printout or fax possibilities for non-EDI trading partners amongst others.

Security

Systems and communications security functions may also be included in the software package. The degree of security required will vary depending on the nature of the business application. Security might range from the basic access control via a simple password to authentication of the parties involved in the business transaction, non-repudiation of messages received, encryption of data or digital signature.

Types of EDI Software

Types of EDI Software can vary from simple converters and stand alone dedicated PC packages to more complex mid-range or mainframe packages and EDI gateways. From the functional aspects, converters will only handle the conversion and management and reporting functions. More elaborate PC workstations will usually include in addition communication and data entry modules and some application interface functionality.

In large organisations where data processing has been decentralised into autonomous business units, EDI gateways enable incoming and outgoing messages to be routed to the appropriate applications through one central point in the organisation. EDI gateways usually operate on mid-range or mainframe computers.

Selecting an EDI Software

Selecting the appropriate software package will require determining which of the software functions are most important or needed, the volume and the speed at which data should be processed now and in the future, commercial viability, service, maintenance and help-desk provided by the software vendor, the cost to list a few. The choice of EDI software will be always limited by the hardware operating system (i.e. AS/400, IBM MVS, UNIX, MS-WINDOWS etc.)

Hardware Analysis

The cost of EDI in terms of computer hardware should not be a major obstacle to the organisation. Of primary concern is the ability to match the system to current and future EDI activities. The costs will largely depend on whether the organisation is implementing EDI reactively or strategically. For reactive implementation a dedicated front-end PC is all that is required. For strategic implementations, a mid range to mainframe computer front-ended by a PC configuration might be more appropriate, depending on the volume of data exchanged, types of applications and level of application integration expected.

LEGAL ISSUES OF EDI

The legal issues raised by the use of EDI include evidential, contractual, and liability issues. The evidential issues of EDI have two distinct aspects. The first considers the questions of admissibility: whether an electronic document is admissible as evidenced in court. The second aspect considers the need to have trade data properly authenticated: the requirement to be signed. Also, since international trade, is based on a series of contracts, performance of these contracts requires an interchange of information. It is inevitable, therefore, that EDI raises certain contractual and liability issues.

Major Legal Concerns

The legal concerns which this wider usage may well bring to the fore include:

- The display of proof that information is authentic (i.e., a means of ensuring that a message is genuine and has not been tampered with) "Authenticity" means that the documents must have not been altered.
- The question of when an acceptance of an offer becomes effective. Is an order which is automatically generated by a malfunctioning computer a binding offer of acceptance? In the normal course of business, documents that are generated or received include: invoices, purchase orders, payment instructions, schedules and manifests. On fulfilling the order the supplier would provide an invoice which would list the supplier's terms and conditions. In legal circles this is known as the "battle of forms". But in EDI-based trade, no physical documents are received, exchanged or generated during the normal course of business.
- The permanent storage of information. Computer records are subject to accidental or malicious corruption or alteration which is extremely difficult to trace, whereas the alteration in paper based transaction can be detected. Therefore, courts have an inherent mistrust of machine-produced evidence, and so the proponent of computer-based records must be able to justify or prove the authenticity of the records. For example, when a floppy disk is presented as evidence in a court, how will the court know what manipulations have occurred?

- The satisfaction of laws which requires certain legal information to be "written" and "signed". Many laws require contracts or commercial documents to be in writing and signed and so preclude contracts, etc., produced or transmitted by electronic communications.
- The appropriating of the risks of errors amongst trading partners and intermediary networks. The recipient, originator and the provider of the communications facility (if applicable) all have different responsibilities to help prevent errors, etc. Who is responsible for a loss if it occurs? The interpretation of government record retention and control requirements for electronic communications. Increased record-keeping requirements, together with wide audit powers, will have a bearing on the ways in which economic records must be kept.

Privacy Concerns

It is likely that a significant amount of data transmitted in EDI messages will contain personal data. It is also possible that an EDI message may include "information which relates to a living individual can be identified from that information (or from that and other information in the possession of the data user)" (U.K. Data Protection Act, 1984, Section 1(3)). There are technical means by which privacy can be ensured, such as, data encryption and system security but the question of which safeguards to use and when to implement them remains unresolved.

The Draft of an EDI Contract

The function of a lawyer in drafting contract between trading partners is to facilitate electronic trading so that the potential benefits of using the EDI can be realised. To do this, he must establish a legal relationship between the client and the trading partners and anyone else involved in the transfer of EDI messages.

EDI SECURITY CONCERNS

EDI messages are no different from any other types of information handled within or between computer systems. The protection of each trading partner site must follow the same physical procedures and organisational rules as other IT systems.

Security begins with the assets of the trading partners as well as any network provider. The can be categorise as physical, for instance computer hardware as personnel, people with access to hardware, and any information stored on the hardware, be it user data or software. Security continues with the creation of EDI Data, data protection techniques applied to the data, and its transmission to the recipient.

EDI Security Threats

The Exchange of EDI messages takes place between the computer systems of importers/retailers and exporters/suppliers. Two security aspects have to be considered:

- Local computer environment
- Communication environment of public or private networks.

Protection of the information within the trading partners organisations is considered to be a separate local issue. Nevertheless, the technical and organisational security on both sides is of great importance. EDI security in this context focuses on the vulnerability of the messages during transmission. The aim is to protect them between leaving the originator and arriving at the recipient.

The Main EDI Security Problems

Loss of Confidentiality

- Unauthorised viewing by tapping into the communication line, other components of the network.
- Unauthorised physical access to the sending or receiving applications or systems.
- Mis-routing by the sender receiver pretending to be a genuine trading partner. This could be simply by misusing an EDI code, or (on the internet) by using the genuine user's internet address.

Loss of Integrity

- The message has been accidentally changed or deliberately modified. This includes deleting, inserting or duplicating or duplicating messages. Modification may result from technical problems, e.g., by telephone line noise.
- The sender or receiver fraudulently claims they did not send or receive the message.

- A third party may pretend to be the sender of a message.

Loss of Availability

- This is particularly relevant for open environments like the internet or interactive EDI.
- Unknown users (hackers) can shut the system service down for genuine users by using up all available computer resources.
- Service disruption due to viruses or junk mail.

IT Countermeasures to Security Threats

Key Requirements

To counter the above mentioned threats, seven EDI security service requirements have been identified. They use one or more methodologies to meet their security object Activities.

Access Control: Protects every EDI system entity (workstation, server, application, network, mailbox service) from illegitimate access by techniques of user authentication.

Message Origin Authentication: Protects the receiver against the sender not being whom he claims to be. Production of a message can be achieved by transmitting a Message Authentication Code (MAC) or by a digital signature of the sender.

Message Content Integrity: Protects against the modification of data. Protection can be achieved by including an Integrity Control Number. This number has no external value but is computed to serve as a check sum. It may be computed from the complete EDI message by using a cryptographic algorithm, such as a Modification Detection Code (MDC). Alternatively, using a MAC will imply content integrity, when the receiver computes the integrity control number of the data he has actually received and compares the result with the value received.

Message Sequence Integrity: Protects against the addition, deletion, loss or replay of messages. The sender must use a Message Sequence Number related to the message flow between him, and the trading partner and the receiver must check it. The sender must request an acknowledgement.

Non-Repudiation of Origin: Protects the receiver from the sender denying ever having sent the message. Protection can be achieved by including a digital signature or MAC. The digital may be verified by using a key. Audit trials are one source of confirmation which could help with many repudiation disputes.

Non-Repudiation of Receipt: Protects the sender from the receiver denying ever having received any messages. Receiver must send an acknowledgement. This could include a digital signature.

Confidentiality of Content: Protects against unauthorised reading and copying of the message. Protection can be achieved by encryption of data using a secret algorithm shared by the sender and the receiver. The trading partners may also implement user authentication and message content integrity. The combination of these countermeasures provides a high level of confidentiality protection.

Different message types require different priorities:

- Access control and Authentication is always required for all messages.
- Integrity is always required for all messages.
- Non-repudiation may be required in the future when the no. of EDI trading partners increases with limited level of trust between the partners.
- Confidentiality is only required for financial messages & for other message types, such as master data, with price information.

EDI Security Guidance

EDI agreements should be expressed explicitly in an 'EDI User and Communication Manual' that is drawn up by the trading partners concerned. This manual should cover the methods and procedures governing the EDI interchange, and from the security point of view it should address:

- Transactions to be transmitted.
- Messages to be exchanged.
- Direction of each message.
- Volume, frequency and time period of transmission.
- Authorised user identification.
- Agreed procedures on:
  - Authentication
  - Commencement and conclusion of message
  - Process disruptions/delays
  - Disaster recovery
  - Stipulated time for acknowledgements.
  - Retention of messages.

EDIFACT Security Developments

Security headers can include:

- The type of security functions provided.
- Whether an acknowledgement is required.

Methodologies

Besides EDI application software related countermeasures such as integrity control numbers, sequence numbers and time stamps, there are five recognised methodologies for fighting possible security breaks. The actions needed depend on the sensitivity of the information transmitted and the assessment of the risk.

In brief, most of the commercial EDI in the distribution industry uses user authentication, acknowledgement and audit trails. In an advanced SCM environment, a digital signature or MAC should be used. Encryption is needed when the risks are considered high, e.g., with FEDI, price or promotional information or when an open environment like the Internet is used to carry the EDI messages.

User Authentication

User authentication aims at unequivocally identifying the sender and the receiver. The simplest protection technique is to employ user IDs and passwords.

- In local workstations and other systems.
- For access to the network.
- For access to the EDI mailbox.

It is also necessary to ensure that messages to be directed to the appropriate user. This can be achieved by using partner lists, either on the network or controlled by your EDI software, or both. Partner lists also help avoid the EDI system to cope with unsolicited messages. Financial EDI payments may have to be submitted from a dedicated workstation secured by special software and physical looks. Smart card readers which contain all necessary authentication information may be used as an additional security interface.

Acknowledgements

Several types of acknowledgements can be given to the sender:

- When the network has placed the message in the recipient's mailbox (delivery notification).
- When the recipient has collected the message from the mailbox (recipient notification).
- The receiving EDI software on the trading partners system sends an EDIFACT functional acknowledgment.
- The receiving applications may send an acknowledgement.

An EDIFACT functional acknowledgement is not necessary as long as a recipient notification from the network exists.

Audit Trails

First of all, EDI audit trails are an operations management tool with activity logs within and between computer systems. They fall into two categories:

- Maintained on the user's site,
- Maintained by a network.

In commercial EDI, they do not contain a complete log of the EDI message content but any reference numbers and control information used in the transmission, such as:

- User IDs.
- Date and time stamps.
- Message header details

This information is used for tracking the message, and makes possible the monitoring and controlling of the EDI operations.

MESSAGE ENCRYPTION

Confidentially of message content can be achieved by encrypting the message so that it cannot be understood by a third party who has accidentally or deliberately gained access to it. In deciding to encrypt a message it is necessary to take into consideration the following:

- Level of confidence between the SCM partners.
- Level of confidence in sharing the same encryption key.
- The existing IT capabilities and resources of the partner.

Encryption requires the use of encryption keys. These may be:

- *Symmetric*: the sender and recipient use the same-shared encryption key.
- *Asymmetric*: both sides have their own private keys, which work together with a known public key which the partners use to encrypt/decrypt their messages. Asymmetric keys need to make use of a respected third party (a certification authority) to hold the public key. An encrypted check sum can be used to show that the EDI message has not been changed or added to in any way.

The choice of the algorithm for the encryption has an effect on the computing power needed in the trading partner's systems. Symmetric encryption requires fewer resources and is therefore more suited when large quantities of data have to be encrypted e.g. on long invoice, payment or master data messages.

PRIVATE KEY SYSTEMS

In private key systems both the sender and the receiver must have the same key. This, however, brings into play the problem of exchanging this private key between the sender and the receiver.

The management of keys must cover the generating, storage, exchanging and changing of private keys. Security relies on both the sender and the receiver, so trading partners have to trust each other over security of the private key. The most well known private key is the **Data Encryption Standards** (DES)

PUBLIC KEY SYSTEMS

In a public key system, a public key is provided to many users and a unique private key is only provided to the sender. Thus the sender only needs to take care about the security of his private key. Public key systems are known as asymmetric because the sender and receiver have separate keys. The distribution and security of the public keys are much easier than when using a private key system.

One problem with this technique is its complexity. A high amount of computer power is usually required for encrypting or decrypting. The main public key system is the RSA system named after its authors Rivest, Shamis and Adelman. The future may see public key systems being implemented in the manufacturing and distribution industry for FED!

THE INTERNET

Security over the internet is a very active area including access to the World Wide Web as well as EDI and electronic payments and all three can be used in combination. Developments can be followed through trade articles, conferences and membership of Electronic Commerce Association as the European Electronic Messaging Associations. (EEMA).

Main Security Concerns

The main security concerns are:

- Lack of accountability leading to problems concerning assured delivery. This is an inherent weakness of the Internet as it consists of the vast number of connected networks and the path through the networks is not predictable.
- Confidentiality and integrity risks are considerably increased because of:
  - Unpredictable routing
  - The increased possibilities of misrouting to the impersonation of the sender or recipient electronic address.
- Availability, which can be affected by a hacker taking up all available resources on a server and thereby preventing access by genuine users.

Counter-measures such as encryption become essential in many cases where they are not essential for EDI traffic over a public or in-house network. Many strides forward are being made in securing EDI traffic, because of possibilities of using the Internet. In cases where a company may want to allow access to the Internet from its network but wishes to restrict it, there are well-established Firewall techniques it should employ

DIGITAL EQUIPMENT CORPORATION: A CASE

Company Profile

Currently operational in 101 countries, Digital Equipment Corporation (Headquatered in Maryland, USA) was founded in 1957. Recognised for product and service excellence, Digital is the leading supplier of high performance, Web-based computing solutions, which help enterprise compete in the global market place. Digital gives its customers a winning Internet advantage through a comprehensive portfolio of Internet solutions, based on award winning systems, advanced networking infrastructure, innovative software and industry applications—including those from its business partners.

Digital's success lies in its ability to deliver internet business solutions, based on high performance and services. The critical components of its growth include:

- Unparalleled experience in delivering the best in high performance enterprise computing platforms based on Digital Alpha and Intel microprocessors. Digital's dual platform strategy offers customers the broadest choice of scaleable systems.
- Three enterprise-strength operating systems—Windows NT, Digital UNIX and Open VMS.
- Leading as an emerging Internet software such as Alta Vista and Millicent.
- Strategic alliances with industry leaders in the field including Microsoft, Oracle and others.
- A world class global service and support organization.

Digital's solutions are the leading choice of customers deploying Windows NT across the enterprise. It offers the enterprise customer, superior value with client server solutions optimized for Windows and Windows NT environments. Digital Unix and Open VMS continue as the most competent of 64-bit enterprise operating systems. Digital brings to the market two decades of Internet experience. It was the first computer company on Internet's predecessor—ARPAnet, and the first fortune 500 companies with a commercial worldwide web server, the first computer vender to use the Internet for interactive ordering. The ALTAVISTA suite of software products from Digital are based on state-of-the-art technology and address the emerging needs for cost-effective and secure connectivity to enhance business productivity within and between dynamic teams.

Along with the best of its partners, Digital combines its best offerings to deliver the best of its information technology solution to its customers. The alliance for enterprise computing between Microsoft and Digital promote Digital's broad range of computing solution, complimented by enterprise software enhancement, establishing Digital as the logical choice for the customer who requires Windows NT based computing solutions. Digital's partnership with Oracle, the industry leading data based company, led to unparalleled performance of the 64-bit Digital Unix enabling very large memory capabilities and applications like data warehousing and Datamart solutions. Digital has significant tie-ups with Computer Associates, Netscape, Informix, Sybase and Software AG, among others.

Digital, with two decades of experience in networking, has an integrating networking product line including the FDDI, ATM, IP switch range GIGA bit Ethernet, which offers customers absolute investment protection. Scaleable from the workgroup range through department servers and into enterprise computing, Digital's award-winning line-up of GIGA switches incorporate high speed switching technology, which was pioneered by Digital. Recently, Digital entered into strategic partnership with Cabletron in a bid to expand its networking capabilities. Digital's role is not restricted to supporting Digital's systems as a vendor service organization. Digital assists its customers in planning, designing, implementing, managing and maintaining software, hardware and networking. Digital's service division aims at providing asset management and a better return on investment by helping users take full advantage of IT resources. In a recent survey conducted by Computer World, Digital has been rated the No. 1 systems integrator in technical performance, project management, business practice and overall satisfaction.

Recently Digital and Compaq have announced the completion of definite merger agreement. This is the largest acquisition in the history of the computer industry, creating a new breed of enterprise leaders and committing to deliver high customer value.

DIGITAL EQUIPMENT (INDIA) LIMITED

Digital Equipment (India) Limited is a subsidiary of Digital Equipment Corporation, U.S.A. It is headquatered at Bangalore, has offices in all major cities and has over 40 customer support locations. Recording a turnover of Rs. 284 crores, Digital India is one of the top five multinational IT vendors in the country. Besides marketing complete range of Digital products and services in India, Digital's ability to provide quality engineering resources across Digital's engineering organisations has resulted in the setting up of exclusive software R&D centers at Bangalore. These expertise centers help in bringing state-of-the-art technology in the Indian market.

Digital India has today become one of the most sought after total solution providers in the field of:

Manufacturing

Digital has partnered with leading ERP software packages like SAP, SSA, ORACLE and BAAN enabling customers to create a cohesive enterprise wide business environment and to manage constant change effectively. Similarly, Digital had partnered with UEC SAIL Information Technology to bring world class technology to the metallurgy industry.

Finance

Digital's solutions for financial services are cost effective, easy to implement and flexible. Ranging from total branch automation at Canara Bank, Punjab National Bank, Corporation Bank. and data warehousing application at Citibank, investor services at UTI, implementing clearing and settlement operations at National Stoke Exchange and many others, Digital's Alpha and Intel based system provide the most conducive IT platform to the financial sector.

Telecommunications

Digital's state-of-the-art technology combined with its services and support capabilities offer a reliable foundation to quickly and cost effectively provide value services in the telecommunications field. Digital has already established itself as a key player in the telecom market, having won 14 telecom circles including Usha Martin, Airtel, Max Cellular, Skycell and RPG cellular, etc. besides providing systems and networking solutions for customer care and billing applications to Alcatel, Hutchison Max, BPL wireless, etc.

Internet

Digital India is the technology provider to VSNL for gateway Internet Access Services (GAIS) providing turnkey solution for country wide internet access. Digital provides complete range of solution for Internet/ Intranet solutions. Digital has set up the first ISP proof of concept in the country at Bangalore.

Continuing in its direction of focussing on sales through channels, Digital has tied up with CMC, CMS, DDE (ORG), Godrej pacific IRIS computers, S. Kumar, etc. Some recent wins with channel partners include a tie up with Nexus and Melstar. Since inception, Digital has internalised the need for customer satisfaction as its prime objective, resulting in its being among preferred service and support provider in the country.

THE DIGITAL EDI SOLUTION

Digital introduced its first EDI software, DEC/EDI in September 1990. Since then Digital has delivered product enhancements, additional EDI software, a range of EDI services and third party programmes designed to meet the evolving business needs of their customers worldwide. DEC/EDI has established itself as an industry leader in providing a client/server architecture that can be deployed in multi-platform environment. Digital offers business a complete portfolio of EDI and Electronic Commerce services encompassing planning, design, implementation, management, training and support.

Digital's EDI Services

Digital offers businesses a complete portfolio of services that encompass EDI and Electronic Commerce planning, design implementation and operational management. Digital has gained a wealth of knowledge and experience implementing EDI and E-commerce solutions internally, working with Digital's trading partners, and working directly with customers.

Key Features and Benefits

- Analyses the business opportunity that EDI can provide and determines the impact of EDI on existing or planned applications.
- Designs and analyses alternative solutions and develops a planned approach for implementing EDI.
- Modifies existing applications or designs and develops new applications to take full advantage of EDI.
- Conducts pilots to assess the impact of EDI and better understand how to implement EDI.
- Provides training to both customers and trading partners on how to trade electronically.
- Implements a full-scale EDI solution.
- Integrates and tests applications that are part of the EDI solution.
- Provides support for the operational EDI environment.
- Evaluates the benefits realised from implementing EDI compared with those predicted or expected.
- Expands and redesigns the EDI solution to link additional trading partners, business documents and applications, and to better link the organizations within the company.

DEC/EDI Software

The DEC/EDI Software is the foundation for Digital's Electronic Commerce offering. The **DEC/EDI server software** provides a range of capabilities including support for multiple trading partner communication options; all national and international EDI transaction set standards (ANSI X12, UN/UDIFACT, TRADACOMS, ODETTE and others); and an advanced table-driven mapper called FileBridge for business application integration. The DEC/EDI server operates in either Open VMS (Alpha and VAX) or Digital UNIX (Alpha) environments.

The DEC/EDI Server consists of the following components which reside on the same node:

- DEC/EDI FileBridge Component
- DEC/EDI Translation Component
- DEC/EDI Communications Component
- DEC/EDI Operational Management Component.

FileBridge Component

This is a table-driven "MAPPER" with a set of data manipulation operators for selected classes of commonly encountered EDI application data. This capability is used to convert data from the existing Business Application file format to that required for Translation into the Trading Partner format as well as formatting data received from the trading partner to the file format expected by the Business Application. The Mapping Process consists of:

Data Specification

The customer defines the application file structure and parameters for mapping between the application file and the EDI standard through a forms—based user interface. FileBridge then generates a conversion table that is used to map data from/to the application file formats.

User Defined Mapping

FileBridge allows fields to be moved directly to the destination or to have built-in mathematical/logical/string operations to be performed upon the fields. The user interface guides the user through this process. The user can define and add code through "hooks" for more complex processing.

Data Definition Extracts

FileBridge also allows for the extraction of Application file record definitions from Oracle[R] CDD/Repository.

The components of FileBridge are:

- FileBridge User Interface (UI): The UI consists of a set of menus and forms that allow users to create, modify and compile FileBridge tables.
- FileBridge Run-time: The Run-time process uses the information in a compiled FileBridge table to map data between the Business Application file and the format required for transmission and receipt to and from trading partners.

Translation Component

The Translation Component takes data that has come from the Business Application and transforms it into the standard EDI formats required for transmission to the Trading Partner. It also takes incoming EDI data that has been received from a Trading Partner by the Communications Component and converts it into the format that is suitable for the Mapper to decode and pass to the Application Client. The Translation Component provides the following features:

- Support for multiple versions of messages (transactions) and EDI standards
- Supports the transmission of 8-bit national character sets
- Compliance checking at Trading Partner level
- Trading Partner specific Code Translation
- User-selection control number generation
- Test transmissions
- Event and Error logging.

An editor is provided to enable users to customise the messages delivered with the product to create new messages for industry specific and Trading Partner specific messages as well as create new messages based on ANSI X12, TDCC, EDIFACT, ODETTE AND TRADACOMS[TM] syntax rules supported by this version of the product. The editor also allows for the creation of new versions of EDI Standard dictionaries. Trading Partner agreement details, including the message formats being used, are recorded in the Trading Partner Profiles. The Translation Component uses the information in the Trading Partner Profiles and the Message Database to perform compliance checking and to check the validity of the transaction.

The Translation Component provides support for the following EDI standards:

EDIFACT has the following characteristics:

- Implements ISO 9735
- Supports character set A, lower case a-z and 8-bit national character sets
- Supports industry-specific transactions adhering to the EDIFACT ISO 9735 syntax.

ANSIX12 has the following components:

- Automatic generation of Functional Acknowledgements and reconciliation of Functional Acknowledgements
- Supports industry-specific transactions adhering to the ANSI XI2 syntax

TDCC has the following characteristics:

- Supports all modes including UCS/WINS
- Automatic generation of Functional Acknowledgements, 997 and 999 and reconciliation of Functional Acknowledgements
- Supports for the following types of messages is not provided:
    - Messages which contain two mandatory instances of the same segment, at the same level, when other segments appear between the two instances (for example, Transaction 994 and its use of the K2 segment)
    - Messages which use EDIFACT enveloping

ODETTE which provides support for Version 2.0 of ODETTE

TRADACOMS which implements the TRADACOMS (October 1989 Update) syntax and messages

Communications Component

The Communications Component supports a range of communications options including OSI. It enables users to send EDI transmissions either through EDI VANs or directly through their trading partners.

The Communications Component provides the following features:

- Flexible scheduling using windows and jobs
- Priority processing to override the scheduler

- User-definable connection and file-based retry limit
- Event and Error logging.

Operational Management Component

An extensive set of management functions is provided to enable the user to easily install, configure and manage an operational EDI system. The Management Component is an integral part of the Application Client, FileBridge, Translation and Communication Components and need not be purchased separately.

The Management Service commands are invoked using DCL. The appropriate form is then displayed on the screen.

The Management Service provides the following facilities:

- *Access Control*
  - Controls user access to the DEC/EDI system
  - User assigned two types of privileges, administration and supervisor
  - Tools are provided to enable the administrator to manage the operational state of the DEC/EDI system
  - The supervisor is provided with tools to monitor the status of documents in the DEC/EDI system
- *Processing Control*
  - Provides end-to-end audit trail of messages status within the DEC/EDI system, including status information from the EDI VAN Services.
  - Provides history files with detailed information to help locate and correct errors
  - Provides messages and transmission file status
  - Time stamps each message status
- *Organisational Control*
  - Provides facilities to monitor the status of the documents in the DEC/EDI system
  - Provides facilities to archive and retrieve documents which have either been processed successfully or cancelled
  - Provides facilities to maintain the list of authorized users
  - Provides facilities to edit and create message format tables.
  - Provides facilities to edit and create Trading Partner profiles
  - Provides a Configuration Verification Program (CVP) to test the linkages between the DEC/EDI Server and the DEC/EDI Client components

DEC/EDI client software integrates EDI into existing business applications. For the end user, the DEC/EDI Command Center and "EDI Cockpit" are available to provide EDI configuration, monitoring and management information. The DEC/EDI Command Center makes it easy to create and maintain DEC/EDI system information, EDI messages and transactions, trading partner profiles, mapping tables and communications options. The "EDI Cockpit" enables business managers to view and analyze trading partner data irrespective of the location or the platform where the applications reside. The DEC/EDI Cockpit, a Graphic User Interface (GUI), running under Microsoft Windows, Windows 98 or Windows NT is available as an option. The Cockpit provides the following functionality:

- A 'system monitor' function providing graphical display of the status of Messages/Transactions. This display can be updated either manually or at a predefined timing interval.
- Access to the DEC/EDI Audit trails at the document and file levels and cross-referencing between documents and transmission files.
- Access to the data contained within the EDI documents.
- Ability to customise the display of the DEC/EDI Audit trails by defining "views".
- Access to the FileBridge Audit trails.
- Access to Error logs.
- Ability to access transaction/message status information from several DEC/EDI servers.
- Ability to reset, cancel or re-send documents or transmission files that are in certain states.

Both end-user clients use Object Broker technology for multi-vendor, enterprise-wide access to the DEC/EDI information.

With concurrent support for multiple clients and a choice of network and communications environments, DEC/EDI enables customers to implement a single, enterprise-wide EDI gateway environment capable of supporting their basic and integrated EDI needs and electronic commerce.

Digital's EDI Services

DEC/EDI is complemented by Digital's EDI services, DEC/EDIs ready to use third-party business applications and inter-operability with their Internet and MAILbus products.

DEC/EDIs ready to use applications solutions such as the SAP AG certified DEC/EDI software and services for SAP R/3 integration are now available for many prominent business applications in the areas of finance, distribution, purchasing, manufacturing, health care, customs/ brokerage and transportation.

When users combine DEC/EDI with Digital's industry leading Internet and MAILbus electronic commerce solutions, the result is the world's most comprehensive set of Electronic Commerce software available today, Electronic Commerce from Digital integrates X.435-based EDI, X.400-based mail, FAX, routing, X.500-based directory services, Internet access via SMTP and more. For organisations that are streamlining their business practices, the Digital electronic commerce solution is a proven platform with worldwide availability.

1. *EDI Orientation Services*

Digital provides customer orientation services to install and tailor EDI software and to provide an understanding of how to implement and manage the EDI environment.

Key Features/Benefits

- Installs the DEC/EDI software correctly on the designated EDI client and server systems
- Provides hands-on training in the configuration and implementation of DEC/EDI
- Tailors the DEC/EDI software to the needs of the customer's EDI environment
- Provides an understanding of DEC/EDI mapping concepts and tools
- Enables utilisation of advanced mapping techniques to address complex tasks and special trading partner requirements
- Provides practical knowledge on EDI system maintenance, monitoring, and problem identification and resolution
- Looks at policies, procedures, documentation and operational practices needed to maximize the EDI technology support for business goals

2. *Packaged EDI Services*

Digital's packaged EDI services take elements of EDI planning, design and implementation and combine them to more quickly and effectively start up the EDI environment and link in trading partners.

Key Features/Benefits

- Sets up the operational DEC/EDI system configured to meet EDI needs
- Demonstrates the use of DEC/EDI to send and EDI document
- Provides instruction in how to manage the operational DEC/EDI system
- Establishes DEC/EDI mapping tables for s selected EDI document and the associated business application file
- Provides advanced DEC/EDI mapping knowledge and techniques
- Assists in setting up the required communications including the link to the selected value added network
- Enables the customer to connect to its first trading partner quickly
- Provides a plan for ongoing EDI implementation activities
- Identifies information needed to add trading partners and EDI documents
- Provides DEC/EDI problem resolution, software updates and documentation.

3. *Customised EDI Services*

For those instances where EDI orientation and packaged EDI services do not fully address EDI needs, Digital provides customised EDI services for EDI planning, design and implementation and management.

Key Features/Benefits

- Provides EDI business needs assessment and planning
- Adds trading partners, EDI documents and transaction sets to existing DEC/EDI implementations
- Installs and provides orientation for additional value-added network and specialized communications connections
- Provides assistance to migrate from another EDI platform to DEC/EDI
- Integrates legacy and DEC/EDI-ready business applications into the DEC/EDI environment
- Manages the DEC/EDI operational environment
- Provides DEC/EDI performance analysis and tuning
- Installs and migrates EDI implementation to X.400/X.435 communications environment

4. *DEC/EDI Integration Services*

Digital offers organisations a complete portfolio of services that encompass EDI and electronic commerce planning, desiga, implementation and operational management. Through customer orientation, packaged services and customised services, Digital delivers EDI consulting and system integration services helping businesses to realise the greatest benefit from their EDI implementation. DEC/EDI services for SAP R/3 provide assistance in setting up EDI within SAP R/ 3, installing and configuring the DEC/EDI integration software, providing an operational SAP R/3 EDI system an conducting training on EDI integration within SAP R/3.

CONCLUSION

EDI is well established as an effective technology to reduce cost and increase efficiency. EDI technologies are approximately the same age as the Internet technology. In the past the technologies have been mutually exclusive, but this is rapidly changing. As the two technological communities begin to merge, and as the business community sees the advantages of this merger, EDI and the Internet will eventually become ubiquitous. EDI users are already seeing dramatic cost savings by moving their traffic from the traditional VAN services to the Internet. Also the Internet can go far beyond EDI in terms of business communications. Some persons feel that EDI never provided the ultimate solution because of the cost and the sophistication required to set it up, the lack of ability to do any spontaneous transactions and the lack of connectivity and proliferation to some customers.

Once the security issues are addressed and the standards are in place, real time EDI over the Internet will replace normal time delay, batch time transactions. However, for the next few years at least, companies planning to conduct electronic commerce will have to conquer several combinations of EDI and Internet.

19

Internet Banking

COMMENTS

Internet Banking, PC Banking, On-line Banking, Electronics Fund Transfer are terms that have come into prominence in recent times because of the rapid growth in the usage of the Internet. This report discusses the subject of Internet Banking along with the related issues concisely.

The concepts of Internet Banking is introduced and explained along with the terms PC Banking, On-line Banking, Direct PC Banking and Indirect PC Banking. Details of the various banking services that can be provided are explained. The likely future services are postulated, based on the services provided by some American Banks. In fact, the service "Build Your Own Bank" (BYOB) provided by Bank of America is explained. The services and opportunities that can and should be provided by PC Banking are explained and discussed along with the potential benefits from each service ('product'). Among the many benefits from Internet Banking that have been discussed, two that have been highlighted and explained are: Measurable and Personalised Banking and Cost Savings. There also a discussion about the flip side of the coin namely about the potential threats of Internet Banking to Banks. The issues discussed under this are:

1. *Security Threats*
2. *Loss of Customer Relationships*
3. *Rapid Technological Advancement.*

Security Issues are discussed later in greater detail, where Data Source Security, Data Transmission Protection and Account Protection Security have been explained well. However, in the discussion under potential threats, the security threats of concern have been mentioned.

Contributed by Meeta Srivastava, Jasber Kaur Dhir, Saurabh Sharma and Joy Samaddar.

The report then goes on to explain the services that occur as a result of Internet Banking. These include:

1. *Better Record Keeping*
2. *Better Access to Relevant Information*
3. *Comparison Shopping Capabilities*
4. *Improvements in Customer Services.*

The discussion continues with the barriers to adoption of online banking and identification of the players in Internet Banking. This includes the Vendors, the End-User Access Software, Network Access Software, Channel Management, Production Applications and Home Banking Applications. There is a brief explanation of Electronic Fund Transfer. Finally, the discussion shifts to the application of Internet Banking in India. All the discussion on Internet Banking has, up to this point, been about Internet Banking in the US. For Internet Banking in India, the example of ICICI Bank has been selected and their banking software "INFINITY" is explained in some detail.

The report presents a balanced view of the shape of things to come in the Banking sector in India.

INTRODUCTION

The rapid growth in the usage of the Internet has created a great deal of interest in the subject of Internet Banking, also known as PC Banking or On-line Banking. Although remote banking has been available in the United States by telephone for nearly 20 years and by personal computer (using direct dial access) for 12 years, the concept of "distance banking" has been slow to gain acceptance in the retail banking marketplace. However, two recent developments have worked together to create a renewed focus on the concept of remote banking. First, the development of software applications utilising graphical user interfaces gave PC's a more broad-based appeal and fuefed the penetration of personal computers into U.S. households. Second, and more recently, the Internet has given this growing group of home computer users an awareness of the PC's ability to help them make more efficient and effective use of their time in communicating with others, obtaining information on items of interest, and making purchases. The growing popularity of the personal computer, a growing number of easy-to-use personal finance software packages, and the realization among consumers that the Internet can be used as a time-saving device have all served to increase the interest in remote banking.

At the current time, PC Banking has made only a small impact on the delivery of bank-related services to retail customers. Estimates of the actual amount of consumers using PC Banking services vary widely. In a piece entitled "On-line Banking by the Numbers" released in its December 1996 issue, The On-line Banking Report (OBR) reported a wide discrepancy in the estimates of current and projected PC Banking usage rates by itself and several leading financial services research companies, including Mentis, Forrester, Faulkner and Gray, Synergistics, Boston Consulting Group, CS First Boston, and Booz-Allen and Hamilton. These differences of opinion reflect the fact that it is much too early to know how much (or little) impact this technology will have on the banking industry. The estimates of online banking users ranged from just 100,000 (Booz-Allen and Hamilton) to as many as 2.5 million (OBR). The On-line Banking Report attributed the disparity in these figures likely had more to do with the varying definitions of "on-line banking" than any other factor. Some industry analysts view on-line banking as online bill paying and/or fund transfers. Because these transactions can also be conducted outside of normal banking channels, OBR ties its definition of on-line banking to "viewing checking account data online":

"An on-line banking user is someone who: Uses a remote terminal (PC, screenphone, Web TV, cable TV, PDA, etc.) to view checking account balance and/or transaction detail, while at home, work, school, or at a public terminal—at least once per month." Regardless of their differences of opinion concerning the number of on-line banking customers in 1996, these industry analysts agree that the coming years will witness explosive growth in the number of U.S. households who will use some form of PC Banking. This belief is due to the emphasis being placed on the technology by the financial institutions themselves, the rapidly expanding number of households with personal computers and Internet access, and the need among consumers for more convenience.

This report seeks to introduce the concept of PC Banking and its impact on the banking industry, and to provide an analysis of the groups involved—the financial institutions currently providing PC Banking services, the customers currently banking on-line, and the companies who design the technology to make PC Banking possible. The report's focus is on the emerging trend of PC Banking via the Internet, the opportunities that this form of PC Banking provides, and its implications for the banking industry especially in India.

PC BANKING: DEFINITION, TYPES AND SERVICES

Definition of PC Banking

PC Banking is a concept that is still in the early stages of development. As detailed in the introduction to this report, even financial service industry analysts do not folly agree on exactly what is meant by "banking on-line". Therefore, to avoid unnecessary confusion, it is important to define "PC Banking" in the context that this term will be used throughout the remainder of this report: PC Banking can be defined as any banking activity which involves the use of a personal computer, some form of software, and a telecommunications circuit to access banking products and/or services from a remote location.

Types of PC Banking: Direct and Indirect

The two ways in which the consumers can access PC Banking services are Direct and Indirect. **Direct PC Banking** is said to occur whenever the banking customer must connect directly to a bank-controlled server with a modem. Once a connection has been established to the bank's server, the customer is able to use bank-provided software to pay bills, check balances, and transfer funds. This type of remote banking has been available in some parts of the U.S. for as many as 12 years. However, because of the relatively small number of home computers in the U.S. market and the more difficult-to-use interfaces available at the time of its introduction, direct PC Banking never gained the widespread acceptance of the banking consumer, and was not actively promoted to the common banking consumer by the financial institutions themselves.

The emergence of Microsoft Windows (and its user-friendly graphical user interfaces) as the PC industry's standard has helped to fuel the growth of the home computer market in the 1990's. This, along with the phenomenal growth of the Internet as a communications, research and purchasing tool, has helped to generate the interest in Indirect (or Internet) PC Banking found in the marketplace today. With **Indirect PC Banking,** customers need neither the bank's proprietary software nor direct access to the bank's server in order to utilise the available services. Instead, the customer needs only a personal computer, Internet access, and a web browser to perform any number of transactions on-line. Most Indirect PC Banking customers use an Internet Service Provider such as America Online to gain access to the Internet. Once connected, the customer can choose a web browser such as Netscape to help locate a bank's web page/From this page, the customer will be able to perform his or her desired transactions.

It is apparent from the above descriptions of the types of PC Banking that the Indirect method, which involves connecting to the Internet, using a web browser, and interfacing with a web site, requires a more complex set of technologies than the direct method of dialing directly into a bank's server via modem. For this reason, it is believed that Direct PC Banking will continue to be the dominant method of PC remote banking access over the next four to five years. The Internet, however, possesses much strength, which will allow it to overcome direct access. As the number of Internet users increase and the necessary technologies continue to evolve, Indirect PC Banking should become the method of choice for both consumers and banks in the coming decade.

PC Banking Services

Banking over the Internet is a concept clearly still in the early stages of development. As the number of financial institutions with an Internet presence continues to grow, so too will the complexity and capability of these sites. The primary services offered to banking customers over the Internet today are account balance inquiries, funds transfer, bill payment, opening accounts, and applying for loans. With the exception of bill payment, these services are those typically provided by banks. Banks have chosen to include bill payment along with these more traditional services for two reasons: First, bill payment is an item normally performed in conjunction with other consumer banking tasks. Second, given the margin pressures faced by commercial banks today, bill payment offers a chance for added fee revenue.

Account Balance Inquiries

Account balance inquiries currently top the list of the most popular Internet banking activities. Because of its relative safety (information flows in only one direction) and ease in programming (coding to handle input from customers is not necessary), balance inquiries were among the first services offered to customers over the Internet. Approximately 50 per cent of those customers engaged in Internet PC Banking activities utilise the account balance inquiry feature.

Fund Transfers

Fund transfers are currently conducted by some 25 per cent of customers banking on the Internet. Interestingly, PNC Bank of Pittsburgh's Managing Your Money, released this past December, was the first Internet banking programme to simplify the process of Intrabank fund transfers. A concern held by many banking executives regarding Internet banking is the potential ease with which funds can be transferred between banks. This has resulted in Internet PC Banking programmes which either do not support this type of transfer, or which make such transfers difficult.

Bill Payment

Bill payment currently accounts for approximately 30 per cent of all Internet PC Banking transactions, and continues to grow in popularity. Having found an area of interest among consumers, bank advertising executives have featured bill payment in a number of advertisements. Nationsbank's television spots for its Internet product, PC Banking, feature persons discussing the time-saving benefits of paying online.

Other Services

Opening accounts and applying for loans online make up a relatively insignificant portion of PC Banking transactions, due to the regulatory compliance demands placed on banks. First, positive identification through the presentation of a picture ID is required in order to open an account. In addition, numerous signatures are required on account and loan documents. Both of these issues make it virtually impossible to utilise the Internet alone to provide such services; information exchanges by mail are necessary as well. Still, given advances in technology and the demand of such services by consumers, banks will continue to explore ways to provide such features as part of their Internet packages.

Future Services

The future services described above provide only a starting point for the types of services which might eventually be offered to consumers via the Internet. As Web technology evolves, a financial institution's service offerings can be further expanded to accommodate the growing and/or changing needs of banking customers. For example *Bank of America* recently announced its plans for a World Wide Web page, which will allow each customer to personalize the bank's web page to meet his/her individual needs. This service, called *Build Your Own Bank* (*BYOB*) will let customers choose the kind of financial information of interest to them, and then see updated information on those topics each time they access the page. Information is available on subjects such as personal finance, business banking, corporate and commercial services, and economic and market data. This data will provide Bank of America customers with financial advice and facts, in addition to the more traditional banking services.

PLAYERS IN INTERNET BANKING: THE BANKS

PC Banking presents banks with new ways to meet many service goals simultaneously, including serving more customers in less time at a lower cost, and providing customers with value-added services that can only be delivered through state-of-the-art technology. In addition, PC Banking offers numerous opportunities, including:

- Expanding the pool of potential customers by removing geography as a barrier;
- Opening new accounts for customers from anywhere in the world, at any time of the day or night Communicating with existing and potential customers more cheaply, whenever they wish to give or receive information;
- Making customers more self-sufficient, giving customer service representatives more time to focus on more difficult customer problems;
- Giving home-based customers more access to services than are currently available at a traditional branch office;
- Allowing customers to pay bills, apply for loans, check account balances, and transfer funds electronically;
- Increasing the number of fee-based services, where banks make a significant portion of their income; and
- Attracting profitable customers.

THE BANKS: POTENTIAL BENEFITS OF INTERNET BANKING

The opportunities listed above provide many potential benefits to banks. Below is a discussion of possible benefits that banks may realize by establishing and maintaining a Web presence.

NEW PRODUCT OFFERINGS

One of the greatest potential benefits is the ability to offer additional products and services to Customers. Bill payment is the most obvious example of a service which was once considered to be very separate from the services traditionally offered by banks, but which now makes a great deal of sense due to the development of both telecommunications and Internet technologies. The business of banking can ultimately be reduced to the transfer of funds between parties, making bill payment a natural extension of the fund transfer services historically offered by banks.

New product offerings such as bill payment provide financial institutions with at least two benefits. First, they increase the possible number of products that any customer can have with his/her financial institution. Many bank executives believe this is vital to prevent customers from switching banks, given the higher costs (both real and perceived) associated with moving these services elsewhere. Second, new product offerings allow banks to invent creative ways to generate increased fee income, an important concept in today's highly competitive and cost-conscious commercial banking environment. PC Internet Banking provides banks with a vehicle for consolidating these products and services in one location—the Web page.

MEASURABLE AND PERSONALISED MARKETING

Another benefit of Internet Banking is the ability to quickly gauge the success (or lack thereof) 'of a bank's marketing efforts. Web technology allows each bank the opportunity to track the number of "hits" to its Web page, which sections of the page (links) are the most popular, and which product/service offerings are purchased as a result of this activity. This stands in sharp contrast to more traditional marketing methods, where it is difficult to determine the impact that each method had on customers.

By maintaining an active PC Banking web presence, banks are able to accomplish at least two things. First, by utilising the information provided by the statistics gathered from the web page, banks can quickly respond to both positive and negative feedback to the products and services offered, and to the messages conveyed on the Web page. This quick response allows banks to become more focussed on meeting the needs of their customers. Second, banks can use this statistical data to segment their markets, tailoring marketing messages to specific customers or groups of customers. By setting up controls to track the areas of its Web page that a particular customer repeatedly accesses, a bank can cause certain advertisements to appear the next time that customer logs into the bank's Internet site. This allows the bank to present those products which they have the highest probability of selling, and allows the customer to receive the information that he/she seeks.

COST SAVINGS

Finally, Internet Banking could provide a means for helping banks reduce the number of branches necessary to service their customers, which could mean significant cost savings for these institutions in the long run. According to a recent study on Internet Banking conducted by On-line Banking Report, there are currently a record 69,902 branches operating in the United States, which equals one branch for every 1,400 households. The branch will undoubtedly continue to be the primary venue for banks to deliver such services as receiving deposits, opening accounts, processing loan requests, and safe deposit box rental. If the number of transactions conducted in traditional branches can be reduced due to Internet Banking, it is possible that banks will be able to reduce the number of branches necessary to service customers, or at the very least, further reduce the number of full-time employees needed to effectively staff each branch. Either scenario could help to reduce overhead costs and to relieve some of the economic pressures facing financial institutions today. Experts agree that any financial benefits to banks derived from Internet Banking are years from fruition, and even then must be measured over a number of years as this delivery channel evolves.

Consider the example of a customer who needs to move $20 from her bank account to her husband's account in order to cover a cheque that he has written. If she visits the branch, this transaction (from teller line to cheque clearing) will cost the bank approximately three dollars. This same transaction, when completed using an ATM, costs the bank approximately 35 cents. If the customer makes this transaction by telephone using the banks touch-tone banking service, the cost to the bank will be between 10 and 15 cents. Finally, if the customer is able to access her bank via the Internet, this transaction will cost the bank 1 penny. A bank providing Internet access can use the savings generated by lower transaction costs to finance capital improvements, improve staff salaries, etc.

THE BANKS: POTENTIAL THREATS OF INTERNET BANKING TO BANKS

In addition to the opportunities it provides, Internet Banking also poses a number of potential threats the banking industry. The following discussion provides a look at some of the more common risks presented by Internet Banking.

Security Threats

Security Threats to the security of both transaction data and the computer systems themselves are commonly cited as the number one concern of banks contemplating the establishment of web sites. Banks (and ultimately, Internet Banking software developers) must take every precaution to ensure that their Web sites provide a safe and secure medium for the thousands of financial transactions which may one day travel over the Internet. By implementing data security measures, such as, public key/ private key encryption to protect data transmission, passwords to protect vital customer information from unauthorised users, and locating the Internet server in a secure location, banks can probably overcome the present threats to data and system security.

In addition to addressing these real security concerns, banks must also overcome the perceived security concerns of many potential Internet customers. According to numerous studies, most consumers are sceptical about the security of confidential transactions over the Internet, even though they will freely give their credit card numbers over the telephone to persons whom they have never met, which poses a much larger potential for fraud. Banks will have to ensure that they adequately communicate the safety precautions taken to their customers in order to gain their confidence.

Loss of Customer Relationships

While Internet Banking provides a means for bank's to expand their potential customer bases by removing geography as a barrier, it also increases the potential number of banking choices for every customer with on-line capabilities. This could serve to impair the ability of banks to maintain customer relationships, because most bank customers cite convenience as their primary reason for maintaining a banking relationship. With other financial institutions on the Internet providing the same level of convenience, it is possible that customers may begin to consider factors such as interest rates or the number of value-added services more closely. Traditionally, banks have sought to sell as many services to each customer as possible, believing that this "entanglement" makes it less likely that customers will move their business when interest rates on deposits are decreased, or when fees on services are increased. Given the time required to visit the bank, close the accounts, and stop automatic drafts and payments, most customers will choose to continue the relationship rather than switch. It is possible that Internet PC Banking may lower the real and perceived costs of switching banks, which would commoditize the industry and further decrease its margins.

In order to address this issue, banks must ensure that their Internet Banking systems are well-designed, easy to use, and secure. In addition, they must continually look for creative product offerings and new system features which keep customers happy and less likely to move their business elsewhere.

Rapid Technological Advancement

The rapid pace of advancement in software and hardware components affects banks just as it does any other purchaser of technology. Just as a computer and the software it contains becomes obsolete shortly after its purchase, an Internet Banking system and the software designed for it may also become obsolete.

The rapid area of technology poses at least two threats to banks which desire a web presence. First, while the figures vary depending upon the source, the capital spent by a large bank on such a system ranges in cost from $25,000 for a glorified billboard to over $1 million for a fully functioning Web site complete with security features. These figures do not include the annual upkeep of such a system. Given the fact that an ironclad industry standard for Internet Banking has yet to emerge, an error in purchasing judgment may leave a bank facing another large expenditure if it turns out that the original system is unable to keep pace with new requirements. This threat is especially troublesome to smaller banks, who may lack the capital to "keep up with the Joneses" of the Internet Banking world. Secondly, even if the system purchased is soundly developed, it must be professionally maintained. Most banks will outsource such tasks to the systems developer, leaving themselves open to ever-increasing charges for maintenance and improvement projects.

PLAYERS IN INTERNET BANKING: THE CUSTOMERS

There appear to be at least four primary needs among consumers regarding their personal/finances: Better record keeping, better access to relevant information, comparison shopping capabilities, and improvements in customer service.

Better Record Keeping

Consumers are faced with increasingly complex financial choices, and have little time with which to consider them. As a result, the ability to keep complete and accurate financial records suffers, and many consumers lose control of their personal financial picture. PC-based financial record-keeping systems, when fully utilised, can help consumers categorise spending in order to better track expenditures for both budgetary and income tax purposes. Linking such financial record-keeping systems to the consumer's financial institution, via direct connection or the Internet, is a natural extension to this process. Such activities as paying bills, balancing the checkbook, transferring funds, and checking balances can be done more quickly and efficiently.

Better Access to Relevant Information

Although not a service commonly provided by banks, access to relevant financial information is an example of a value-added service that most consumers would appreciate. Electronic capabilities such as the Internet allow banks to provide relevant, concise and personalized financial information to customers whenever they are ready to receive it.

Comparison Shopping Capabilities

Consumers often choose banks based on proximity to the home or business, not necessarily on which provides the best products and services for each particular customer. As a result, consumers are often unsure if they are receiving the best deal. The Internet removes geographical barriers that would otherwise limit choice, and provides consumers the opportunity to shop for the best rates on deposits and loans.

Improvements in Customer Service

Banking industry consolidation in the 1990's has had a negative impact on customer satisfaction levels. Larger, less personalised banks generally trade individualised customer service for business volume, and the result is more time spent in the drive through, at the teller window line, or on the telephone. Internet banking provides customers the ability to secure effective "do it yourself" access to accounts, along with other services, at their convenience.

THE CUSTOMERS: BARRIERS TO ADOPTION OF ON-LINE BANKING

Despite the growing number of home personal computers, the increasing need for convenience among banking customers, and the potential of the

Internet, some consumers are still resistant to the idea of banking on-line. Some are concerned with privacy and security issues, while others simply do not see the benefits of PC Banking. Most persons simply do not feel that they have enough information to make an assessment of its features and benefits. Current PC users believe that the Internet will never be secure enough to trust for financial transactions.

While these statistics pose a large obstacle to the growth of Internet Banking, they are by no means insurmountable. If banks are to overcome these obstacles, however, they must make efforts to communicate the benefits of online banking in general, as well as the safety and benefits of banking on the Internet in particular.

PLAYERS IN INTERNET BANKING: THE VENDORS

The tremendous interest in both direct and indirect PC Banking has fueled an influx in the number of information technology firms developing the necessary hardware and software applications.

The Vendors: Product Segments

There are several specific products which fall under the umbrella of the larger PC Banking industry. Remote PC Banking technology can be segmented into five groups: End-User access software, Network access, Channel management, Production applications, and Home banking applications. A brief description of each group and its subsegments is provided below.

End-user Access Software

End-User access describes the software which provides the consumer with access to personal financial management or electronic banking services. Personal financial management software provides complete financial management capabilities including direct PC Banking Vendors which provide personal financial management software include Microsoft (Managing Your Money), Intuit (Quicken), and Meca. Electronic Banking software, on the other hand, provides customers with specific services (balance inquiries, fund transfers, etc.) through direct access to the financial institution using software provided by the bank. Electronic banking software is still offered by most large banks (Citibank, Chase, etc.), although it is slowly being replaced by the indirect Internet access.

Network Access Software

Network access describes the software which gives consumers indirect access to PC Banking services by providing Internet access. Once on the Internet, customers can transact business with their banks through the World Wide Web. In addition to the growing number of local access providers, companies such as MCI, Sprint, AOL, and CompuServe are actively involved in the network access market.

Channel Management

Channel management describes the technology and services which manage the point of customer contact and each interaction with the bank's core production system. This includes channel access (providing customers with a connection to the bank's system), security measures, the unique applications within the channel (each bank providing customers with slightly different services), and sending messages to the core production system.

Channel Management can be thought of as the "backbone" of PC Banking, since it joins the customer interface (via Internet or direct access) with the bank's other systems, allowing these transactions to be properly accounted for. Vendors involved in Channel management include Five Paces, Edify, Braun Simmons, and a number of small, emerging firms.

Production Applications

Production applications describe those applications which receive transaction data from the channel management technology and actually post these on-line transactions to a bank's core production system in real time. Because of the necessary integration between production applications and channel management systems, many of the same vendors deal in both market segments. CFI Proservices and Braun Simmons are two of the more well-known vendors in this segment.

Home Banking Applications

Home banking applications are specialized interfaces, which enable banks to provide bill payment services to customers.

SECURITY ISSUES

As banks go on-line, security is a big issue today. There is a dual requirement to protect customers privacy and protect against fraud. Banks put in place industry standard security measures like firewalls and authentication services. These measures are supplemented by a proprietory security solution. The emphasis is on confidentiality of the transaction as much as protecting against unauthorised access. With security being such a raging issue banks are spending a lot of money. The Development Bank of Singapore spent $ 2 million as its total set-up costs.

To explain how banks scrutinise their transactions, we give an example of Virtual Financial Services Inc. Virtual financial services offers financial institutions a home banking and bill payment solution that allow customers to access their funds via the Internet while maintaining the strictest security measure possible. They have developed a three-tiered security policy that governs all *htmlects* of the *Internet* home banking and bill paying service. Each security policy governs a unique *htmlect* of the customer's session and transactions with the credit union. These policies are made up of software and hardware solutions and procedures. When combined, these policies from a rigid security implementation that enables customers to conduct business with the credit union via the Internet with an extremely high degree of security. The security policy consists of:

- Data Source Security
- Data Transmission Security
- Data Protection Security

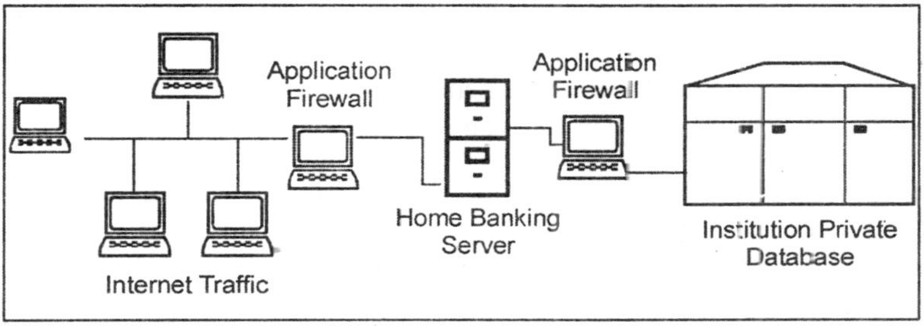

Data Source Security

At no time does anyone from the outside world have access to a financial institution database via the Internet. Any request for data must pass through two distinct validation and control centres, the Application Firewall and the WWW home-banking server. Each request and answer is logged at each stop through the firewall. Any suspicious activity is logged and causes an alarm.

Firewall Segmentation helps to segment the banks and network and internet. This firewall can be instituted by combination of hardware and software that focuses on protocols, types of traffic traversing the firewall, frequency of both successful and unsuccessful attempt to pass through the firewall and other methods of assessing internet traffic. A secured audit trail is an essential part of any firewall installation.

Any incoming request from the internet is the first captured by the application firewall and validated. The Application Firewall stops all incoming traffic. The Application Firewall then makes a request to the WWW Home Banking Server mimicking the request made by the customer on the Internet as long as that request is a valid secure HTML request. The WWW Home Banking Server will only take requests from the Firewall. The WWW Home Banking Server then makes a request to the Firewall speaking its own language on a secret port for a secret address. The Firewall again validates this request and talks to the credit union steps, the credit union database then receives the request for data.

All ports and IP addresses used behind the firewall are private and cannot be routed via the Internet. Therefore, no communication is ever possible between the internet and the credit union's private database.

Real Time Security and Auditing

Through the use of blanketed network monitoring, the Application Firewall stands guard over the entire network, notifying the system administrator of any attempts of unauthorised access or hacking. Every transaction is audited and contains all network information. This allows the Application Farewell to act as an effective phone tap and tracing tool.

Data Transmission Security

All transmissions via the Internet between the user and the credit union are protected by Secure Socket Layering (SSL), SSL utilises authentication and encryption (also known as public key encryption) provides for:

- Server Authentication (thwarting imposters)
- Privacy Using Encryption (thwarting eavesdroppers)
- Data Integrity (thwarting vandals)

Public key encryption is a technique that uses a pair of asymmetric keys for encryption and decryption. One is called the public key, and one is called the private key. The public key is made public by distributing it widely. The private key is never disturbed and is always kept secret. When data is encrypted using public key, it can only be decrypted using the private key. Conversely, when data is encrypted using the private key, it can only be decrypted using the public key. The generation, transmission and storage of these keys are called key management.

Key management is the core element in maintaining the security of this type of cryptosystem because compromising the key will result in the compromise of all the existing uncrypted data. A message encrypted with a 40-bit RC4 takes on average 64MIPS-year to break. In other words, a 64-MIPS computer needs a year of dedicated processor time to break the message's encryption.

Account Protection Security

Account Set-up Protection

No customer can have access to his accounts via the Internet Home Banking and Bill paying Service unless he specifically signs up for this service through his credit union. This process requires that the customer sign two forms (one for Home Banking and one for Bill Paying) in order to be allowed to use the service. The set-up process can only be performed by financial institution personnel. The customer will receive confirmation of this set-up in the mail along with his credit union generated account number and Personal Identification Number (PIN). Only after receiving his account number and PIN will the customer be able to access the service.

Account and PIN Validation Protection

To access the Internet Home Banking and Bill paying service, a customer must enter correct account number and PIN for that institution. The account number that the customer enters is not his account number in the CIF for the all of his other financial function. This account number is a nine digit number generated by the credit union. The credit union may change this number if any situation arises that would deem it necessary to do so. The PIN is a four-digit number. This number is initially generated by the institution. The customer has the option of changing the PIN whenever he wishes via the Internet Home Banking and Bill Paying service. The combination of a nine (9) digit account number and a four (4) digit PIN gives the customer and credit union over 9 trillion possible combinations for access.

Account and PIN Retry Protection

As stated before, to gain access to the Internet Home Banking and Bill paying service. A customer must give correct number and PIN. These fields are hidden on the computer screen when they are entered by the customer (when a customer types in these fields, only asterisks '*' appear). If a hacker tries to arbitrarily guess at an account number and PIN by writing a programme that tries all possible combinations or by just manually typing in combinations, this action will be detected immediately by the Application Firewall, and he will be denied access to the entire system by his IP address or range of IP addresses. He can then be traced back to his source

In the event that someone knows a customers account number, and this person tries to guess at an incorrect PIN, entry to the account can only be unlocked by financial institution personnel. The institution may ask a customer to come into a branch and show a picture ID to have his access restored. The institution may also choose to give the customer a new account number and PIN.

Programme Feature Set Protection

The Internet Home Banking and Bill paying Service does not allow for simple fraudulent use. A customer can only view his account balances and allowable transactions are reversible transactions. Customers are not allowed to transfer money to other accounts.

Virtual Financial Services developed this three-tiered security policy to guard against the unauthorised use and access of customer's data at the credit union. By separately securing the credit union data source, the data while in transmission, and the members' account, they have been able to allow their customers to access their funds via the Internet 24 hours a day, 7 days a week. With this kind of technology security may not be a big problem. It may actually be safer than phone banking, but user perception is just opposite. User co-operation is also essential to maintain a high level of security. For example, a user must remember to log off the system once they have completed their banking session.

ELECTRONIC FUNDS TRANSFER

Electronic funds transfer is the pre-authorised transfer of funds from one bank account to another. Funds are transferred into your bank account or to the bank account you direct the funds to go to on the due date. EFT can be used for recurring amounts (the same amount each period) or for nonrecurring amounts (different amounts). EFT can also do a one-time only funds sweep.

EFT eliminates the ongoing inconvenience of accepting, processing and handling monthly payments by cheque, cash or other paper instruments. Designed to dramatically improve the customer service and reduce the inherent cost of processing paper transactions, EFT is the means by which pre-authorised debits and credits are electronically transmitted from a customer's account to the business account.

Each day millions of electronic transfers are transmitted through Automated Clearing House (ACH). The ACH is the electronic banking network through which all electronic commerce is conducted in the US. Over 95 per cent of the nation's banks belong to the ACH including the Federal Reserve.

How does EFT work?

- The bank inputs all its customers data into the database of the software and indicates the time of the month that they are to have payments made or withdrawn
- Or the bank can download the information from its computer directly in to the software.
- Two business days before the due date the bank transmits the information on to the ACH PC processing company.
- On the due date, the money will be transferred into the corresponding bank's account and all the transferred data will be transmitted back into the bank's software.

The Major Benefits of EFT

- Timely payment—improves cash flow by having payments in on time
- Less paper work—reduces paper work
- Reduces follow—up work-reduces follow-up calls and reminder notices
- NSF control—allows you to react much faster to NSF (non-sufficient funds) payments.

There are two types of EFT as follows:

Debit EFT: Funds are transferred from a customer's account to the merchant account for the payments of goods and services.

Credit EFT: Funds are transferred from the merchant account to the customer account.

There are a number of companies providing Windows software for the complete support of all EFT transactions.

Electronic Banking Systems (EBS) is a company dedicated to the electronic funds transfer-EFT industry. In fact EBS has become a leader in marketing of the EFT systems, softwares, and products. EBS offers the following services:

1. *Direct deposit*: is the direct payment of salary into an employees current or savings account by an employer. This is much more efficient, less costly and less time-consuming than paper checks.
2. *Cheque transaction*: is the changing of a paper Cheque at the point of sale. The benefit is that the paper Cheque is no longer processed thus saving processing fees, deposit preparation, banks runs, lost or stolen Cheque etc.
3. *Electronic Cheque Representment*: is the electronic collection of non-sufficient funds Cheque using Netstar representment software. The benefit is that the process time is 6 business days vs. 10 days or more for paper representment.

BANKING THROUGH THE INTERNET IN INDIA (ICICI)

ICICI has been the first Indian bank to harness the net for providing information about the various products and services and also distributes the forms for the same. As the realisation dawned about the net's potential, since the domestic customer base was still at an embryonic stage, the NRI community became the main focus area.

ICICI site which started off with disseminating information about the bank, its products and services has now graduated to providing information about NRI deposit scheme. The site also provides for downloading of account opening forms.

INFINITY is the Internet Banking Service from ICICI Bank. Infinity provides convenient remote banking facility. It provides various options to get information about all your accounts at any branch of the Bank, over Internet. You can view your account information like account details and statement. It provides Financial modelling tools, which help you to simulate various deposit options and arrive at an optimum combination. You can communicate with the Bank using a messaging facility. Functionality planned in the next phase would permit you to transfer funds between your accounts.

The most important feature of Infinity is that it provides a highly secure environment and ensures that your financial transactions are carried out safely and with confidentiality. You can also register for Infinity, by filling the registration form

WHAT CAN YOU DO WITH INFINITY

1. Account Information

Infinity gives you the following options to retrieve your account information efficiently and effectively.

- *Quick View*: View the last 8 transactions in your account.
- *Query*: Enquire on transactions in your account based on the criteria like amount range, date range, cheque number-specified by you.
- *Schedule*: View the payment schedule in case of loan and deposit accounts.
- *Details*: View additional information about your account.

2. Mail

You can send and receive e-mail to and from the bank. You can use this option to ask any question you may have about your accounts.

3. Request

You can send instructions such as Cheque Book request, stop payment instruction, to the bank-just by filling up and submitting the related form.

4. Customise Infinity

Infinity allows you to customise it according to your preferences. You can specify your viewing preferences such as specifying your primary account, date and amount formats you want. You can even set nick names for your account(s). Also use this option to change your password.

5. Personalised Content

Infinity provides customised content exclusively for its users. Current channels include

- Personal finance containing views of experts in the field of personal finance and taxes.

- Select articles on electronic commerce, information technology.
- Life-style, articles on various places of tourist attraction.

6. Funds Transfer

Own account transfers through the Net is now brought to you by ICICI Bank via INFINITY-Internet Banking service. The funds transfer facility is an extremely convenient service which allows you, whether a domestic or NRI Infinity User, to transfer money by choosing a source account and a target account across any of our branches. And, currently the service is available free.

SECURE ENVIRONMENT OF INFINITY

Infinity provides a highly secure environment for carrying out your banking activities on the Internet.

A multi-layered security architecture comprising of firewalls, filtering routers, encryption and digital certification will provide you the confidence that your account information is protected from unauthorised access.

Firewalls and filtering routers ensure that only legitimate Internet users are allowed to access the Infinity system. Encryption techniques used by the Bank (including the sophisticated public key encryption) would ensure that the privacy of data flowing between the browser and the Infinity system is protected. Digital certification procedures provide the assurance that the data you receive is from the Infinity system.

You can access Infinity only by using your User Id and Password.

Infinity can be accessed by only using SSL-enabled browsers. The software works with Microsoft Internet Explorer version 3.0 (or later) and Netscape's browser version 3.0 (or later) which come with SSL support.

You yourself provide one of the systems's most important safety features by taking the following precautions:

- Protect the confidentiality of your Infinity login name and password.
- Make sure no one can see the account login name or password you're entering when you log on to Infinity.
- Logout of Infinity before moving on to other Websites.
- Do not write your Infinity login name or password.
- Do not leave your login name and password such that someone sitting at your computer could see them.
- Never reveal your Infinity login name and password to anyone (no representative of ICICI bank will ever ask you for your infinity password).

- Notify ICICI Bank immediately if you notice any unusual account activity.
- Keep all documents that include your account information in a secure location.

What is Infinity?

ICICI Bank's Infinity is an on-line link to your ICICI Bank accounts through the Internet ***(www. icicibank. com)***.

Infinity provides you convenient facilities like:

- Details of transactions and balance information about your accounts as at the end of the last business day
- Download of all your account information in QIF or text format
- Dispatch Mail and Instructions to ICICI Bank
- Transfer of funds between your own ICICI Bank accounts (in a later version).

The Equipment and Software Needed to Access Infinity

If you have a PC connected to the Internet, you will be able to use Infinity. For optimal performance and security, the recommended browsers are:

- Microsoft Internet Explorer 4.x or later
- Netscape Navigator 4.x or later

It will not be possible to access Infinity using text-based browsers like Lynx.

How to Register

You can register on-line by filling up the appropriate registration form.

When to Start Using it

After the application form is received duly signed by you, the mailer is despatched to your correspondence address. It contains your Login name (i.e. user-id), Login password and Transaction password. You will have to confirm having received the mailer by signing and sending back the acknowledgement slip. You can send it back directly to infinity Centre at Mahalaxmi, Mumbai or through ICICI Bank's branch network. For security reasons, the Infinity account will be activated only after ICICI receives the customer's acknowledgement.

During the first login attempt, it is mandatory to change both, the Login and Transaction password.

The Cost of Using Infinity

Individual account holders: Absolutely FREE.

For corporates, the rates will be decided on a case to case basis.

Which Services are available in Infinity?

The Bank plans to implement the service in phases. In the first phase, it was possible to view the Infinity Demo to get an idea about the functionality of the package. The Demo is still available.

In the second phase you can see your account information using Quick View, Query, Details and Schedule on all your accounts with the bank. You will be able to do send and receive messages to and from your Account Manager at the Bank. You will also be able to send cheque book requests, stop payment instructions and some other activities using standard forms available online.

In the third phase, additional facilities like funds transfer between your accounts would become possible.

The schedule of implementation of each of the phases will be announced on the Web page of the Bank at the appropriate time.

Use of Internet Infinity Overseas

Infinity can be used from any country in the world provided you have access to the Internet and the equipment/software mentioned earlier. However, please be aware that operations in your accounts with ICICI Bank and the Infinity service are subject to the relevant laws of India and not any other country.

Locking of the User Id

User id and password make your KEY to the Bank. To protect against any unauthorised access to your accounts, your login is automatically disabled after 5 consecutive login failures. In case your access is locked, you will have to inform the Bank. The Bank will then enable your User Id and send you a new password by courier after verification.

When the Customer Forgets the Password

You just have to inform ICICI Bank by way of a duly signed letter. The Bank will then issue a new password and send it by courier. Upon its acknowledgement, ICICI Bank will activate your account with new password. During first login attempt, it is mandatory to change the password.

THE SUCCESS OF INFINITY

A quantitative way of judging the success of Web site is by measuring the "hit" count or exposure count. However, the true measure would be the number of responses which the site has encountered. Calculating on such parameter, one can confidently assert that it is banking and finance that have managed to evoke a quantitative response.

Fine-tuning Website has paid off ICICI Bank in many ways. During the first 13 months of existence of the web site, the ratio of NRI deposits as percentage of total deposits of the bank grew from 4.5 per cent to more than 10 per cent. Presently about 80 per cent of bank's customers come from the internet. This is particularly significant since ICICI bank does not advertise in any US or European publication.

On a qualitative basis, 4700 messages are registered in Guest book of the site. On the flip side, the bank has also been able to achieve significant savings in the form of mailing and courier costs. The bank's on-line presence has also got glowing compliments in Indian e-mail discussion groups.

LEGAL ISSUES

The impact of changing technology on banking activities is not yet felt due to the fact that the mechanisation and automation in the banking industry is at a relatively less mature state when compared to its counterparts abroad and also because we are still to come to terms with the comparative advantages in working on information based systems in place of traditional banking systems.

Computerisation in Banks

In the initial stages, computerisation in many banks was undertaken mainly to deal with the complex problems of interbranch reconciliation. However, it was the Rangarajan Committee which set the pace of Computerisation of banks in the country.

The Rangarajan Committee on 'Computerisation and Mechanisation in Banking Industry'

The 1984 Recommendations

The report of the committee, submitted in the last quarter of 1984 recommended that the process of mechanisation and computerisation should encompass activities at the branch; regional/zonal and head office levels. The Committee suggested installation of mainframes at the central/ head offices for processing data received from regional/zonal offices. The objectives include orderly storage of data in meaningful formats for easy retrieval, analysis of data to exercise control and audit, generation of reports for management policy formulation and evaluation of performance. This would help banks in housekeeping and providing slightly better service to customers. It had inter-alia recommended the mechanisation of cheque clearing, introduction of national clearing and computerisation of forex transactions.

The 1989 Recommendations

Against the background of the experience gained in the eighties and the task of the banking system, the Reserve Bank of India set up in September 1988 another committee under the chairmanship of Dr.C.Rangarajan, the then Deputy Governor to draw up a perspective plan of computerisation of the banking industry 1990-94 and consider other allied issues: The objective for the committee were:

1. To suggest modalities for implementing on-line banking specially at the branch level.
2. To indicate the application areas (like Electronics funds Transfer, E-mail, message switching) relating to the use of BANKNET and SWIFT.
3. To suggest guidelines for installation of ATM's and other self service machines by the banks.
4. To assess the requirements of trained personnel and the need for training on computerisation and to recommend suitable arrangements.
5. To suggest modalities for phased introduction of bilingualisation in computers used by banks.
6. To make any other recommendations which are incidental or related to the above terms of reference.

The recommendations of the Committee were:

(a) Computerisation of branches, administrative offices, networks, ATM's, teller systems, Signature storage and retrieval systems
(b) On-line terminals at the corporate customer sites
(c) Customer authentication through machine readable cards, credit cards, credit clearing and office automation.

Recently, the RBI appointed a committee on 'Technology Issues in the Banking Industry' for action-oriented steps and solutions on different aspects of computerisation such as remittance facilities, screen-based reporting of transactions in Government securities, clearing operations etc. In December 1994, a committee on Technological Issues was formed under the chairmanship of W.S. Saraf to look into the issues relating payment systems, cheque clearing settlements in Banking. The major recommendations were:

- Setting up of Electronic Funds Transfer (EFT) systems. The B-ANKNET communications may be the carrier.
- The ultimate goal of EFT is to facilitate funds transfer between two bank branches.
- For operationalising EFT scheme, banks may install the necessary computer and communication infrastructure (a PC/AT, a printer, a modem and direct telephone line) at their service.
- Steps may be initiated by RBI to enact the suitable legislation on the lines of EFT Act 1978 in the USA and Data Protection Act 1984 in the UK.

The committee also studied the issues relating to legal framework necessary for large-scale computerisation in the Banking industry to make it more technologically competitive, certain legal restrictions imposed by the RBI Act 1934 and Bankers Books Evidence Act and the proposed Acts, namely the Electronic Fund Transfer Act, and computer Misuse Act had to be modified, since they restricted the full technological development in this sector. The RBI is now framing regulations under Section 58 of the RBI Act, 1934, to implement the Electronic Funds Transfer System.

BOTTLENECKS WITH REGARD TO IMPLEMENTATION IN INDIA

Banking through the Internet is undoubtedly a promising technology. It will go a long way in improving the quality of the services offered by banks. Its adoption and success in the developed world has opened the doors for banks in developing countries like India to follow suit.

The next challenge for banking would obviously be conducting the entire banking on the net. Nevertheless, there are hurdles towards its adoption in India.

Today most of the Banks are bound by infrastructure and policy restrictions. Though now with the process of privatisation of Internet services and the entry of more internet service providers, the infrastructural limitations would probably be minimised to a large extent.

However, any additional computerisation of bank operations has also been held up due to electricity and telecom bottlenecks, and lack of funds. Since most bank-branches function from rented premises, they have to first obtain the landlord's permission to get additional electric and telephone connections. All this also requires money. And, squeezed by high capital adequacy norms and recent liquidity crunch, banks rarely have cash to spare. Above all, in today's scenario legal issues for on-line transactions are also a key challenge ahead.

20

Electronic Modes of Payment and Related Issues

COMMENTS

This report discusses certain aspects of electronic commerce, dealing with various modes of payment. The report goes into the historical background of money and is interestingly called "The History of Money". The manner in which the use of the money system began is described along with the advantages of a money economy. Continuing with background information, the report goes on to introduce and explains the subject of cryptography and encryption procedures-issues such as public and private key encryption, digital signatures and issues that make E-money possible. The different ways on which cryptography is used by different entities are explained

The report then concentrates on the issue of the digital cash system. The key elements of a digital cash system are discussed and certain properties—identified as mandatory properties—are singled out. Similarly, some desired elements and optional properties are also identified. The discussion then moves on to the issue of implementation of the E-Money system. The two distinct types of e-money are made out: identified e-money and anonymous e-money, also known as digital cash. There are two varieties of each type of e-money, on-line e-money and off-line e-money. These basic explanations having been completed, the accent shifts to products of certain companies that re providing the basic software that is being utilised to further the e-money revolution. These include ***CyberCash Inc.*** *that provides secure credit card payment system and* ***DigiCash,*** *which is the company behind the e-cash system and the* ***Mondex*** *project. The Mondex project is explained as to how it uses smart cards to ensure its cryptographic security. It is pointed it that Mondex*

Contributed by Priya Chandra.

replaces physical cash and is more auditable and, therefore, provided Central bankers with more accurate information about the behaviour of monetary aggregates and potentially more controls. Digital money has several obvious advantages-for users, for the issuer, for Banks and for the Retailers. These advantages are enumerated for each pf the parties involved

The discussion then moves on to other issues related electronic to electronic modes of Ss payment. These issues are:

1. *Financial Issues*
2. *Legal Issues*
3. *Market Access Issues*

Each issue is discussed rationally and reasonably. For example, a major financial issue may be a possible uncontrolled growth of E-cash systems. Similarly, in case of legal issues, cryptography, security and privacy are issues that may become important and regarding market access, the rising of E-cash could foster a have and a have-not society. These queries have been posed and possible solutions discussed. The note concludes with a discussion on "The double spending problem".

INTRODUCTION

In these days, electronic commerce seems like the upcoming and exploding prospective market and as a consequence, a great opportunity to make new and remodelled business in a global environment. Companies are changing dramatically as they strive to lower organisational overhead, improve customer service while reducing costs and build a worldwide presence independent of the actual location of the company offices. The convergence of commerce, networking computing and money are relevant basics for the foundation of a global electronic marketplace.

With the unprecedented growth of the Internet, the standards for electronic commerce and electronic forms of money are still evolving. No particular governmental money unit has set a foothold in this new field. The monetary landscape is wide open and no restrictions prevent private financial institutions from establishing private currencies. In the past, governments issued currency and retained a monopoly over its creation. The existing rules governing currency can be used while paying with credit card information over the Net.

Important key questions need to be resolved regarding the issue of electronic commerce and payment systems for this upcoming global marketplace. How will electronic commerce be transacted? What will the mercantile process look like in an electronic environment? How will businesses present themselves and how will transactions be executed in the on-line markets of the future?

Also, a lot of concerns arise with the use of electronic forms of money in the above-mentioned on-line environment. What are the prospective institutions for regulating electronic commerce and payment systems? What forms and characteristics of payment instruments—electronic cash, electronic checks, credit/debit cards and others—will consumers prefer? How will on-line goods and services be paid for? Can the government, especially the federal bank of issue and traditional currency, remain immune to these trends? These key questions need to be solved before a breakthrough regarding new payment methods can be achieved.

Because of the dynamic and fast-paced evolution in this business sector, the answers to these questions are still vague and unclear. New opportunities and challenges in business arise for individuals, firms, non-banks and banks. Safe and simple electronic payment systems are expected to catalyze sales for on-line products and services.

THE HISTORY OF MONEY

The history of money until now has been a process of climbing up layers of abstraction.

The concept of money developed for two reasons:

(i) as a "medium of exchange",
(ii) as a "standard of value" route.

Things began with *barters:* Objects were exchanged directly. If one wanted a sack of grain from another then one would have to have something that the other wanted...perhaps one makes ploughs and the other needs a plough. On the other hand, perhaps one does not have a plough, but knows a blacksmith who has a plough who needs some storage jars and the former has some storage jars... This can get very complicated very quickly, so a *medium of exchange* was required.

The first thing to come along was token-based. The earliest forms of tokens were objects that were generally agreed to carry value themselves. Thus, precious stones and shells formed the early currency. Later, specific coins were minted in precious metals. The tokens themselves were valuable. The next evolution of the system involved tokens that themselves carried no value. A pound coin is not worth [sterling] 1 because of the metal that it is made from, but because we have all agreed that it is worth a pound. Paper notes are the same. The value carried by this kind of token is *conventional,* a matter of consensus. The actual circulating media of exchange, e.g., paper money and cheques, need not themselves be the standard of value, but they must be expressed in terms of the standard of value in the unit of account, whatever the government may define that to be.

Once the tokens become detached from "real" value, the next obvious step is *notational* money. The canonical example of notation is the cheque. The cheque is not worth anything, neither is it a token of value. It is a piece of notation that is directly tied to value stored elsewhere (usually in a bank account). However, notational money is still tied to actual value stored somewhere. The next step after that is *credit*-based systems, with the credit card. Here it is possible to make a transaction that is not even directly tied to value stored anywhere. It is indirectly tied to value in that when it is used, one undertakes to become liable for the amount of the transaction.

The choice of the money system is an act of government by legislation or decree. In this system, there is a standard money and a unit of account; all other moneys are multiples or subsidiaries of the standard unit. Provision is made for converting any type of money into any other. Coinage has become a state monopoly, and government further controls the money supply directly by legal-tender laws that define the moneys which a creditor must accept when offered by a debtor in settlement of a debt. At the moment, most electronic transactions use notational systems (electronic funds transfers) and credit systems (credit card payments over the net). The next stage, and this is the real innovation of electronic money, is to lift actual value tokens into the electronic domain and come with electronic systems of payment!!!!

ADVANTAGES OF A MONEY ECONOMY

- In today's money economy, increasing specialisation is promoted by greater ease of exchange. The resulting economising of skill and effort is one of the great blessings that money has conferred on mankind.
- A second aspect, of particular importance in a capitalistic society, is the automatic direction of the productive enterprise by the price system, which establishes the relative values of all goods and services and thereby largely determines what economic activities shall be pursued.

- The third great benefit derived from the use of money is the availability of money loans whereby the producer may pay for his materials, machines, and labour before his own products reach the market. Large-scale productive enterprise could hardly function without available credit to support its development and sustain it when sales are low.

INTRODUCTION TO CRYPTOGRAPHY

Public-key cryptography and Digital signatures make e-money possible. The banks and customers would have public-key, encryption keys. Public-key, encryption keys come in pairs. A private key is known only to the owner, and a public key, made available to everyone. Whatever, the private key encrypts, the public key can decrypt, and vice-verse. Banks and customers use their keys to encrypt (for security) and sign (for identification) blocks of digital data that represent money orders. A bank "signs" money orders using its private key and customers and merchants verify the signed money orders using the bank's widely published public key. Customers sign deposits and withdraws using their private key and the bank uses the customer's public key to verify the signed withdraws and deposits.

Cryptography is used in most digital money schemes in many different ways:

- Authenticate the user to the bank.
- Money can be generated by banks, represented by a long number.
- Nobody except the bank can alter the number representing the value, without the "bill" losing its value.
- Send data safely over a public phone line or the Internet. An eavesdropper cannot use, read or understand that data.
- Digital signatures.

An encryption system is a procedure, which takes the original message (plaintext) and a small piece of information arranged in advance between sender and receiver (the key) and creates an encoded version of the message (the cipher text).

KEY ELEMENTS OF A DIGITAL CASH SYSTEM

Confidence, A Key Element

Commodity money circulates because the stability of the society gives ground for confidence that the money will continue to be accepted habitually in exchange for goods and services. Similarly, the credit instruments of any government, bank, corporation, firm, or individual will circulate more or less widely in proportion to public confidence in its promises to pay. Likewise, a possible introduction of a digital cash system will require a period of long-term trust.

Mandatory Properties

Security

The transaction protocol must ensure that a high-level security is maintained through sophisticated encryption techniques. The communication can be made secure using private key encryption, whereas the authentication of sender and receiver can be accomplished using digital signatures. On-line verification can prevent double-spending, or other off-line techniques must be used.

User-friendly

The digital cash should be simple to use from both the spending perspective and the receiving perspective. Complicated systems are difficult to administer and raise the failure rate due to errors of the user. Simplicity leads to mass use and mass use leads to wide acceptability.

Portable

The security and use of digital cash should not be dependent on any physical location. The cash can be transferred through computer networks and off the computer network into other storage devices. Digital wealth should not be restricted to a unique, proprietary computer network

Two-way and Transferability

The digital cash should be transferable to other users. Essentially, peer-to-peer payments are possible without either party required to attain registered merchant status as with today's card-based systems. This is a very important element if digital cash is to replace ordinary cash. Think of situations where you hand money personally to someone else, e.g., as a gift, as charity, or as a tip. It is important to waiters to have some means to supplement low pay. There are also other person-to-person payments for which only cash is regarded appropriate, like payments to children, friends, colleagues or neighbours.

DESIRED ELEMENTS

Off-line Capable

The protocol between the two exchanging parties is executed off-line, meaning that neither is required to be host-connected in order to process. Availability must be unrestricted.

Divisible

The cash must be tongible so that a reasonable portions of change can be made. The smaller, the better to enable high quantities of small-value transactions.

Infinite Duration

It should maintain its value until lost or destroyed provided that the issuer has not debased the unit to nothing or gone out of business. Without this property, people would want to keep all their money in bank accounts or cash and only use digital cash if they have to.

Wide Acceptability

Digital cash should become well-known and accepted in a large commercial zone. Primarily a brand issue, this feature implies recognition of and trust in the issuer. Even with several digital cash providers displaying wide acceptability, Alice should be able to use her preferred unit in more than just a restricted local setting.

No Secret Algorithm

The security of digital money should not depend on the secrecy of the algorithm used to manufacture or transfer money. Thus, giving everybody the possibility to verify and check the correctness and strength of the used cipher, such a system will inspire overall confidence. At the same time, everyone is able to write software for all electronic devices that can be found in such a system, making manufacturing monopoly and special security measures obsolete.

OPTIONAL PROPERTIES

Anonymity

This ensures the privacy of a transaction on multiple levels. Beyond encryption, this optional intractability feature of digital cash promises to be one of the major points of competition as well as controversy between the various providers. Transactional privacy will also be at the heart of the government's attack on digital cash, because it is that feature which will most likely render current legal tender irrelevant.

If digital currency were fully untraceable, it would open up opportunities for abuse that are not available to criminals now. In the physical world, money is bulky. In the physical world, it is possible to follow people, so a kidnaper can potentially be caught if the currency is marked, if the money was being observed on location, or if the serial numbers were recorded. Fully anonymous cash might allow opportunities for counterfeiting and fraud.

On the other hand, traceable digital cash would give enormous power to the government and the credit agencies. Currently, credit card charges can be monitored and compiled to create a portfolio on you and all of your purchases. Cash is free from this threat. With non-anonymous electronic cash, there is the fear that ALL transactions will be recorded, and that consumers will lose all purchasing privacy.

Already, public and private sector organisations acquire extensive personal information and exchange it amongst themselves. Individuals have no way of knowing if this information is inaccurate, outdated, or otherwise inappropriate, and may only find out when they are accused falsely or denied access to services. New and more serious dangers derive from computerised pattern recognition techniques: even a small group using these and tapping into data gathered in everyday consumer transactions could secretly conduct mass surveillance, inferring individuals' life-styles, activities, and associations. The automation of payment and other consumer transactions is expanding these dangers to an unprecedented extent.

Unit-of-value Freedom

Normal cash is issued by the government, but digital cash does not need to be. It could be denominated in market-determined, non-political monetary units.

Such a transition to a privately-operated digital cash system would require a period of brand-name recognition and long-term trust. Opportunities would abound for almost anyone, but in reality the greatest advantage currently goes to the on-line shopping malls and the large merchant sites on the Internet, such as Open Market, Internet Shopping Network, and Net Market. For it is this group that will directly influence the payment channel between consumer and merchant through their extensive contact with both. And, this influence could be used to their advantage to build preference for their "site" through money issuance, in much the same way that various forms or coupons build customer loyalty and guarantee repeat visits.

Other potential unit providers include Internet service providers, bulletin board system operators, content publishers, card-based payment networks, and well-known manufacturer or service companies. They all share in common the existence of an extensive base of on-line customers. As the new digital cash providers, international brand names, such as, Coca-Cola, Microsoft and IBM, find themselves in an enviable position to capitalise immediately on their global name recognition.

IMPLEMENTATIONS OF E-MONEY SYSTEMS

Varieties of Cash

Normal cash is issued by the government, but digital cash does not need to be. It could be denominated in market-determined, non-political monetary units. Such a transition to a privately-operated digital cash system would require a period of brand-name recognition and long-term trust. Opportunities would abound for almost anyone, but in reality the greatest advantage currently goes to the on-line shopping malls and the large merchant sites on the Internet, such as Open Market, Internet Shopping Network, and Net Market. For, it is this group that will directly influence the payment channel between consumer and merchant through their extensive contact with both. And this influence could be used to their advantage to build preference for their "site" through money issuance, in much the same way that various forms or coupons build customer loyalty and guarantee repeat visits.

Other potential unit providers include Internet service providers, bulletin board system operators, content publishers, card-based payment networks, and well-known manufacturer or service companies. They all share in common the existence of an extensive base of on-line customers. As the new digital cash providers, international brand names, such as, Coca-Cola, Microsoft, and IBM, find themselves in an enviable position to capitalise immediately on their global name recognition.

In general, there are two distinct types of e-money:

- Identified e-money, and
- Anonymous e-money (also known as digital cash).

Identified e-money contains information revealing the identity of the person who originally withdrew the money from the bank. Also, in much the same manner as credit cards, identified e-money enables the bank to track the money as it moves through the economy. Anonymous e-money works just like real paper cash. Once anonymous e-money is withdrawn from an account, it can be spent or given away without leaving a transaction trail.

There are two varieties of each type of e-money:

- On-line e-money, and,
- Off-line e-money.

On-line means you need to interact with a bank (via modem or network) to conduct a transaction with a third party. Off-line means that you can conduct a transaction without having to directly involve a bank. Off-line anonymous e-money (true digital cash) is the most complex form of e-money because of the double-spending problem.

CyberCash

The CyberCash Inc. provides a range of commerce solutions from credit card based payment systems to secure micro payment systems. The main CyberCash transaction system is based around secure credit card payments. This takes advantage of the existing major electronic payment systems and simply integrates the vendor-side software as if it were a physical POS terminal, with CyberCash servers acting as a gateway between the vendor on the Internet and the banks' secure financial networks. The transaction can be described in seven steps:

1. The customer, having placed an order with the vendor, receives the invoice.
2. The customer uses the CyberCash wallet to pay, which generates an encrypted payment sent to the vendor.
3. The vendor strips the order from the packet, digitally signs the payment advice and sends it to the CyberCash server. The vendor cannot see the customer's credit card number, thus reducing the possibility of merchant fraud.
4. The CyberCash server takes the transaction off the Internet, uses dedicated hardware to decrypt it, reformats the message and forwards it to the vendor's bank.
5. The vendor's bank then forwards the transaction to the customer's bank who send an approval or denial back to the vendor's back.

6. That code is then sent back to the CyberCash server.
7. CyberCash then send the approval or denial code back to the vendor.

Steps 1,2,3 and 7 take place on the Internet and involve a combination of public key and symmetric key cryptography. Steps 4 and 6 take place over dedicated lines. Step 5 takes place on the existing financial networks. CyberCash claims a complete transaction time of 15-20 seconds.

- Because the CyberCash system makes extensive use of the existing secure transaction networks it is very attractive to the major banking houses.
- In addition, because it does not require vendors to establish new banking relationships, it is attractive to them.

The system is not economic for small payments because of the use of credit cards as the grounding financial instrument. Also there is a question mark over its scalablity due to the use of central servers at CyberCash to interface between the Internet and the secure financial networks. CyberCash hint at franchising these services to third parties, and in that case the problem will be ameliorated.

The next table shows the CyberCash visibility profile:

Table 20.1: Visibility of CyberCash Transactions

	Seller	*Buyer*	*Date*	*Amount*	*Item*
Seller	Full	None	Full	Full	Full
Buyer	Partial	Full	Full	Full	Full
CyberCash	Full	None	Full	None	None
Bank	Full	Full	Full	Full	None
Observer	None	None	Full	None	None

The CyberCash group are in the process of adding two new payment options:

- A cash-like system "CyberCoin", and
- A cheque-like system "CyberCheque".

These will be implemented as extra payment options in the existing CyberCash software wallet.

The CyberCoin system is designed to be an extension to the basic CyberCash wallet. It is claimed to be a micro-payment system, although its stated lower bound on payments is 25cents, which contrasts with the usual notion of a micro-payment as a sub-cent transaction. Part of the reason that this lower bound is so high is that the system uses public key in combination with symmetric key encryption. Also, merchants are charged on a per-transaction basis, which will also serve to keep the lower bound higher than it might be otherwise. In addition, the system is not implementing coins in the strict sense of the term, as it is a notational rather than a true token system. Value is never held on the customer's PC, the money stays within the existing banking network and what the customer has on their computer is a legal record of the money. This has the advantage that, should the customer's computer crash, the money is unaffected. However, CyberCoin does not support peer-to-peer transactions. Finally, the system is not anonymous, and transactions can be tracked.

DigiCash

DigiCash are the company behind the e-cash system. As the name implies this is designed to be a cash like system. It is "software only" mode. It uses public key cryptography and is intended for use over the Internet, including email. The cash is stored locally, but requires that the user have an account with a digital bank, from which they then withdraw cash. The system is coin-based, therefore, requiring clearing of the coins by the issuing bank. The following quote is of interest in connection with this: like banknotes, e-cash can be withdrawn from and deposited to transaction demand deposit accounts. And like banknotes, one person can transfer possession of a given amount of ecash to another person. But unlike cash, when a customer pays another customer an electronic bank will play an unobtrusive but essential role.

E-cash is capable of supporting full anonymity with its protocols; however, the system actually implements offers one-sided anonymity; when clearing a transaction the payee is identified by the bank, but the payer remains anonymous unless they chose to reveal their identity. In addition, like the CAFE system, coins can be signed to detect the buyer should they double spend. This partial anonymity is cited as a deterrent to the criminal use of the system. The e-cash system works in the following manner:

1. A user connects to an e-cash issuer and purchases electronic coins to some value.
2. These coins are then downloaded onto the user's local machine. These tokens can be made anonymous via the use of the cryptographic technique of blind signatures.

3. The user can then spend those coins by simply transmitting them to a vendor over the network.
4. The seller must then present the tokens to the issuing bank for verification; in order to insure the transaction is valid, the vendor must get the tokens validated immediately. It is at this stage that *double spending* is detected: the DigiCash servers maintain a list of all coins that have been spent in a database; each transaction is checked against this list; and if the coin is already known then a case of *double spending* has been detected.

However, this-payer anonymity is not without problems. Given that it is achieved by the bank blind-signing coins generated out with the issuer's control and then later checking that the deposited coin has, in fact, been signed by the bank, the following scenario could occur. When the bank itself generates all serial numbers for their coins (as is the case with the NetCash system) then it can be sure about *double spending.* However, suppose two separate users, A & B, both by chance generate the same base serial number and get the bank to blind-sign it. Then, two legitimate transactions could look like a case of *double spending.* Although this is relatively unlikely, the chances of it occurring at some point increase as the amount of anonymous currency in the system increases. More importantly, allowing arbitrary user-generated serial numbers for coins imposes certain storage overheads in the detection of *double spending*

The general lack of atomicity in the basic transaction arises if the transfer of the tokens from the buyer to the seller is not completely clean: under such a situation, it would be possible for both (or neither) parties to believe that they had legitimate access to the token. Clearly, this particular lack of atomicity causes consistency and isolation to be violated, along with durability. The possible limit in scale arises because of the need to verify tokens during the transaction—this is a potential network bottleneck, but one that any on-line system will have to resolve. Similarly, the interoperability problem is not particular to DigiCash.

Table 20.2: Visibility of E-cash Transactions

	Seller	*Buyer*	*Date*	*Amount*	*Item*
Seller	Full	Partial	Full	Full	Full
Buyer	Full	Full	Full	Full	Full
Bank	Full	None	Full	Full	None
Observer	Partial	Partial	Full	None	None

Mondex Project

This is a hardware device. The Mondex project is at a much more developed stage, including manufacturers making and selling the necessary equipment to engage in Mondex transactions, and a number of successful pilot schemes.

The system uses special-purpose hardware on smartcards to ensure its cryptographic security. This is the case in the Internet transactions—there is simply a Mondex compatible card reader attached to the computer; when a transaction is required, the computer talks to the card through the interface. This means that the Mondex system is able to support transactions both in the real world and in the virtual world. An important point about Mondex transactions is that value can only move from one Mondex card to another, and can only be stored on Mondex cards. This obviously makes the system highly proprietary.

Because Mondex employs state-of-the-art hardware technology, which would need to be duplicated in any attempt at forgery, it is assumed that such fraud would be uneconomical. Further, Mondex plans to issue regular and "sporadic" upgrades to the chip, so that any successful forgery would rapidly be rendered obsolete. In addition, Mondex plans to use transaction sampling to maintain a picture of the flow of value, and to use this to identify fraudulent and illegal use of Mondex value before any such forgery investment could be recouped.

Clearly, for this to be possible, the Mondex system cannot be anonymous. They say:

> "To the extent that Mondex replaces physical cash and that Mondex is more auditable than physical cash, it will provide central bankers with more accurate information about the behaviour of monetary aggregates and potentially more control."

ADVANTAGES OF DIGITAL MONEY

For Users

One of the benefits for the user is increased convenience. Increased convenience also derives from the freedom of individuals to obtain their card computers from any source, to use whatever hardware or software they choose, and to interface with communication systems wherever they please. This permits card computers to be adapted to the requirements of sophisticated, naive, and handicapped users alike.

Handicapped people nowadays cannot access most ATM machines, and even other people refuse to use ATM machines due to various reasons. Some of them think they are too complicated to use or they cannot remember their personal identification number, and others just do not like to interact with a machine. They prefer to talk to a human being.

Using digital cash, people might choose never to actually see their digital pseudonyms or to be concerned with other implementation details. The user is also protected against the bank's refusal to honour a legitimate note, since nobody is able to counterfeit the bank's digital signature on the note. Another important benefit for the user is improved security. By requiring a password akin to the PIN (personal identification number) now used for bank cards, the electronic wallet could safeguard itself from abuse by thieves by making encrypted backup copies of its contents. A replacement card could then recover these contents if the original electronic wallet were lost.

At the same time, abuse of a lost or stolen card computer by another individual would be very difficult without the owner's secret authorising number. The card would require the authorising number, which might typically be about six digits long, before allowing any transactions. A reasonably tamper-resistant device within the card computer could, for example, read fingerprints, or the like, to prevent use by anyone but the card owner accept a special authorising number that the owner could use in case of duress to trigger a pre-arranged protective strategy permit only the current owner to reset the card for a new owner, to prevent its use as a replacement by a thief. True anonymous digital cash would also provide unconditional intractability. The "blinding" carried out by the user's own device makes it impossible for anyone to link payment to payer. But users can prove unequivocally that they did or did not make a particular payment, without revealing anything more, if they need to. And, of course, the dangers deriving from computerised pattern recognition techniques mentioned earlier would be non-existent.

For the Issuer

There would be no more need to find new security techniques to print bank notes. Printing costs would also disappear. Another aspect is also that the system provider does not need to supply user organisations with tamper-resistant terminal equipment for each entry point, any more than he needs to supply card computers to individuals. Thus, user organisations can supply their own terminal equipment wherever they please and take advantage of the latest technology.

For Banks

Digital cash offers reduction in cost for banks in several ways. First of all, processing effort might be lower than with existing post-pay cards since single transactions need not be authorised on-line, debited from the customer's account or printed for the customer. This will be particularly true for small amount of payment. Furthermore, the current card system requires widely trusted, tamper-resistant devices at all points of entry to transaction systems. Such a requirement implies substantial initial agreement, outlay, and commitment to design, and can be expected to result in technology that is outdated when systems come into widespread use.

With the security measures built into the electronic wallet mentioned above, fraud costs and costs for clarifying disputed transactions could be reduced. Nowadays, card fraud is a very important problem. The same argument applies also to card counterfeiting and forged bank notes. Other costs linked to normal cash are costs for printing and mailing statements to customers, or the costs for handling cash and for cash dispensers. These could be reduced with home banking possibilities, i.e., transactions that can be made using your personal computer at home. If the digital cash system implemented were anonymous, it would also reduce the sensitivity and the quantity of consumer data in the hands of the bank and, by the same token, it reduces their exposure to incidents that might incur legal liability or hurt their public images.

For the Retailers

First of all, stored value payments should be a lot faster than post-pay payments. Infra-red payments might also be faster than cash payments, resulting in a saving of time. But costs could also be reduced. Nowadays, retailers must pay a fee of 2 to 7 percent of the purchased amount to the credit card company. The fees for digital cash transactions are likely to be smaller than for today's cards because of smaller operating costs for the issuer.

There would also be less vandalism, since with infra-red interfaces, there are no slots or contacts. Ticket vending machines could use electronic receipts rather than printers. This would reduce maintenance costs. In general, infra-red terminals should be cheaper than stripe or chip card terminals and unlikely to wear off as fast. The back office costs for counting, storing and transporting cash would also fall away.

RELATED ISSUES

Financial Issues

A number of different types of electronic payment systems (EPS) are developing that can support international commerce over the Internet. *A major issue is a possible uncontrolled growth of E-cash systems. Such a monetary explosion could undermine bank and government-controlled money systems, giving rise to a confusing and inefficient Babel of competing systems.*

What should be Done to Solve it?

From a long-term perspective, government guidance is needed on issues that the marketplace alone may not be well equipped to address. A technical study group should be formed on the international fora, that will bring the full range of government interests to examine the major financial issues raised by the development of electronic money services. Its function should include guiding policy approaches toward electronic money in different countries, also providing an exchange of information and views among supervisors of financial institutions, central bankers, and law enforcement officials.

For the success of the above initiatives, it is imperative to seek private sector input to inform policy development. It will be critical for public sector officials to collaborate with industry leaders to ensure that governmental efforts fully appreciate and understand the developments and trends unfolding in the marketplace.

Legal Issues

"Uniform Commercial Code" for Commerce Conducted on the Internet

It will be a lot harder for the government to control E-cash than real cash. Money laundering and tax evasion could proliferate in stateless E-money systems. If the monetary system were anonymous, criminals could use untraceable cyberdollars to hide assets offshore.

That is why the American government is trying to make the public key/private key system that is used by DigiCash and others illegal. The government would prefer for everyone to use their Clipper chip, which is similar to the above method, except that they can also tap in "if necessary".

In France, cryptography is completely forbidden by the law. Obviously this is currently an area of heated debate.

What should be Done to Solve it?

To encourage electronic commerce, the government should support the development of both a domestic and global uniform commercial legal framework that will validate, recognise, enforce, and facilitate electronic commercial transactions worldwide.

Participants in the market place—including consumers, businesses, financial institutions, and on-line service providers—should define and articulate most of the rules that will govern electronic commerce. To enable such private entities to perform this task and to fulfil their roles adequately, however, governments should foster the development of harmonized domestic and international rules and norms which will serve as the legal foundation for commercial activities in cyberspace.

Security and Privacy

Safety is the major problem. What if the chosen cryptographic algorithm is not safe? Most algorithms used in these monetary systems have been around for many years already (DES [DP92J, RSA [RSA 78]), and some of the best cryptology experts tried to break them using sophisticated methods without success. Although, chances that this might happen are very small, still one can never exclude the possibility that someone invents a new brilliant way to attack a given algorithm.

Another weak spot is the users' personal hardware (e.g., the electronic wallet) and his copy of the software. Only complete physical security can guarantee that nobody can ease a user of some money, businesses and consumers must feel comfortable that the Internet can function as a secure and reliable commercial medium, and that safeguards are in place to ensure the reliability of the Internet.

The installation in the bank must be perfectly secure too

What should be Done to Solve it?

Efforts to protect the privacy of personal data and to protect intellectual property on-line are futile unless there are technologies and policies in place that ensure the security and integrity of on-line information.

Because cryptography is an important tool for providing computer security, a great deal of attention and effort has been devoted to developing policies to promote the development and use of effective encryption products which encode both stored data and electronic messages. Unfortunately, strong encryption not only enables law-abiding citizens to protect better their trade secrets and personal records, it also can be used by criminals and terrorists to hide their activities and thwart legally-authorised investigations.

Market Access Issues

Consumer Access

While technology in the service of customers is a worthy goal, it may not be realised for all. Few of the poor own computers, and fewer still can pay the monthly Internet access charges required to carry out electronic banking. While computers are available in public libraries, many people refrain from doing their banking there—although, consumers might use telecommunication centres.

It is important to consider how electronic money products will affect non-users. New electronic money products may prompt the use of pricing strategies that further segment financial markets and raise prices paid by low-balance customers. For example, tying no-fee checking to high-value SVC cards could raise the prices on other services, pushing low-income, low-balance customers out of the banking system.

The rising of E-cash could also foster a have and have-not society: Those with PCs would have ready access to the new technology, while those without, many of them low-income consumers, would not.

What should be Done to Solve it?

The Governments should pursue international agreements to eliminate regulations which restrict on-line service providers from reaching end-users on reasonable and non-discriminatory terms and conditions. Genuine market opening will lead to increased competition and improved telecommunications infrastructure, and to lower prices and increased and improved services.

User Acceptability

Fewer people can understand the technology behind digital money and thus it does not inspire confidence. Conventional money, on the other hand, does not require any profound knowledge in order to use it.

With digital cash, with numbers stored somewhere in a pocket computer, the notion of money is somewhat lost. Sometimes we use the physical tangibility of cash in order to hand money personally to someone else, e.g., as a gift, as charity, or as a tip. This cannot be done with digital cash. It might be difficulties to learn to get a feeling of "how much" something costs.

Technical Standards

To ensure the growth of global electronic commerce over the Internet, standards will be needed in areas, such as:

- Electronic payments;
- Security (confidentiality, authentication, data integrity, access control, non-repudiation);
- Security services infrastructure (e.g., public key certificate authorities);
- Electronic copyright management systems;
- Electronic catalogues;
- Video and data-conferencing;
- High-speed network technologies; and
- Digital object and data interchanges.

Premature standardisation, however, can "lock in" users to outdated technology. Standards also can be used as a non-tariff barrier to trade, in essence to "lock out" non-indigenous businesses from a particular national market.

CONCLUSION

The convergence of money, communication and personal computers represents one of the booming markets in the future. As the most relevant precondition for establishing a new market, an innovative form of electronic money, integrated in an effective electronic transaction system of micro-payments, has to be installed. A real break-through will come when electronic forms of money are valued like the value of any other existing currency. In the past, governments proclaimed their currency legal tender, but how can an international global currency be handled?

To date, private companies, especially financial institutions with experience in risk management, take over this role on the Internet. Firms compete among each other as to who will establish the most accessible and secure system to exchange information, services, products a and currency!!! Currently, there are various approaches to payments systems. The most innovative one is a token-based payment system where individuals make purchases with digitised money or electronic checks.

Another mechanism using smart cards or debit cards has been developed and currently is in the pilot phase. The third system is the already known credit card-based payment system with transferring information based on a WWW-based encryption or a third-party authorisation number via the Internet. As a result, current approaches or solutions are not mature enough to have a world-wide and robust system against attacks and capable of handling a large number of transactions at an acceptable price.

The institutions of the private sector will compete among each other on the basis of established trust, fees and value-added services, but also with government as a counterpoint in this fast-paced evolution. Agreements between banks, financial service providers and non-banks regarding allocation of risk and splitting of fees for their offered services will be needed. The Internet is a new virtual world with increasing rate of "citizens" and a new world demands new standards and also-a new standard of value.

Whether the global market will adopt an electronic payment system and if so, which one will be the favourite is still not predictable. For now, appropriate institutions and responsible organisations are still watching the trends and evolution of the pilot projects and trials—to see what the most likely path of the transition to digitised money will look like. Electronic commerce and its involved payment methods are still vague and remain a frontier town—a land of opportunities, for sure, but also one of tremendous risks regarding the indicated problems.

APPENDIX

THE DOUBLE-SPENDING PROBLEM

Since e-money is just a bunch of bits, a piece of e-money is very easy to duplicate. Since the copy is indistinguishable from the original, you might think that counterfeiting would be impossible to detect. Obviously, real e-money systems must be able to prevent or detect double spending.

Online e-money systems prevent double spending by requiring merchants to contact the bank's computer with every sale. The bank computer maintains a database of all the spent pieces of e-money and can easily indicate to the merchant if a given piece of e-money is still spendable. If the bank computer says the e-money has already been spent, the merchant refuses the sale. This is very similar to the way merchants currently verify credit cards at the point of sale.

- Off-line e-money systems detect double spending in a couple of different ways. One way is to create a special smart card containing a tamper-proof chip, called an Observer (in some systems). The Observer chip keeps a mini database of all the pieces of e-money spent by that smart card. If the owner of the smart card attempts to copy some e-money and spend it twice, the embedded Observer chip would detect the attempt and would not allow the transaction. Since the Observer chip is tamper-proof, the owner cannot erase the mini-database without permanently damaging the smart card.
- The other way off-line e-money systems can handle double spending is to structure the e-money and cryptographic protocols to reveal the identity of the double spender by the time the piece of e-money makes it back to the bank. If users of the off-line e-money know they will get caught, the incidence of "double spending" will be minimised (in theory). The advantage of these kinds of offline systems is that they do not require special tamper-proof chips. The entire system can be written in software and can run on ordinary PCs or cheap smart cards.

It is easy to construct this kind of off-line system for identified e-money. Identified off-line e-money systems can accumulate the complete path the e-money made through the economy. The identified e-money "grows" each time it is spent. The particulars of each transaction are appended to the piece of e-money and travel with it, as it moves from person to person, merchant to vendor. When the e-money is finally deposited, the bank checks its database to see if the piece of e-money was double spent. If the e-money was copied and spent more than once, it will eventually appear twice in the "spent" database. The bank uses the transaction trails to identify the double spender.

Off-line anonymous e-money (sans Observer chip) also grows with each transaction, but the information that is accumulated is of a different nature. The result is the same, however. When the anonymous e-money reaches the bank, the bank will be able to examine it's database and determine if the e-money was double spent. The information accumulated along the way will identify the double spender. The big difference between off-line anonymous e-money and off-line identified e-money is that the information accumulated with anonymous e-money will only reveal the transaction trail if the e-money is double spent. If the anonymous e-money is not double spent, the bank cannot determine the identity of the original spender nor can it reconstruct the path the e-money took through the economy. With identified e-money, both off-line or on-line, the bank can always reconstruct the path the e-money took through the economy. The bank will know what everyone bought, where they bought it, when they bought it, and how much they paid.

REFERENCES

www.digicash.com
www.cybercash.com

21

Security on Internet

COMMENTS

The Internet is the mother lode that has spawned a whole host of recent advances like e-commerce, Electronic Data Interchange (EDI) and several other similar developments. It is, therefore, necessary to consider security on the Internet. The report explains the Internet briefly and simply, and explains why security is an important consideration for a network, the Internet being a set of networks. The sources of security problems for the Internet are identified. They are:

1. *Access Points*
2. *Mis-configured Systems*
3. *Software Bugs*
4. *"Insider" Threats.*

Next, the report spends time in explaining the vulnerabilities of the system. These require identifying the assets and identifying the threats. The assets usually include the hardware, the software, data, users, documentation and supplies. The threats are more varied and consist of:

- *Unauthorised Access*
- *Disclosure of Information*
- *Denial of Service*
- *Ease of Spying and Spoofing*
- *Inherent Problems with TCP/IP Protocols*
- *Wide-open Network Policies.*

Some TCP/IP services have been identified as high-risk and should be blocked and certain network services have been identified as requiring special attention from the point of view of network security. These include

---

Contributed by Anchal Jain, Niti Guglani and Prerna Gupta.

Authentication, Gateways, Modem Pools, Public Access Systems and System Security Tools. Effective Security Tools have been identified and explained. These are:

- *Firewall*
- *Confidentiality including Encryption and Privacy Enhanced Mail*
- *Origin Authentication*
- *Information Integrity*
- *Limiting Network Access through Gateway Routing Tables and Router Packet Filtering*
- *Authentication Systems.*

Examples of Authentication Systems that have been discussed include First Virtual Accounts, DigiCash, CyberCash and Secure Electronic Transaction protocol (SET). All these authentication systems have been discussed and their strengths and weaknesses have been analysed. In addition, there is a discussion of Problem Reporting Tools. These include McAfee Total Virus Defence, Total Network Security, Sniffer Total Network Visibility and McAfee Total Service Desk. The paper concludes with elaborate and well-reasoned recommendations for new as well as existing Internet connections, including general security precautions.

INTRODUCTION

The "Internet" is that set of networks and machines that use the TCP/IP protocol suite, connected through gateways, and sharing a common name and address spaces. Many organisations and individuals are looking at the Internet as a means of expanding their research interests and communications. Consequently, the Internet is now growing faster than any telecommunications system thus far, including the telephone system.

New users of the Internet may fail to realise, however, that their sites could be at risk to threats such as intruders who use the Internet as a means for attacking systems and causing various forms of computer security incidents. Consequently, new Internet sites are often prime targets for malicious activity, including break-ins, file tampering, and service disruptions. Such activity may be difficult to discover and correct, may be highly embarrassing to the organisation, and can be very costly in terms of lost productivity and damage to data.

When discussing electronic commerce on the Internet, usually the first concerns/that come to mind are its lack of security, reliability and accountability. The ability to encrypt data (cryptography) is one of the essential tools in the quest for secure communications on the Internet. New and existing Internet users need to be aware of the high potential for computer security incidents from the Internet and the steps they should take to secure their sites. Many tools and techniques now exist to provide sites with a higher level of assurance and protection.

The component networks are inter-connected at various points to provide multiple routes to destinations and a high level of overall service to users. Many academic business, and government organisations are connected to the Internet. In 1993, over five million users were connected, with roughly half being business users. The Internet provides several types of services, including terminal emulation and remote system access (telnet), file exchange (ftp), electronic mail (smtp), and a number of other services for information exchange.

In deciding whether to use TCP./IP or OSI, it is recommended agencies procure Open Systems Interconnection (OSI) networking products for new network implementations and for multi-vendor information exchange. At the same time, it is practical for agencies to consider connections to the Internet for the purpose of exchanging e-mail and files with the large number of existing Internet sites.

SOURCES OF SECURITY PROBLEMS

Access Points

Access points are typically used for entry by unauthorised users. Having many access points increases the risk of access to an organisation's computer and network facilities. Network links to networks outside the organisation allow access into the organisation for all others connected to that external network. A network link typically provides access to a large number of network services, and each service has a potential to be compromised.

Dial-up lines, depending on their configuration, may provide access merely to a/login port of a single system. If connected to a terminal server, the dial-up line may give access to the entire network. Terminal servers themselves can be a source of problem. Many terminal servers do not require any kind of authentication. Intruders often use terminal servers to disguise their actions, dialling in on a local phone and then using the terminal server to go out to the local network. Some terminal servers are configured so that intruders can TELNET in from outside the network, and then TELNET back-out again, making it difficult to trace them.

Mis-configured Systems

Mis-configured systems form a large percentage of security holes. Today's operating systems and their associated software have become so complex that understanding how the system works has become a full-time job. Often, systems managers will be non-specialists chosen from the current organisation's staff. Vendors are also partly responsible for mis-configured systems. To make the system installation process easier, vendors occasionally choose initial configurations that are not secure in all environments.

Software Bugs

Software will never be bug-free. Publicly-known security bugs are common methods of unauthorised entry. Part of the solution to this problem is to be aware of the security problems and to update the software when problems are detected. When bugs are found, they should be reported to the vendor so that a solution to the problem can be implemented and distributed.

"Insider" Threats

An insider to the organisation may be a considerable threat to the security of the computer systems. Insiders often have direct access to the computer and network hardware components. The ability to access the components of a system makes most systems easier to compromise. Most desktop workstations can be easily manipulated so that they grant privileged access. Access to a local area network provides the ability to view possibly sensitive data traversing the network.

Understanding Vulnerability of Your System

There are two elements of a risk analysis:

1. Identifying the assets
2. Identifying the threats.

For each asset, the basic goals of security are availability, confidentiality and integrity. Each threat should be examined with an eye to how the threat could affect these areas.

Identifying the Assets

One step in a risk analysis is to identify all the things that need to be protected. Some things are obvious, like all the various pieces of hardware, but some are overlooked, such as the people who actually use the systems. The essential point is to list all things that could be affected by a security problem.

1. *Hardware*: boards, keyboards, terminals, workstations, personal computers, printers, disk drives, communication lines, terminal servers, routers.
2. *Software*: source programmes, object programmes, utilities, diagnostic programmes, operating systems, communication programmes.
3. *Data*: during execution, stored on-line, archived off-line, back-ups, audit logs, databases, in transit over communication media.
4. *People*: users, people needed to run systems.
5. *Documentation*: on programmes, hardware, systems, and local administrative procedures.
6. *Supplies*: paper, forms, ribbons, magnetic media.

Identifying the Threats

Once the assets requiring protection are identified, it is necessary to identify threats to those assets. The threats can then be examined to determine what potential for loss exists. It helps to consider from what threats you are trying to protect your assets,

The following are some threats to security:

Unauthorized Access

A common threat that concerns many sites is unauthorised access to computing facilities. Unauthorised access takes many forms:

Use of another user's Account to gain access to a system: The use of any computer resource without prior permission may be considered unauthorised access to computing facilities. The seriousness of an unauthorised access will vary from site to site. For some sites, the mere act of granting access to an unauthorised user may cause irreparable harm by negative media coverage. For other sites, an unauthorised access opens the door to other security threats. In addition, some sites may be more frequent targets than others; hence the risk from unauthorised access will vary from site to site.

Disclosure of Information: Another common threat is disclosure of information. Determine the value or sensitivity of the information stored on your computers. Disclosure of a password file might allow for future unauthorised accesses. •

Denial of Service: Computers and networks provide valuable services to their users. Many people rely on these services in order to perform their jobs efficiently. These services are not available when called upon, a loss in productivity results.

Denial of service comes in many forms and might affect users in a number of ways. A network may be rendered unusable by a rogue packet, jamming, or by a disabled network component. A virus might slow down or cripple a computer system. Each site should determine which services are essential, and for each of these services determine the Affect on the site if that service were to become disabled.

Ease of Spying and Spoofing: The vast majority of Internet traffic is unencrypted and therefore easily readable. As a result, e-mail, passwords, and file transfers can be monitored and captured using readily available software. Intruders have been known to monitor connections to well-known Internet sites for the purpose of gaining information that would allow them to crack security or to steal valuable information. This information sometimes permits intruders to spoof legitimate connections, i.e., trick system security into permitting normally disallowed network connections.

Inherent Problems with TCP/IP Protocols: The TCP/IP protocol suite, as implemented in the Internet, does not contain provisions for network security. A number of the TCP/IP services, e.g., rlogin, rsh, etc., rely on mutually trusting systems and are inherently vulnerable to misuse and spoofing. Ironically, some of these services, e.g., nis and nfs, are widely used to coordinate local area network security and to distribute system resources among other local systems. If the vulnerabilities in these services are exploited, security on the local area network could be badly compromised.

Wide-open Network Policies: Many sites are configured unintentionally for wide-open Internet access without regard to the potential for abuse from the Internet. Some systems still employ password-less guest accounts or anonymous ftp accounts that can be written to, without restriction. Others keep sensitive information on network-accessible systems where it can be easily read. The vast majority of sites permit more TCP/IP services than they require for their operations and do not attempt to limit access to information about their computers that could prove valuable to intruders.

In recent years, a number of security problems with the Internet have become apparent. Newspapers have carried stories of high-profile "cracker" attacks via the Internet against government, business, and academic sites. Crackers often roam the Internet with impunity, covering their tracks by moving from system to system. Intruders have been known to use systems illegally to exchange copyrighted software, to obtain sensitive information such as business secrets, and in general to cause mischief. System administrators are often unaware of break-ins and unauthorised users until they learn by accident.

A number of factors have contributed to this state of affairs:

Complexity of Configuration: Many sites connect systems to the Internet with little thought to the complexity of system administration and the increased potential for abuse from the Internet. New systems often arrive "out of the box" with network access controls configured for maximum, i.e. "least secure" access. The access controls are usually complex to configure and monitor. As a result, controls that are accidentally Mis-configured can result in unauthorised access.

Network Service Policy: The first step is to create a policy that details what types of connectivity will be permitted. If, for example, e-mail is the only service required, then other forms of access such as telnet and ftp can be restricted and overall risks reduced. Eliminating the TCP/IP services that are not needed will help to provide a simpler, more manageable network environment.

The following is a partial list of TCP/IP services that should be restricted or blocked from passing through a site's Internet gateway:

Table 21.1: High-Risk TCP/IP Services

DNS zone transfers	Leaks names of internal systems
tftp	Intruders can read password file
RPC (eg., NIS, NFS)	Intruders can read/write files
rlogin, rsh, etc.	Relies on mutually trusting systems
X windows, OpenWindows	Intruders can monitor users" sessions
Telnet, ftp, smtp	Can be abused, access should be restricted to selected systems.

Authentication Systems: That can be accessed from the Internet or via modem should require strong authentication such as provided by smart cards and authentication tokens. These systems use one-time passwords that cannot be spoofed, regardless of whether the passwords are monitored.

Gateways: Secure gateways, or firewalls, are highly effective for improving site network security. A secure gateway is a collection of systems and routers placed at a site's central connection to a network whose main purpose is to restrict access to internal systems. A secure gateway forces all network connections to pass through the gateway where they can be examined and evaluated. The secure gateway may then restrict access to or from selected systems or block certain TCP/IP services or provide other security features. A simple network usage policy that can be implemented by a secure gateway is to provide access from internal to external systems, but little or no access from external to internal systems (except perhaps for e-mail).

For small sites, a simple router with packet-filtering capability may serve as an effective gateway. This type of router can typically restrict access to selected systems and services by examining each packet according to a sequence of filtering rules. A more flexible and robust approach is to combine the router with other systems capable of logging network access and restricting access and services on a finer basis. Some secure gateways implement proxy services that require all ftp or telnet connections to be first authenticated at the gateway before being allowed to continue.

Without a secure gateway, a site's security depends entirely on the collective security of its individual systems. As the number of systems increases, it becomes more difficult to ensure that network security policies are enforced. Errors and simple mistakes in one system's configuration can cause problems for other interconnected systems.

Modem Pools: Unrestricted incoming and outgoing modem pools (including PABX systems) can create backdoors that would let intruders get around the access controls of secure gateways. Modem pools need to be configured to deny access to unauthorised users. Systems that can be accessed from modem pools should require strong authentication such as one-time passwords. Modem pools should not be configured for outgoing connections unless access can be carefully controlled.

Public Access Systems: Public access systems, such as, anonymous ftp archives are often prime targets for abuse. Such systems, if mis-configured to allow writing, can permit intruders to destroy or alter data or software, which can prove highly embarrassing to the organization.

System Security Tools: The existence of a secure gateway does not negate the need for stronger system security. Many tools are available for system administrators to enhance system security and provide additional audit capability. Such tools can check for strong passwords, log connection information, detect changes in system files, and provide other features that will help administrators to detect signs of intruders and break-ins.

EFFECTIVE SECURITY TOOLS

Firewall

A firewall is a system or a group of systems that enforces an access control policy between two networks. Though the actual implementation varies, it consists of two mechanisms: one, which exists to block traffic and the other to permit traffic.

Configuration

Conceptually, there are two types of firewall:

- The network level
- The application level.

Network-level firewalls generally make their decisions based on the source, destination address, and ports in individual IP packets. A simple router is an example of the "traditional" network level firewall, since it is not able to make particularly sophisticated decisions about what a packet is actually talking to or where it actually came from. Network-level firewalls are generally hosts running proxy servers, which permit no traffic directly between networks, arid which perform elaborate logging and auditing of traffic passing through them.

Application-level firewalls can be used as network address translators, since traffic goes in one "side" and out the other, after having passed through an application that effectively masks the origin of the initiating connection. Application-level firewalls tend to provide more detailed audit reports and tend to enforce more conservative security models than network level firewalls.

Protection

Generally, firewalls are configured to protect against unauthenticated interactive "logins from "outside" world. Some firewalls permit only e-mail traffic through them, thereby protecting the network against any attack other than an attack against the e-mail service. More elaborate firewalls permit passage only from the inside to the outside while blocking all incoming traffic. By plugging the firewall, the user is protected against any type of network-borne attack. Firewalls cannot protect against attacks that do not go through the firewall. Firewalls cannot protect very well against things like viruses.

There are too many ways of encoding binary files for transfer over networks, and too many different architecture and viruses to try to search for them all. In general, a firewall cannot protect against a data-driven attack-an attack in which something is mailed or copied to an internal host where it is then executed.

Firewall can be used to enhance your site's security in a number of ways. The most straightforward use of a firewall is to create "internal site", one that is accessible only to computers within your own local area network. If this is what you want to do, then all you need to do is to place the server INSIDE the firewall:

```
other hosts
  \
server <—> FIREWALL <—> OUTSIDE
  /
other hosts
```

However, if you want to make the server available to the rest of the world, you will need to place it somewhere outside the firewall. From the standpoint of security of your organisation as a whole, the safest place to put it is completely outside the local area network:

```
other hosts
  \
other hosts <—> FIREWALL <—> server <—> OUTSIDE
  /
other hosts
```

This is called a "sacrificial lamb" configuration. The server is at risk of being broken into, but at least when it is broken into it does not breach the security of the inner network. It is not a good idea to run the WWW server on the firewall machine. Now any bug in the server will compromise the security of the entire organisation.

There are a number of variations on this basic set-up, including architectures that use paired "inner" and "outer" servers to give the world access to public information while giving the internal network access to private documents.

Confidentiality

Confidentiality, the act of keeping things hidden or secret, is one of the primary goals of computer security practitioners. Several mechanisms are provided by most modern operating systems to enable users to control the dissemination of information. Depending upon where you work, you may have a site where everything is protected, or a site where all information is usually regarded as public, or something in-between. Most sites lean toward the in-between, at least until some penetration has occurred. Generally, there are three instances in which information is vulnerable to disclosure: when the information is stored on a computer system, when the information is in transit to another system (on the network), and when the information is stored on back-up tapes.

The first of these cases is controlled by file permissions, access control lists, and other similar mechanisms. The last can be controlled by restricting access to the back-up tapes (by locking them in a safe, for example). All three cases can be helped by using encryption

Encryption

Encrypted security tools are used to ensure the privacy, integrity and authenticity of messages, files, transactions and data traffic. Cryptography is frequently used in electronic mail (e-mail) and financial transactions.

Different groups of people have dedicated themselves to promoting privacy in these areas. Some, like banks and software companies, such as, Netscape, are spending millions on software development to design unbreakable systems. Some civil libertarian ones like the cyberpunks@toad.com!

The afore-mentioned security concerns regarding electronic commerce are shared by both consumers and suppliers alike since, today, the greatest use of commerce across the Internet is consumer-to-business as opposed to business-to-business.

Strong network security is extremely hard to implement and requires eternal, almost obsessive vigilance to maintain. However, there are many companies that do conduct business over the Internet. These companies see the Internet as a means of providing a global presence at very low costs.

There are many forms of encryption tools used by the Internet community today. Most businesses employ a form of public key encryption to affix a digital signature to messages, to encrypt messages, and to initiate encrypted sessions that will be carried out with private key encryption. Public key encryption uses four keys, or mathematically generated numbers, to encode and decode information. Both the sender and the recipient have a unique public key and private key. Public keys can be freely distributed; private keys should not be shared with anyone.

Here is how it works. The sender encodes the message using his or her own private key and the recipient's public key. When the message arrives, the recipient decodes it using the sender's public key, and a private key. By decoding the message with those two keys, the recipient confirms the identity of the sender, and the sender is assured that the message can only be deciphered by the intended recipient.

Other forms of cryptography and some of the related tools used are listed below. These are:

- E-mail Encryption
- Pretty Good Privacy (PGP)
- Privacy Enhanced Main (PEM)
- Voice Encryption
- PGPFone
- Nautilus
- Disk or File System Encryption
- Secure File System (SFS)
- CryptDisk.

Encryption works by encoding the text of a message with a key. In traditional encryption systems, the same key was used for both encoding and decoding. In the new public key or asymmetric encryption systems, keys come in pairs: one key is used for encoding and another for decoding. In this system, everyone owns a unique pair of keys. One of the keys, called the public key, is widely distributed and used for encoding messages. The other key, called the private key, is a closely-held secret used to decrypt incoming message. Under this system, a person who needs to send a message to a second person can encrypt the message with that person's public key. The message can only be decrypted by the owner of the secret private key, making it safe from interception. This system can also be used to create unforgeable digital signatures. Most practical implementations of secure Internet encryption actually combine the traditional symmetric and the new asymmetric schemes. Public key encryption is used to negotiate a secret symmetric key that is then used to encrypt the actual data.

Since commercial ventures have a critical need for secure transmission on the Web, there is very active interest in developing schemes for encrypting the data that passes between browser and server. There are several types of commercially available encryption packages in both hardware and software forms. Hardware encryption engines have the advantage that they are much faster than the software equivalent, yet because they are faster, they are of greater potential benefit to an attacker who wants to execute a brute-force attack on your encrypted information. The advantage of using encryption is that, even if other access control mechanisms (passwords, file permissions, etc.) are compromised by an intruder, the data is still unusable. Naturally, encryption keys and the like should be protected at least as well as account passwords.

Information in transit (over a network) may be vulnerable to interception as well. Several solutions to this exist, ranging from simply encrypting files before transferring them (end-to-end encryption) to special network hardware which encrypts everything it sends without user intervention (secure links). The Internet as a whole does not use secure links, thus end-to-end encryption must be used if encryption is desired across the Internet.

Privacy-Enhanced Mail

Electronic mail normally transits the network in the clear (i.e., anyone can read it). This is obviously not the optimal solution. Privacy-enhanced mail provides a means to automatically encrypt electronic mail messages so that a person eaves-dropping at a mail distribution node is not (easily) capable of reading them. Several privacy-enhanced mail packages are currently being developed and deployed on the Internet.

Origin Authentication

We mostly take it on faith that the header of an electronic mail message truly indicates the originator of a message. However, it is easy to "spoof", or forge the source of a mail message. Origin authentication provides a means to be certain of the originator of a message or other object in the same way that a Notary Public assures a signature on a legal document. This is done by means of a "Public Key" cryptosystem.

A public key cryptosystem differs from a private key cryptosystem in several ways. First, a public key system uses two keys, a Public Key that anyone can use (hence the name) and a Private Key that only the originator of a message uses. The originator uses the private key to encrypt the message (as in DES). The receiver, who has obtained the public key for the originator, may then decrypt the message.

In this scheme, the public key is used to authenticate the originator's use of his or her private key, and hence the identity of the originator is more rigorously proven. The most widely known implementation of a public key cryptosystem is the RSA system. The Internet standard for privacy enhanced mail makes use of the RSA system.

Information Integrity

Information integrity refers to the state of information such that it is complete, correct, and unchanged from the last time in which it was verified to be in an "integral" state. The value of information integrity to a site will vary. For example, it is more important for military and government installations to prevent the "disclosure" of classified information, whether it is right or wrong. A bank, on the other hand, is far more concerned with whether the account information maintained for its customers is complete and accurate.

Limiting Network Access

The dominant network protocols in use on the Internet, !P (RFC 791), TCP (RFC 793), and UDP (RFC 768), carry certain control information which can be used to restrict access to certain hosts or networks within an organisation. The IP packet header contains the network addresses of both the sender and recipient of the packet. Further, the TCP and UDP protocols provide the notion of a "port", which identifies the end-point (usually a network server) of a communications path. In some instances, it may be desirable to deny access to a specific TCP or UDP port, or even to certain hosts and networks altogether.

Gateway Routing Tables

One of the simplest approaches to preventing unwanted network connections is to simply remove certain networks from a gateway's routing tables. This makes it "impossible" for a host to send packets to these networks. This approach is commonly taken in "firewall" systems by preventing the firewall from advertising local routes to the outside world.

Router Packet Filtering

Many commercially available gateway systems (more correctly called *routers*) provide the ability to filter packets based not only on sources or destinations, but also on source-destination combinations. This mechanism can be used to deny access to a specific host, network, or subnet from any other host, network, or subnet.

Authentication Systems

Authentication refers to the process of proving a claimed identity to the satisfaction of some permission-granting authority. Authentication systems are hardware, software, or procedural mechanisms that enable a user to obtain access to computing resources. At the simplest level, the system administrator who adds new user accounts to the system is part of the system authentication mechanism. At the other end of the spectrum, fingerprint readers or retinal scanners provide a very high-tech solution to establishing a potential user's identity. Without establishing and proving a user's identity prior to establishing a session, your site's computers are vulnerable to any sort of attack.

Some authentication systems are:

(i) *Kerberos* is a collection of software used in a large network to establish a user's claimed identity.

(ii) Smart Cards Several systems use "smart cards" (a small calculator-like device) to help authenticate users. These systems depend on the user having an object in their possession. One such system involves a new password procedure that require a user to enter a value obtained from a "smart card" when asked for a password by the computer. Typically, the host machine will give the user some piece of information that is entered into the keyboard of the smart card. The smart card will display a response, which must then be entered into the computer before the session will be established. Another such system involves a smart card which displays a number which changes over time, but which is synchronised with the authentication software on the computer.

First Virtual Accounts, DigiCash, CyberCash, SET are all schemes that have been developed to process commercial transactions over the Web without transmitting credit card numbers or other confidential information.

First Virtual System

The First Virtual scheme is designed for low—to medium-priced software sales, fee-for-service information purchases, and other types of "intangible" merchandise that can be delivered over the Internet. It is not designed for the purchase of tangible items like computer hardware or dishwashers. Before making purchases with the First Virtual system, the consumer first signs up for a First Virtual account by filling out an on-line application form at First Virtual's site. The consumer then completes the process by telephone. During the sign-up procedure he provides his credit card number and contact information, and receives a First Virtual personal identification number (PIN) in return. Thereafter, to make purchases at participating on-line vendors, the user provides his First Virtual PIN in lieu of his credit card information. First Virtual will later contact him by e-mail, giving him the chance to approve or disapprove the purchase before his credit card is billed. The consumer pays a one-time charge of $2.00 to open a First Virtual account. There are no additional charges, and no special software is required on the user side of the connection.

Merchants wishing to accept First Virtual payments must open up an account with First Virtual for a one-time processing fee of $10.00. First Virtual will provide the merchant with simple software for validating users' PIN numbers and notifying First Virtual when a purchase has been made. It's straightforward to integrate this software into a "shopping cart" style CGI script. In addition to the one-time fee, First Virtual charges the merchant a transaction fee of $0.29 per transaction, plus 3 per cent of the transaction price.

Digicash

Digicash, a product of the Netherlands-based Digicash company, is a digital cash system that works something like the familiar phone card or subway token. Under this system, users purchase "CyberBucks" from a bank that supports the Digicash system. CyberBucks can be purchased remotely via credit card or wire transfer. These CyberBucks, in actuality a series of specially-encoded serial numbers, can be used just like real cash: used as payment for tangible merchandise, exchanged in return for intangible information, or traded among individuals. At any time, you can deposit CyberBucks in a bank and receive "real" currency in return.

The software that supports Digicash prevents the CyberBucks from being forged or spent more than once. Like true cash, Digicash can be spent anonymously. You do not have to authenticate yourself in order to spend or receive Digicash, and its use leaves no audit trail. This is in contrast to credit-card based systems like CyberCash and SET in which every transaction leaves a paper trail that can be used to gather information about the consumer's spending habits. In addition, Digicash can be used to transmit money safely between peers, allowing ordinarily people to sell merchandise and services across the Internet without involving the banking system.

Digicash requires special software to be installed on both the consumer's and merchant's computers. It is currently available for Windows NT, Windows 95, and some Unix systems. Since the system has only recently come out of its pilot phase, it is not yet widely adopted, but this situation is expected to change.

CyberCash

CyberCash, a product of the CyberCash Corporation, uses specialised software on the merchant and customer's sides of the connections to provide for secure payments across the Internet. For a consumer to make CyberCash payments, he must first download a free piece of software called the "Wallet" from CyberCash's Web-site and initialise it with payment information and personal identifying information. Payment options currently include credit card numbers and bank accounts numbers. The wallet stores this information, in encrypted form, on the user's personal computer.

When a user goes to purchase an item from a CyberCash-accepting merchant, the wallet software pops up and requests the user to select the payment system. He may choose to charge the purchase against a credit card, in which case the charge will appear on his next credit card statement, or against his bank account, in which case the sum will be debited immediately. Software installed on the merchant's side of the connection validates and records the transaction by connecting to a server maintained by CyberCash, a process that takes 10-15 seconds. The Wallet maintains a record of each transaction, allowing the user to rapidly review his purchases and check them against his credit card and/or bank statement. It is available for many platforms, including Macintosh, Windows 95 and Windows NT.

The system uses strong cryptography to prevent transaction information from being intercepted by unauthorised third parties. Further, because actual credit card account numbers are never recorded on the merchant's server, there is no chance that credit card numbers can be stolen by individuals who have broken into the merchant's computer system.

For a merchant to accept CyberCash payments, he must open an account with a bank that supports the system. A CyberCash account is similar to the checking account that is set up for mail order houses that accept conventional credit card orders and has similar fees: a one-time fee of approximately $100 to create the account, a monthly fee of approximately $15 to keep the account open, and a transaction fee of 2-3 per cent of the purchase price of each transaction. The exact fees are set by the regional banks themselves, so the numbers will vary somewhat. There are currently several hundred banks that support CyberCash accounts, and the number is growing steadily.

After setting up a CyberCash account, the merchant must install software called the "Electronic Cash Register" on his Web server. This software is launched when the user presses the "pay" button in a shopping cart script (or equivalent), and takes over the transaction, creating an entry in a transaction log that the merchant can feed into his order entry and shipping systems. The Cash Register is downloadable free of charge and available for many platforms, including Windows NT and Unix.

The main advantage of CyberCash over Digicash is that it provides the customer with the same degree of consumer protection that credit cards provide. If a merchant fails to deliver a product or delivers unsatisfactory merchandise, the customer can appeal to the credit card company. Disadvantages include the loss of anonymity that accompanies any credit or debit card transaction, and the inability of CyberCash to handle peer to peer payments. In addition, the fees charged to merchants for processing credit card transactions make CyberCash impractical for small purchases, such as, "play per pay" on-line video games. This last problem, however, has been addressed by CyberCash's recently introduced "CyberCoin" system, which allows the customer to pay a lump sum to CyberCash in advance and then make small purchases against this sum.

Source Electronic Transaction (SET)

SET, or Secure Electronic Transaction protocol, is an open standard for processing of credit card transactions over the Internet created jointly by Netscape, Microsoft, Visa and Master Card. The main justification for SET is interoperability. By adhering to the standard, one vendor's software will be interoperable with any other vendor's software. To address the high potential for fraud on the Internet, the SET standard uses a complex system of certifying authorities to vouch for the identify of every party in the transaction: customer, merchant, card-issuer and merchant's bank are all identified by signed, unforgivable certificates. To address privacy concerns, the transaction is separated in such a way that the merchant has access to information about what is being purchased, how much it costs, and whether the payment is approved, but no information on what payment method the customer is using. Similarly, the card-issuer has access to the purchase price, but no information on the type of merchandise involved. Despite these precautions, however, SET does not provide the complete anonymity that the Digicash system has.

SET requires specialised software on both the customer's and merchant's side of the connection. On the customer's side, at least, the software can be delivered transparently in the form of Java Applets and/or ActiveX controls. Two SET-compliant products are currently in the offing. Microsoft Merchant is an Internet commerce system offered by the Microsoft Corporation. Based around Microsoft's Internet Information Server, Microsoft Merchant offers online credit card validation services using the Verified service. In addition, Microsoft Merchant provides such frills as catalogue management, shopping cart scripts and sales tax calculations.

First Data—This is an optional module for the Netscape Secure Commerce Server, which provides secure credit card submission, validation and processing. You can add other modules to provide automatic catalogue and inventory management, shopping carts, and corporate database interfaces. In its current incarnation, Live Payment uses a precursor to SET that is similar, but not identical, to the standard.

SSL, SHTTP, etc., are all proposed encryption and user authentication standards for the Web. Each requires the right combination of compatible browser and server to operate, so none is yet the universal solution to the secure data transmission problem.

Secure Socket Layer (SSL) is the scheme proposed by Netscape Communications Corporation. It is a low-level encryption scheme used to encrypt transactions in higher-level protocols such as HTTP, NNTP and FTP. The SSL protocol includes provisions for server authentication (verifying the server's identity to the client), encryption of data in transit, and optional client authentication (verifying the client's identity to the server). SSL is currently implemented commercially on several different browsers, including Netscape Navigator, Secure Mosaic, and Microsoft Internet Explorer, and many different servers, including ones from Netscape, Microsoft, IBM, Quarterdeck, Open Market and O'Reilly.

SHTTP (Secure HTTP) is the scheme proposed by Commerce Net, a coalition of businesses interested in developing the Internet for commercial uses. It is a higher level protocol that only works with the HTTP protocol, but is potentially more extensible than SSL. Currently, SHTTP is implemented for the Open Marketplace Server marketed by Open Market, Inc on the server side, and Secure HTTP Mosaic by Enterprise Integration Technologies on the client side.

Open Market Web Commerce System

Open Market, Inc., also offers an on-line commerce system. In the Open Market scheme, Open Market itself acts as the credit card company, handling subscriptions, billing and accounting. The scheme is integrated into its Open Marketplace Server, and requires a secure browser that supports the SHTTP or SSL protocols. Open Market's products are geared mainly to large corporations, banks, and service providers who wish to set up virtual "malls", and are priced accordingly.

Keeping up-to-date

Sites need to be aware of other resources and information that will permit them to update site security, as new vulnerabilities are discovered and as new tools and techniques to improve security become available. In particular, sites need to know who to contact when trouble arises.

Vendor Support

Several system vendors now regularly distribute software update notifications and related security information *via* e-mail. Customers need to contact vendors and determine whether they can receive such information.

Incident Handling Teams

A number of vendors, other businesses, and government-affiliated organisation have created computer security incident handling teams. These groups typically provide assistance in determining whether an incident has occurred and how to correct any vulnerabilities that were exploited.

Problem Reporting Tools

Auditing

Auditing is an important tool that can be used to enhance the security of your installation. Not only does it give you a means of identifying who has accessed your system (and may have done something to it) but it also gives you an indication of how your system is being used (or abused) by authorised users and attackers alike. In addition, the audit trail traditionally kept by computer systems can become an invaluable piece of evidence, should your system be penetrated.

Tools

COPS is a security tool for system administrators that checks for numerous common security problems on UNIX systems. COPS is a collection of shell scripts and C programmes that can easily be run on almost any UNIX variant. Among other things, it checks the following items and sends the results to the system administrator.

The challenge of security has caught the network professionals in a state of inadequate preparation without the training, resources or expertise. Fast-evolving technologies, based on open systems and extensive connectivity, has ushered in risks of intrusions and information compromise. Ever-increasing tribe of hackers are using various techniques to gain access to corporate information, resulting in huge monetary losses to corporate. According to "Trends in Intellectual Property Laws", a study for the American Society for Industrial Security (ASIS), the losses from intellectual property thefts for US-based companies world-wide are estimated to be $24 billion annually,

To protect today's networks which are based on global communications models, firewalls and perimeter protection tools are inadequate. Corporate networks require a mature level of protection from technologies that in addition to actively monitoring the networks also allow for immediate protective action to be taken. With the threat to network integrity and the cost of managing networks increasing exponentially, it is imperative for corporates to choose an integrated solution, which will significantly lower the cost of network management and protect it from every possible security threat. Net Tools from Network Associates is one such tool. Net Tools is built around two integrated, scaleable suites: NET TOOLS SECURE and NET TOOLS MANAGER.

Net Tools Secure protects the enterprise from viruses, hackers, thefts, lost data and threats to data security while **Net Tools Manager**, a network management and-service desk solution, improves network efficiency and performance.

Net tool Security Includes: McAfee Total Virus Defence—a multi-tired approach to virus protection at the client, server and internet Gateway and PGP Total Network Security-which combines the latest security products with desktop encryption software and management tools.

Net Tool Manager includes: Sniffer Total Network Visibility-a comprehensive suite of products and services for network fault and performance management and McAfee Total Service Desk, which combines robust help desk technology with electronic software distribution, automatic inventory and network management technology.

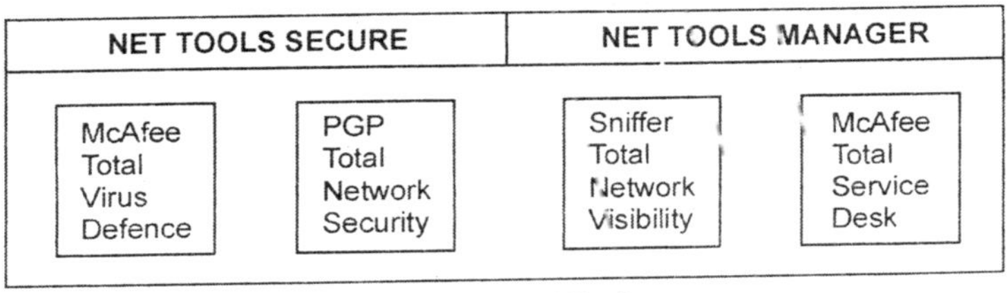

Fig. 21.1: Net Tools

McAfee Total Virus Defence

Recent studies have shown that viruses cause more financial losses than any other security threat. Corporate networks, which in the recent past had to be protected from the attack of viruses through floppy disks alone, now require virus protection from multiple points of entry.

"Network Associates" Anti Virus Emergency Response Team (AVERT) provides 24 hours support for enterprise customers. Virus signature files are updated on an hourly basis on the web as new strains are discovered.

Dr. Solomon's Anti Virus Toolkit which is capable of detecting and identifying over 19000 known computer viruses is being made available as part of "Network Associates" Total Virus Defence Suite and Net Tools Secure.

Total Network Security

This offers an integrated solution which address the security needs, enterprise wide, from desktop to the internet (security issues like such as electronic sabotage, and espionage on the internet, intranet and extranet).

It protects all digital assets of the company through encryption, authentication, intrusion protection and firewall technologies. The product's ease of use and scalability along with centralised manageability and policy-based administration makes it ideal for both large and small companies.

Sniffer Total Network Visibility

With the explosive growth of the internet, not only have networks mushroomed in size, their complexity and bandwidth have also increased considerably. Also the increasing user needs and ever changing technologies have fast outpaced the resources and personnel required to manage them. In this scenario, the need for automated tools and services like Sniffer

Tool Network Visibility that helps prevent, find and fix network problems have become more essential than ever. It provides corporates with end to end intelligence for optimising performance and preventing problems.

McAfee Total Service Desk

With the networked management world rapidly converging with the centralised help desk world, corporates need a total service desk solution that will help them capitalise on this emerging trend, like for instance McAfee's total service desk, which provides pro-active network management and robust help desk technology as a single integrated solution.

RECOMMENDATIONS FOR NEW AND EXISTING INTERNET CONNECTIONS

New and existing Internet sites need to take strong and specific measures to improve computer security. These measures include creating a TCP/IP service access policy, using strong authentication, and using a secure Internet gateway that can implement network access policies. It is important for an organisation to have a sound policy for network security.

Before deploying an effective security mechanism, consider the following:

1. What resources are you trying to protect?
   - CPU, files, storage devices phone lines, etc...
2. Determine the host-specific security measures needed. Password protection, file encryption, firewall, etc.
   - Determine who the computer systems must be defended.
   - Determine the likelihood of a threat.
   - Implement measures to protect network resource.
3. Consider the corporate budget when planning for Internet Security.
4. Design a Security Policy that describes your organization's network security concerns. This policy should take into account the following:
   - Network Security Planning
   - Site Security Policy
   - Risk Analysis
   - What you need to protect
   - What you need to protect it from

- How to protect it
- Estimating the risk of losing the resource
- Estimating the importance of the resource

5. Consider the following factors to determine who will grant access to services on your networks:
   - Will access to services be granted from a central point?
   - What methods will you use to create accounts and terminate access?
6. Design and Implement Packet Filter Rules.
7. Ensure your Firewall has the following properties:
   - All traffic from inside to outside, as well as outside to inside must pass through the firewall.
   - Allow only authorised traffic as defined by your corporate security policy be passed through the firewall.
   - Ensure the firewall is immune to penetration.
8. Educate users about password protection:
   - Educating users not to use passwords that are easy to guess.
   - Ensuring the password lengths are adequate.
   - Running a password guesser.
   - Requiring the use of a password generator.
   - Always using a mixture of upper- and lowercase characters.
   - Always using at least one or two non-alphanumeric characters.
   - Never using dictionary words or ones spelled backwards.
   - Never using a portion or variation of a proper name, address or anything that could obviously identify you (the user).
9. Security-related organisations play an integral role in the development and deployment of Internet technologies. Keep abreast of the latest in security-related activities by visiting their Web sites.

The following aspects should be kept in mind while framing a network policy. They are:

Organization Issues

The goal in developing an official site policy on computer security is to define the organisation's expectations of proper computer and network use and to define procedures to prevent and respond to security incidents. In order to do this, aspects of the particular organisation must be considered.

First, the goals and direction of the organisation should be considered. Second, the site security policy developed must conform to existing policies, rules, regulations and laws that the organisation is subject to. Therefore, it will be necessary to identify these and take them into consideration while developing the policy. Third, unless the local network is completely isolated and standalone, it is necessary to consider security implications in a more global context. The policy should address the issues when local security problems develop as a result of a remote site as well as when problems occur on remote systems as a result of a local host or user.

There are a number of issues that must be addressed when developing a security policy. These are:

1. Who is allowed to use the resources?
2. What is the proper use of the resources?
3. Who is authorised to grant access and approve usage?
4. Who may have system administration privileges?
5. What are the user's rights and responsibilities?
6. What are the rights and responsibilities of the system administrator *vs*. those of the user?
7. What do you do with sensitive information?

General Security Precautions

If you are a Webmaster, system administrator, or are otherwise involved with the administration of a network, the single most important step you can take to increase your site's security is to create a written security policy. This security policy should succinctly lay out your organisation's policies with regard to:

- Who is allowed to use the system
- When they are allowed to use it
- What they are allowed to do (different groups may be granted different levels of access)
- Procedures for granting access to the system
- Procedures for revoking access (e.g., when an employee leaves)
- What constitutes acceptable use of the system
- Remote and local login methods
- System monitoring procedures
- Protocols for responding to suspected security breaches.

This policy need not be anything fancy. It need only be a succinct summary of how the information system works, reflecting your organisation's technological and political realities. There are several benefits, to having a written security policy:

1. You yourself will understand what is and what is not permitted on the system. If you do not have a clear picture of what is permitted, you can never be sure when a violation has occurred.
2. Others in your organisation will understand what the security policy is. The written policy raises the level of security consciousness, and provides a focal point for discussion.
3. The security policy serves as a requirements document against which technical solutions can be judged. This helps guard against the "buy" first, ask questions later" syndrome.
4. The policy may help bolster your legal case should you ever need to prosecute for a security violation.

For Web servers running on Unix and NT systems, the following are some general security precautions to take:

1. Limit the number of login accounts available on the machine. Delete inactive users.
2. Make sure that people with login privileges choose good passwords. The Crack programme will help you detect poorly chosen passwords.
3. Turn off unused services. For example, if you do not need to run FTP on the Web server host, get rid of the ftp software. Likewise for tftp, sendmail, gopher, NIS (network information services) clients, NFS (networked file system), finger, systat, and anything else that might be hanging around. Check the File Manager for a list of servers that may be lurking. De-activate any that you do not use.
4. Remove shells and interpreters that you do not absolutely need. For example, if you do not run any Perl-based CGI scripts, remove the Perl interpreter.
5. Check both the system and Web logs regularly for suspicious activity. The programme Tripwire (Unix), and Internet Security Scanner (Unix & NT) are helpful for detecting this type of activity:

22

Planning and Implementation of ERP

COMMENTS

The report looks at the ERP phenomenon critically and clinically. Enterprise Resource Planning (ERP) and its objectives are explained and it is pointed out that 'ERP is the finest expression of the inseparability of Info Tech and business' The history of ERP is traced and it is pointed out that today's ERP systems are an evolutionary development from MRP II (Manufacturing Resource Planning) systems. Indeed, a tabular comparison between the two highlights the development and the gains that have occurred as a result of this evolution. The lack of penetration of ERP in India is discussed and the reasons for this situation are pondered over while considering the benefits of ERP. This is also looked at from the point of view of the main industry sectors to which the existing users belong. The question "Why ERP", that is often asked by traditional managements, is asked and answered from a holistic point of view. The possible benefits from ERP are enumerated and it is pointed out that there is a suitable time in the life-cycle of an organisation that it should go the ERP way. At a latter stage in the report, an analysis is made of the role BPR (Business Process Re-engineering) and ERP and whether it is necessary to go through BPR before venturing into ERP. BPR is explained and it is explained that BPR and ERP should dovetail into one another, for best results.

The report shifts its focus to the issue of implementation of ERP. It is pointed out that the issue of implementation of ERP is not a technology decision—as is the popular notion—but a decision based on business needs and benefits. The report explains at great length the issues that need to be addressed for proper implementation of ERP. The

---

Contributed by Abhishek Shukla, Anuj Kumar Agarwal, Gaurav Srivastava and Richa Singhal.

implementation of ERP is a project by itself and after analysing the factors that make the implementation successful; the report lays down a step-by-step methodology for successful implementation. In fact, different methodologies are suggested and discussed, highlighting the pros and cons of each and pointing out the snags to beware of at each step.

The discussion on ERP does not end with successful implementation. In fact, for the implementation to be successful, the organisation should not have any post implementation blues. The post implementation expectations and fears are listed and steps to overcome them are suggested. A brief analysis of the major ERP vendors is done and the strengths and perceived weaknesses of each are talked about. Some statistics about users are also given. There is a note on the topics that are often taken up when an ERP is to be evaluated. A tabular SWOT analysis of ERP is presented to help decision-making. There is a note on the likely future development. The report concludes with the presentation of two case studies of ERP implementation.

INTRODUCTION

What is ERP?

Enterprise Resource Planning or ERP systems which are coming into vogue are built with a vision to provide businesses with an integrated information system. These systems implement business processes within the organization to achieve synergy in operation across various business units. The challenge for ERP systems is to set-up and integrate information resources across geographically spread business units to enable optimisation across the organisation. ERP is basically the planning of the 4M's—of an enterprise's resources—Man, Money, Materials & Machines to their best synergistic value. It helps an organisation in addressing needs like reduced cycle time, customer focus, sharing information seamlessly across the enterprise globally. ERP integrates the entire enterprise starting from supplier/vendor to the customer.

Objectives of ERP

- Provide support for all variations of best business practices.
- Enable implementation of these practices with a view towards enhancing productivity.

- Empower the customer to modify the implemented business processes to suit their needs.

ERP is thus the finest expression of the inseparability of Infotech and business but it must be remembered that there is no magic wand for success—using IT does not mean setting up computers to manage jobs ERP helps link together disparate elements of an organisation so that whenever an "external pull" takes place, the enterprise is able to adjust to it immediately and effectively. This proactive adaptability of an enterprise is known as enterprise-wide integration and this is what ERP helps achieve. The trend of function-oriented businesses to process-driven entities requires ERP packages which help it not only at the IT level but also at the application level. ERP suppliers thus provide for all the functional areas' processes in a modular manner helping to encompass the entire organization.

HISTORY OF ERP

Information systems technologies evolved from mainframe-based computing through the client-server era to where we are now heading—the Internet era. These distinct phases are marked by parallel development in hardware technologies. Client-server era began when computing power delivered at desktop machines increased manifold and matched mainframe computing power. These technologies limited the availability of information services to users within an organisation. The Internet era has brought in the ability to deliver information around the globe. This is made possible with advancements in communication infrastructure.

From a historical perspective, today's ERP systems have evolved from MRP II(Manufacture Resource Planning) systems focusing on planning and scheduling activities. However, increasing demand for improving business effectiveness to be competitive world-wide made it essential for ERP to ensure all the functions and business processes are seamlessly integrated, so that they work in harmony. Apart from the functional scope, the primary difference between MRP II and ERP is in software architecture as seen in Table 22.1. ERP systems are platform independent, supported by client-server architecture, DBMS technology and graphic user interface capabilities. Due to the evolution background, initially ERP had its place only in manufacturing organisations. Today, it has universal applications irrespective of nature and scope of business.

Table 22.1: MRP II and ERP

MRP II	*ERP*
➢ Assumes a static nature of the enterprise and fits the system to the enterprise. It automates existing processes based on the existing workflow in the enterprise. ➢ Enables high degree of integration with other automated processes in the organisation ➢ Problems of MRPII 1. Heavily dependant on available hardware and software platforms in the companies 2. This is an automation solution and in many cases may not be a business solution. ➢ Generally manufacturing organisations gain from this ➢ Is an expense.	➢ Runs in tandem with a fresh relook at the existing business processes in an organisation. ➢ Runs in close association with BPR, thus initially questions the value-adding potential of the particular process, and then goes on to put some system and procedures in practice. Thus, ERP is the process for the entire organisation regardless of size, number of plants and locations. ERP does not blindly automate ➢ Takes into account not only the company's internal issues but also factors in the external environment in terms of competition, time to market and so on. ERP is considered as a next generation of post-automation business solution ➢ Includes non-manufacturing and even service organisations ➢ Is an investment.

ERP IN INDIA

For a country as large as India, with the potential being what it is, the market penetration of ERP systems remains abysmally low. The irony of it is visible. India is a country that boasts of five decades of domestic manufacturing, easy availability of skilled workers and English being the accepted business language. It is also a country where the first MRP-II/ ERP systems made their entry over a decade ago. And yet, the market penetration of ERP is estimated at 6 per cent! And if it were not for the last two years, when the market is estimated to have grown by 75 per cent, this percentage would have been even smaller. And just how small is indicated by the fact that in absolute value terms, the market is expected to be around Rs. 400 crore very soon. There are several reasons given for this. Most notable among them are:

- ERP systems are only for large organisations
- ERP systems cost too much
- ERP service and support are not up to standards
- ERP systems do not provide true benefits.

ERP and Large Organisations

Though there are no formal data available, going by market feel, there are as many (if not more) small and mid-sised companies that invest in ERP solutions as large corporations. More to the point is the fact that the investment in ERP systems as a percentage of annual sales is more or less constant, showing that smaller companies invest fewer Rupees in installing ERP systems. The mid-sized customers have varied reasons for investing in ERP. Some of them are:

- need to crash the 'order-to-cash' cycle time
- need to optimise finished goods distribution
- need to close the books in a stipulated time
- need to give sales people accurate dispatch information.

The common denominator, therefore, was that an ERP system linked the organisation and provided the capabilities to achieve business objectives. So, in actuality, there is no correlation between the size of an organisation and its readiness for an ERP system. Hence, it is not truly an impeding factor in so far as ERP systems penetration is concerned.

ERP and High Cost

Certainly, ERP systems do not come cheap but the investment needs to be weighed against benefits and the opportunity cost of not implementing. Most companies which have done this exercise say that without their ERP system their very competitiveness would be under threat. Unfortunately, ERP vendors, in their anxiety to create largest market space for themselves, responded by slashing prices to correct this anomaly. In fact, a better part of 1997 and 1998 was spent by vendors trying to outdo each other by exploring new depths in product pricing. In consequence, losing out on the opportunity of educating Indian corporates on correctly evaluating the need for ERP.

ERP and Lack of Back-up

The feeling is that the there are not enough people out there who know enough and more about these systems. So, they fear that there may not be service and support both during and post implementation. However, over the years, ERP vendors have created a network of distributors and system integrators, while simultaneously increasing the number of consultants. Also, newer technologies and improvements in communications have given ERP vendors unconventional, yet effective, methods of servicing customers. For instance, most ERP vendors host a Web-site where software patches to the most oft reported problems are listed. Customers need only download the desired 'fix', install it, and run. In fact, most vendors are well geared to sell, implement and service customers. Yet even these capabilities have not brought a rush of new ERP customers.

ERP and Benefits

Recently in an ERP seminar, a company using an ERP system shared its experience. The presenter, the General Manager for Materials, described the ERP system thus: "It is a damsel during evaluation, transforms into an elephant during implementation, and finally, a dinosaur after some time." Clearly, this organisation did not reap the benefits of the ERP system. Does it mean that this is the case rather than the exception? No, simply because an implementation has to go through a formal review, make out an accurate estimate of the cost and a list of realistic benefits. Companies which fail to do these often find their ERP implementation not succeeding. The above, mentioned company did not go through the above process and paid for it. Unfortunately, such misadventures get amplified and lead to the wrong conclusions about ERP systems.. India, is a first-time user market. Computer applications are not a pervasive way of doing business. Even today state excise authorities (with jurisdiction over liquor and tobacco) refuse to accept excise returns in any format other that manual registers. EDI is not an established data-interchange practice in trade. Electronic payments are not fully legal. One estimate puts India's PC penetration at 0.7 per one thousand people. Historically, organisations in India did not invest in commercial information systems. Rather, the in-house EDP team wrote applications to handle departmental functions like financial accounting, payroll and finished goods tracking. The cash outflow to acquire such systems was only the cost of development tools. The cost of labour, time overruns, poor quality, multiple rewrites were not factored into the total cost. So when the ERP systems entered, companies made the mistake of comparing the cost of 'in-house' systems and the ERP systems. Implementation of an ERP system should be treated as a project, which means an appraisal should be done before management sign-off. Companies should keep in mind that the perceived benefits, commitment of management time, roles and responsibilities, are some aspects that need to be reviewed, agreed and communicated prior to the start of the project. The non-pervasiveness of information technology and the absence of a deep-rooted IT culture is one chief factor inhibiting deeper penetration of ERP systems in India. In effect, what all the above points indicate is that while there are problems in each, they are not the real cause of low market penetration. In some ways, one could say that misconceptions about ERP systems and lack of awareness as being two major factors for the low penetration of ERP systems in India.

Main industry segments who use ERP packages

Seventy-five per cent of users are from the manufacturing sector out of which 60 per cent are in the discrete segment e.g. Machine tools, switch gears etc. The automobile industry is the largest followed by electronic and machine tool industry. This is because these industries have large operating size and hence require a lot of co-ordination and planning for the optimum utilization of resources. 15 per cent are in the process segment, e.g., Pharmaceuticals, chemicals etc.

WHY ERP?

Managers need to take an integral look at their companies and realise that functional areas like marketing, logistics and financial accounting cannot be separated. Thus, every entrepreneur has to remember that he gets up every day for one thing: "To make money". In this environment, concepts such as "technology push" are obsolete. "market pull" is valid the customer dictates the company.

Therefore, when challenges faced by a company become more complex, managers who are very specialised begin to generally lose control. It is here that integration becomes a primary goal. As each specialist manager knows only a part of the company, the information flow is sub-divided into different sub flows. This information is too diverse and even redundant in some cases. And when information needs to be consolidated at the top management/corporate level—things, more often than not, go wrong. This is where ERP comes in, as seen in Figure 22.1.

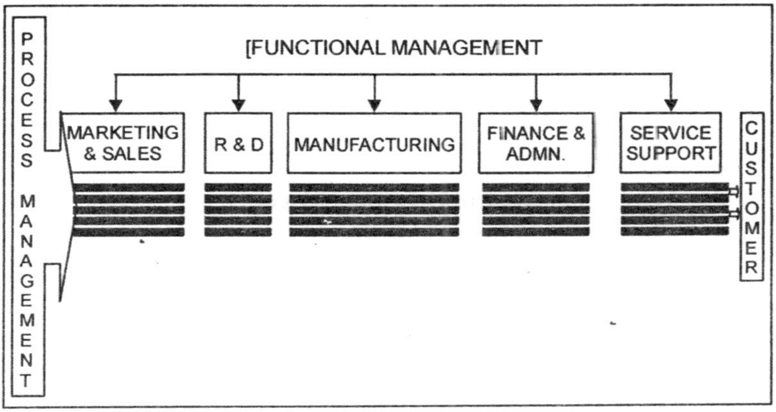

Fig. 22.1: Process Based Management

It is, thus, an opportunity for business reengineering. The organisation and the people are restructured and reorganized to give the organisations various benefits. Most of all is the organisation's capability to meet global competition; this is greatly enhanced with the coming in of ERP. It facilitates learning the best practices, which helps reap good benefits in the future.

BENEFITS OF ERP

- ERP helps in integrating the various functional spheres in an organisation—a link through the entire supply chain, aimed at adapting best industry and management practices for providing the right product, at the right place, at the right time, at least cost
- ERP generally follows Graphical User Interface (GUI) approach. This enhances the usability of ERP systems
- Bringing together people who work on shared tasks within the same enterprise or in their dealings with suppliers and customers
- Executive time available for more productive causes
- Organisations ensure a smoother flow of information at all levels and between all parts of the organisations
- To access up-to-date information
- Enterprise replace legacy systems that are expensive to support and maintain
- Work flow integrates business processes

- Improved process yields and better capacity utilization's. It also uncovers the hidden capacities
- Better cost identification and management
- Quicker response to market change
- Reduced down time
- More efficient use of warehouse space
- Improved operations flexibility
- Reduction in inventory
- Diversion of manpower-routine activities to profitable ones
- Elimination of unproductive activities.

WHEN TO GO THE ERP WAY

Based on a study done by KPMG for the magazine *Dataquest* (February 15, 1998), a framework has been suggested as to whether an organisation needs ERP, BPR or an IT solution aimed at automation. The framework proposes that an organisation can map itself on a grid with "pressures to perform" on the Y-axis and organisational "readiness to change" on the X-axis.

Readiness to Change

BPR and ERP entail change albeit in varying degrees. Unless an organisation on its own feels the need to change, prescribing a strong dose of BPR or ERP can only result in strong side effects. The organisation's readiness to change can be measured on the following parameters:

- Top Management Initiative
  Is the top management pro-change or anti-change? Does it encourage innovation?
- Organisation culture and management style
  Is it family owned or professionally managed? Are the employees empowered, the decision-making participative or not?
- Relationship with workplace
- Level of job motivation? Current skill-sets of the employees? And what are the new ones required?
- Financial health
  Profitability and cash position are very important, as a cash strapped organisation cannot possibly implement BPR or ERP.
- Legal shackles
  The above is mostly valid for governmental or semi-governmental organisations.

Pressures to Perform

Business is always defined from 'outside in'—by customers, competitors and complementors. If the business environment is in a flux, an organisation must respond and change in order to maintain or elevate its performance level. The pressures to perform can be external or internal:

External Pressures

These include factors like:

- Competitive positioning
- Structural changes in the Industry and changes in regulatory structures
- Threat of ending spending privileges
- Shortened product life-cycle
- Shortened technology life-cycle
- Danger of over supply or challenge to create demand.

Internal Pressures

These Include the following:

- Increase in business complexity
- Changes in strategic aspects like product portfolio and market segment
- Increase in workforce turnover and other labor problems.

The parameters for pressures to perform can be evaluated in terms of market-share/value leadership, share price, industry trends like mergers and acquisitions, profitability, product obsolescence rate, manpower attrition rate and so on. Pressures to perform lead to need for change. So the grid translates into cross-matching the need and willingness to change for an organisation.

The Grid

If Both are Low or Moderate

The organisation can afford to work with custom developed systems for its various functions. The main thrust is automation of work and the organisation can think in terms of linking these localised systems as a step toward better performance.

If Both, the Need and the Willingness are High

Then BPR in its true radical sense is the most logical option.

High Pressures to Perform

These organisations can opt for an ERP solution with strategic focus on cost cutting, if they are not ready for change. The ERP package will thus help them to cut wasteful expenditure and inefficient buffers built in their supply chains. The package will let them maximise their current resources and thus become lean to compete.

Higher Readiness to Change but Lower Pressures to Perform

These organisations can strategically use the ERP package to widen the competitive gap between them and their competitors. The ERP package thus helps them to have better control over business activities and endows them anticipatory skills via pro-active decision making capabilities. Organisations can build a solid information structure and focus on elevating their business performance. An organisation can, therefore, focus on its core competencies and can groom them more

The rating of parameters on this grid is a tricky issue. A lot needs to be done in this regard. Apart from some readily quantifiable parameters like market share or profitability or share price, others have to be looked at in more qualitative terms. The mapping exercise, as seen in Figure 22.3 should be done in close consultation with the top management of an organization, with individual perceptions playing an important role. Then again outside consultants can also be beneficial.

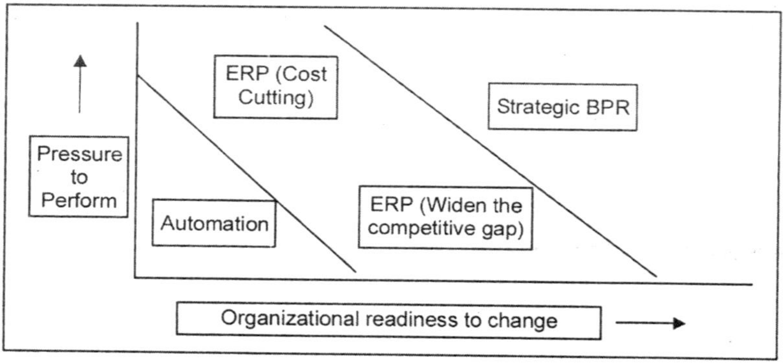

Fig. 22.2 : The grid

IMPLEMENTATION OF ERP

Pre-Implementation

Know, Prepare and Communicate

The popular notion is that implementation of an ERP is a technology decision. The fact is, it is not. In fact, it is a decision that ideally should be based on business needs and benefits. Ideally speaking, the CEO and the apex management must be more than clear as to what they expect from the entire initiative. Unfortunately, though, what happens in actuality is none of the above. Very often, one finds the CEO and CIO speaking different languages and, worse, are at different wavelength. John Donovan of the MIT puts it succinctly. Says he, "as CEOs struggle to get their e-mail to work, the CIOs bombard them with opaque talk of bytes and ram."

The Business Case and Its Realization

The CEO must insist upon benefits from the ERP team that would be felt as a one-time measure and those on a recurring basis. The costs of the entire initiative must also be calculated on a capital and a revenue basis. The system must be positioned as a tool and an enabler to achieve specific benefits. These desired outcomes, in turn, should be articulated in a business case.

For instance, a CEO could ask, during the course of appraisal, the following for a clear understanding and communication of needs and expectations:

1. Create a base case of annual savings from cost cuts that could be made without the ERP system in place.
2. Create an ERP case of annual savings that could be made with the ERP. This should include savings that do not depend on ERP.
3. Subtract the base case savings from the ERP case savings on annual basis, and calculate the NPV of the residual cash flow. A positive NPV will indicate that you should proceed with the deployment of the ERP.
4. If step 3 produces a positive NPV, conduct a sensitivity analysis to ensure that the business case is strong enough to withstand time overruns and cost overruns.
5. Back allocate all ERP system deployment costs to individual business units so that they can factor them into their planning. Ensure each unit is held responsible for producing the promised savings.

Setting Metrics and Assigning Responsibility

Getting an enterprise system up and running is only the beginning. To ensure that the system will meet the organisation's business objectives, the CEO and the apex management must establish key metrics and assign key process owners specific responsibility for achieving the benefits. The CEO must also assign individual responsibilities for achieving the benefits. A company's metric could be achieving better quality (less scrap, fewer returns, lower warranty costs, etc). Others may set faster product development times, improved customer satisfaction, more accurate order fulfillment or faster delivery times as benefit goals.

Apart from the hard benefits that he is likely to earn and a case for ROI, the CEO should mould also look at the security of his investments and the rate of obsolescence. One cannot expect to be technology-proof in the current state of technological boom, but, all the same should ensure that it does not become a losing proposition either. One way to do it is by ensuring that the product in a few years will be supported by evolving standards A case in point is the argument on proprietary versus open systems. Selection of the IT framework, though highly technical and subjective, must be delved into by the CEO for a comfort on future investments (in the long term). Very often, CEO's comment on the CIO's approach to capital expenditure on IT systems as "one more plaything for my IT guys". Sadly, this has been proved true quite often. The lesson for the CIO to learn is that they need to provide a clear IT roadmap for the enterprise to the top management.

Choosing the Approach

A very critical decision here is whether one should go the big-bang way or the incremental route. In weighing this, a lot has to do with culture and politics of the organisation. In some companies, lots of small successes are needed to build up the case. Here, the organisation might opt for the incremental or the pilot and rollout approach. The CEO and the apex management's decision on this subject are of utmost importance here.

A subject that should be addressed early in the stages of the ERP initiative is the approach to re-engineering. The top team must understand that enterprise packages are tailorable, but only within limits. The ERP applications already have a gamut of best-practices incorporated by design. They also assume standardisation of business processes across the enterprise. A sound approach is to develop a philosophy at the outset of the project about the level of re-engineering and package-changes needed, and then communicate that philosophy to the project team. The best-known method is to re-engineer processes concurrently with implementation, up to a point.

Continuous Communication

Deployment of the ERP is a tricky job as it involves a whole gamut of people, all interacting or rather trying to communicate at the same time. A whole lot of personal agendas surface during this period, for instance, operational and tactical staff, whose interests were earlier trampled by the IT staff, try to settle scores. Hence, it becomes very imperative that the key stake-holders in the initiative agree upon the project plan at the beginning of the project. The project plan should also clearly mention dates for steering committee meetings, etc. And these meetings must be used by the CEO or the CFO to communicate the original vision, the strategies and objectives of the project. Continuously re-emphasising the message helps in reducing the chances of "straying off" from the target objectives as well as the target dates. The success of the ERP initiative, on the softer side, can be attributed to two things. Firstly, in all the communication about the project, a practice of tying central messages and specific department objectives and needs back to the overall company. Secondly, a regular mix of efforts to include everything from conducting workshops, publishing newsletters and holding focus groups to organising lunch-time discussions and travelling road-shows each designed to suit different stages across the implementation life cycle.

You have to explicitly, candidly and constantly communicate the business case and realities including goals timetable and expectations. The execution and the timing of the communications plan is extremely important. Consider a case where the communication pitch in an ERP implementation reached a crescendo by calling the end-users and stake-holders in an inaugural user-training workshop on a Saturday morning. Members wore T-shirts and caps with logos of the project insignia to demonstrate their commitment to the new system. The entire effort went berserk when most users and stake-holders complained of the system being useless as the names of the suppliers and the customers the users saw were that of "The ABC Shoppe" or "XYZ Corp". The transactions would work perfectly as it would do in the real world, but the master data had still not been uploaded and test data were being used.

Forming the Team

Selecting the right project leader is as important as selecting the right package. Team leaders should know the business well and have cross functional experience. They should also be politically savvy, have credibility within the organisation and be good communicators, and, of course, be from the business side and not the IT management.

The team is also as important as its leader. The apex management must make sure that the best and the brightest join the team. The members must have some cross-functional knowledge. The member must have some benchmarking experience and above all, must be a team player. A lot of good people do not like to be pulled out of their jobs to serve on an ERP team. The challenge is to convey not only how important the teams' work is to the company but the invaluable career enhancement benefits that accrue to each member as a result of his of her involvement. If attracting the best and brightest people in the organisation to the team is important to successful implementation, so is keeping them.

Change Management and Culture

Enterprise systems are all about the enterprise and not about systems

So, a top management champion can play a critical role in effective change management and implementation of an enterprise system. However, a lot of people feel that the CEO is not always the best choice. The higher up the organisation you go, the more risk you run that the champions' involvement will be delegated.

The support of a champion does not replace the need to sell to other members of top management the benefits of the system. It has to be done early and often by using multiple approaches. All senior managers in the organisation should be able to support the enterprise system. Every manager should be able to clearly explain to any employee who asks why the company is doing what it is doing. The explanation provided must be reasonably consistent across the company.

Because of the range and magnitude of alterations that must be made, ERP requires fundamental shifts in assumptions, beliefs, and day-to-day behaviours throughout the organisation. The biggest leap is the necessity for each person to have a basic understanding of the overall organisational system rather than just their piece of it. In each case, organisations find that just as they begin to address the demands they anticipated, they are engulfed by a second wave of implications that they had not anticipated. Unless the organisation is prepared to come to a dead stop while putting the new system in place, people must maintain performance levels while preparing to shift to the new system. The level of effort required continue to escalate throughout the process, and typically does not diminish for a substantial period after the conversion. When these demands are placed on top of an already-high workload, the resulting strain leads to decreased output, increased turnover, delays in work in process, and negative attitudes towards the organisation. If issues related to increased workload are not managed effectively, and resistance is driven underground, the resulting problems can be very costly to the organization in terms of quality and productivity.

To summarise, one can clearly say that to make sure an ERP initiative is a sure-shot success, one must:

1. Realise the business benefits
2. Recognise the cultural implications
3. Get adequate support from senior management
4. Plan and manage the implementation effectively
5. Plan for high impact and high resistance.

Steps for ERP Implementation

1. Take a fresh look at business processes of the organisation
2. Pick out the core business processes
3. Find the information required of these processes
4. BPR: Business Processes in a few problem areas in the organisation are redesigned
5. Prototyping: Package is customised and demonstrated to the users/clients
6. Designing and developing enhancements, work around and reports
7. Transitioning of existing data from the legacy system
8. Finally, productive operation of the system.

Implementation Methodologies

Working of a Methodology

A methodology divides the entire implementation into smaller and more manageable groups of activities. A top-drawer methodology should encompass all the components that go on to provide a complete business solution, rather than merely implementing the software package. It should ideally incorporate the following five tracks which run throughout any ERP implementation life-cycle:

- Business process analysis

- Change management
- Technical infrastructure development
- Programming for enhancing package features
- Project management.

These five tracks would incorporate activities spread typically across four broad sections or stages which are a part of any software implementation: analysis, design, construction and implementation Let us now look within each of these four stages and explore some of the key activities that would typically be covered within these phases.

Analysis

Analysis should start off with strategy alignment, which should align the project with the strategic goals of the enterprise. The aim of this exercise should be to arrive at a set of critical success factors that should henceforth be the 'success-or-failure' criteria for the implementation effort. This phase should also see the determination of the implementation strategy, the sequence in which the ERP package will be implemented and rolled out to various business units. Some of the choices that you have are:

- *Big Bang:* where all modules of the ERP package are implemented at one go across all business units of the organisation.
- *Pilot Approach:* where all modules are implemented across some key business units and then roll the system out to other offices.
- *Phased Approach:* where implementation is phased out according to the modules of the package, for instance, finance implementation, followed by materials management, etc.

A recent trend in this area has been the 'business model' approach, which is a combination of the big-bang and the pilot approach. In this scheme, a global prototype is developed, followed by a geographical roll out across all business units. 'As-is' analysis would typically follow strategy alignment. Here, the existing business processes of the organisation is studied and documented. Simultaneously, an evaluation of the existing technical set-up of the organisation should be carried out. While the functional teams (the finance or the logistics team, for example) conduct as-is analysis, the project management should look to finalise a detailed project plan and a quality plan for the project. This phase can be referred to as the project work plan development phase.

This is followed by 'to-be' analysis by the functional team. This phase is critical, as the exercise involved is quite tricky and needs the involvement of experienced resources. The team needs to carefully marry the existing processes with the package functionality and industry best practices—the goal is to devise an optimal process that can be supported by the package. During this phase, the functional teams should also map the envisioned processes to the package features, and come out with gaps, special reports, forms or interfaces with existing system that will continue to function in parallel with the ERP package. It is very important to appreciate that an ERP package comes with standardised functionality. You are expected to 'configure' the package to follow your 'to-be' business processes by setting parameter values in tables. However, the inherent assumption in implementing an ERP is that you will not change the basic way in which the package is structured, i.e., you will not change the source-code of the package. Customisation or code change is a dreaded word in the world of ERP implementation and should always be desisted from, however strong the temptation is to make the package work exactly the way your legacy system used to. Indiscriminate customisation can lead to a system that it impossible to upgrade; most ERP vendors, such as SAP, often withdraw their product warranty if their source code is customised.

Design

Design mainly consists of configuring the ERP package as per the analysis specifications, and developing design specification for custom code objects. Custom objects are programmes that are written to augment standard functionality that the ERP package provides. The design stage will start off with the functional teams developing the organisation structure to support the target processes they have designed. During this phase, called the key data structures, the organisation tree with business units like company code, factories, sales offices, branches, warehouses, etc., is developed. In some cases, the organisation structure is treated as a part of the 'to-be' phase. However, it is often advisable to first nail down the target processes, and then design a hierarchy to support them, rather than the other way round.

Then comes the prototyping. The prototyping phase will see the ERP package being configured through parameter-settings to reflect the organisation structure and the 'to-be' processes designed earlier. This phase is really the heart of the project. While the functional teams configure the system, the custom development team will study the gaps identified during the 'to-be' phase, and develop specifications in consultation with the functional teams. Once the gap analysis is complete, the development team will start designing the modules that will augment the standard functionality provided by the package.

During implementation, the number of gaps that you decide to approve for development should be minimised. This is where most projects slip—too many gaps identified too late in the day proves their undoing. Custom code design will primarily focus on the following objects:

- Custom reports to meet client-specific requirements
- Interfaces to existing legacy applications which need to exchange data with the ERP system
- Data conversion routines to load historic data
- Modifications to the package programmes (which is normally recommended against).

Construction

Construction deals mainly with development of custom code objects, such as, reports, forms, interfaces with legacy applications, and data migration routines. This stage also involves end-user training and integration testing. The construction stage primarily involves custom code construction and unit testing by the programming team. It is imperative for the functional team to start conduct end user training after prototyping. It is better for this training to be 'process driven' with people trained only on that part of the package which figures in his or her daily task list.

Subsequent to this, the system should be put through a thorough test on an simulated environment resembling the production environment. This shows the integration of the prototyped system and the custom developed objects.

Implementation

Implementation deals with setting up the production environment and making sure that hardware/software/networking infrastructure is in place. Once everything is ready, the ERP implementation goes 'live'. Typically, a period of hand-holding is required after go-live, and this is achieved by forming a helpdesk which is 'on-call' to assist end-users. The implementation stage performs the last few critical activities necessary for system live-processing. These mainly involve setting up the ERP front-end on individual workstations, configuring printers, testing the connectivity and promoting the tested system to production.

Post-Implementation

Post-implementation blues are a reality. It could be kept to the minimum by taking preventive measures, starting with communicating the message across the organisation, cutting across functions and hierarchy. It is very well recognised that technological advances of any nature bring change. Enterprise Resource Planning is no different. The introduction of ERP systems has created a mini revolution in the functioning of industries. Like problems faced with any new technology, ERP implementation and usage has its own set of unknowns to be tackled. These have to be proactively handled to derive maximum benefits. Needless to add, there are mechanisms for putting in place a continuous improvement programme.

Expectations, Fears, Reality

To start at the beginning, many post-implementation problems can be traced to wrong expectations and fears. The expectations that corporate managements have while choosing an ERP are no less than a miracle. Of course, some of the blame for this have to be laid at the doorstep of the ERP vendors and their pre-implementation sales type.

A few of the popular expectations are:

- An improvement in processes
- Increased productivity on all fronts
- Total automation and disbanding of all manual processes
- Improvement of all key performance indicators
- Elimination of all manual record keeping
- Real time information systems available to concerned people on a need basis
- Finally, total integration of all operations.

ERP implementation also engenders a host of fears. Some of them are:

- Job redundancy
- Loss of importance as information is no longer an individual prerogative
- Change in job profile
- An organisational fear of loss of proper controls and authorisation
- Increased stress caused by greater transparency

- Individual fear of loss of authority
- Balancing the expectation and fears is a very necessary part of the implementation process.

Finally, a successful ERP implementation is not the end of the road as far as change is concerned, and continuous improvements are required to reap the benefits. One of the most obvious ways to balance the expectations, fears and reality is during the process of implementation.

Life After Implementation

Effective use of ERP is a direct result of steps taken at the time of implementation towards preparing the organisation. Change integration has to be necessarily embedded in the task list for any ERP implementation. The main tool for this is the process of communication in all forms—written, oral, workshops, meetings, etc. The process should start quite early, by educating all layers of the management on the particular ERP product, its relevant functionality, limitations and benefits. Also, at the start of the project, the Critical Success Factors (CSF) for the company as .a whole should be listed. This should be drilled down to CSFs for respective functionalities or departments. From these CSFs, performance measures required to address these CSFs should be culled out. The numeric figures against these performance measures can be classified as the key performance indicators. The process of firming up the above is usually done through workshops. This has to be completed before the processes to be configured on the ERP are drawn up.

Envisioning the processes to be configured on an ERP is the critical portion to ensure user buy-in during the post-implementation phase. There are various ways this could be done, but what is important is the following should be borne in mind. The important end-users should be involved in evolving the process. This should be done keeping the ERP functionality in mind. The Key Performance Indicators (KPI) derived from the organisational goals and CSF should be kept in mind too. Having evolved the processes while the configuration, construction and implementation are in progress, the organisation needs to ready itself for the post implementation period. Some of the tasks that are to be performed are:

- Develop the new job descriptions and organisation structure to suit the post-ERP scenario
- Determine the skill gap between existing jobs and envisioned jobs

- Assess training requirements, create and implement a training plan
- Develop and amend HR, financial and operational policies to suit the future ERP environment
- Develop a plan for workforce logistics adjustment.

Post-Implementation Blues

While the above checks would take care of most post-implementation blues, certain problems are bound to be encountered. The major task is to monitor **the Key Performance Indicators** and take the correct business decisions to improve them. In most Indian organisations, however, these indicators may be non-existent before the implementation. So the immediate task is to set attainable goals. However, this may be unrealistic to be achieved in the first go. The more realistic path would be to have a stretched target to be achieved in phases. Similarly, certain KPIs, though existing in the system, are better monitored and controlled after the ERP system attains maturity.

The other major problem faced is that, more often than not, for reasons of data, transfer or just to be safe it is decided that the legacy systems run for a period of time. The users having a choice, many a time, display resistance to change. The only panacea to this is a strong management resolve to insist on implementation of the system. Even with all the preparations during the implementation, during post-implementation, there will need to be course correction many times. Simply because,

- A change in the business environment requires a change in the CSFs, resulting in a new or changed set of KPIs necessitating reconfiguration
- A review indicates a need for change in some process
- Vision changes in the ERP and improvements in hardware and communication technology necessitate changes
- New addition to the business requires extra functionality.

The international trend is to outsource the activity of maintenance and upgradation to enable the company concentrate on its core business activity. Correcting its course can be done by going in for an ERP audit, which is an emerging trend. This audit could be general in nature or very specific. One of the specialised areas is to evaluate the security, authorisation and controls. An audit could be triggered either by a perceived inadequacy in terms of return on investment or by a simple desire to improve existing systems. To conclude, investment in an ERP system is substantial for any organization. While implementation itself is a challenge, the ultimate test is in proper usage. This can be ensured by integrating the business objectives with the ERP functionalities during the implementation stage. The limitations of an ERP must also be recognised to get the right expectation. A periodic independent audit would be a proper mechanism for an organisation to ensure that they get the best return on their investment.

ROLE OF BPR IN ERP

The discussions around whether BPR precedes ERP or the other way around continues to confuse corporates. In order to meet the new challenges, Indian executives have been experimenting with the new management philosophies and tools like Business Process Reengineering or BPR, Enterprise Resource Planning or ERP, Total Quality Management or TQM, Just In Time or JIT, etc. Most of these philosophies and tools, imported mainly from USA and Japan, have met with only partial success, despite the big claims made by many global consultants operating in the country. Although some of these philosophies are conflicting to each other (e.g., BPR and ERP), there is a big confusion in the industry about similarities and dissimilarities among those.

What is BPR?

"BPR is the *fundamental rethinking* and *radical redesign* of processes to achieve *dramatic improvement* in critical, contemporary measures of performance such as cost, quality, service and speed." This has a few important key words, which need clear understanding. Here, dramatic achievement means to achieve 80 per cent or 90 per cent reduction (in, say, delivery time, work in progress or rejection rate) and not just 5 per cent, 10 per cent reduction. This is possible only by making major improvements and breakthroughs, and not small incremental changes (like those in TQM or suggestion schemes).

Radical redesign means BPR is *reinventing* and not *enhancing* or *improving*. In a nut-shell, a *'clean-suite approach'* of BPR says that 'Whatever you were doing in the past is all wrong', do not get biased by it or reassemble your new system to redesign it afresh. *Fundamental rethinking* means asking the question 'why do you do what you do?', thereby eliminating business process altogether if it does not add any value to the customer. There is no point in simplifying or automating a business process, which does not add any value to the customer. A class example is that of asking for an invoice from the supplier for payment when the company has already received and accepted a particular quantity of materials physically and at an agreed price. Receiving, processing, and filing of invoices add no value to customer and makes only the supplier unhappy for delayed payments. Thus, BPR aims for major transformation of the business processes to achieve *Dramatic Improvement*. Here, the business objectives of the enterprise (e.g. profits, customer-satisfaction through optimal cost, quality, deliveries, etc.) are achieved by 'transformation' of the business processes which may, or may not, require the use of the IT.

What is ERP?

Unlike BPR, ERP does not seem to have a well-documented and accepted definition. However, one formal definition is: "An accounting oriented *Information System* for identifying and *planning* the *enterprise-wide* resources to take, make, ship account for *customer orders*". The key words involved here are the *'Information System'*, which indicates that ERP is primarily an information technology-oriented system which provides immense help in *planning* the *enterprise wide* resources (which may also mean multi-location manufacturing units, multi-location warehouses besides multi-location procurement sources-covering main resources of man, machine, material and money).

ERP is primarily an Information and Communication Technology (ICT)—driven philosophy. In the past, most of the poor decision-making process (which led to low utilisation of men, machine, material and money resources) was due to non-availability of information at the right time and inaccuracy (or multiple sources) of information. ERP provides a technology platform to bridge this gap. Information is accurate, available in time and is cheap. Thus, the business objectives of the organisation (profits, customer satisfaction, etc.) are achieved by sharing the accurate and instant information throughout the enterprise. This was possible due to innovation in information and communication technology (e.g. client/ server technology, object-oriented programming (OOP), Internet and Intranet, etc.)

The Confusion about BPR and ERP

It is clear from the fundamental definition that BPR and ERP are quite different from each other. While BPR focuses more on the transformation aspect of the business processes using innovative business concepts, ERP focuses on the automation aspect of the business processes using ICT to achieve the same objective. An example will make it clear. If a large corporate fertilizer plant has two strong functional departments (namely, Operations and Maintenance) located far from each other, there will be considerable delay in sending a work order from operation department to maintenance department in case of an equipment breakdown. This delay would result in low availability of the equipment, hence higher costs and longer lead times.

An ERP can make the delay almost zero, where work order generation and communication is automatic and instant, thereby reducing the delay of attending it, increasing the availability of the equipment and decreasing cost. A BPR, however, may result in drastically changing the business process, where there are no separate departments for operation and maintenance. Additionally, maintenance (at least 95 per cent of the cases, except where exceptionally high expertise is required) may also be required to be carried out by the operations man. This would not only make him more responsible for operations, but also eliminate the need for generating a work order for maintenance department. Of course, this will lead to multi-skills concepts, etc. This BPR would also increase the availability of the machines (due to eliminating delays), reduce cost, etc. and may not need automation at all. Thus, the approach of BPR and ERP can be quite different to achieve the same goal and more availability of equipment and lesser cost.

However, with time, BPR has become more and more dependent upon the availability of ICT; on the other side, ERP has been upgrading itself in terms of adapting various proven good business processes in the industries. Many of the business 'transformations' are possible because of availability of ERP with high ICT. Similarly, many of the good proven industry-wise (say, automotive, electronics, pharmaceutical, etc.) 'transformed' business processes have been built in as business templates in ERP solutions.

Finding Common Grounds

For instance, many ERP solutions have graduated from APICS Definition and presently include many features like human resource management modules, supply chain modules (linking suppliers and customers), product data management, electronic data interchange, engineering change management, multi-currency and multi-location plants facilities, which is allowing many drastically transformed business processes to be adopted quickly.

Comparison between BPR and ERP

- ERP users in definite, but not necessarily radical, changes across the whole organisation unlike BPR.
- ERP does not force an organisation to become process-centric; it helps to streamline processes within functions.
- ERP and BPR both require extensive change management and committed leadership to succeed.
- ERP cannot substitute a BPR exercise in terms of its radical gains but is considerably less risky and painful.
- The most important point is that both ERP and BPR exercises should not be divorced from a business strategy, as they are not just a solution to an existing problem but also contribute to long term capabilities of an organisation.

Most consultants take advantage of this overlap and confusion. They talk whatever is convenient at the point of sale to puzzled Indian executives in order to clinch the deal and meet their own business interests. The following benefits may be listed towards customer's major five expectations as in Table 22.2.

Table 22.2: Middle Capability of BPR and ERP Towards Customer

S. No.	*Customer Expectations*	*Benefits from BPR*	*Benefits from ERP*
1.	Quality/Reliability	Explicitly	Implicitly
2.	Cost	Explicitly	Implicitly
3.	Delivery Time	Explicitly	Explicitly
4.	Flexibility (customisation, delivery)	Implicitly	Explicitly
5.	After-Sales Service	Implicitly	Explicitly

When it comes to overcome the 3C Crisis (Customer, Competition, Change), BPR and ERP are complementary to each other. But questions which top management may ask are:

- Whether to transform the business processes first through BPR and then adopt ERP or
- First adopt an ERP solution with world-class proven practices, and then keep on doing re-engineering or.
- Adopt them together, business process module-wise, one after other, till the complete business is covered.

The arguments and counter-arguments for these 3 routes are presented in the Table 22.3 below.

Table 22.3: Arguments, Counter Arguments for BPR, ERP Sequencing

Route	*Argument*	*Counter-Argument*
BPR before ERP	Why customise a BPR on ERP which may not be required at all after BPR? (e.g.) maintenance work order generation process mentioned earlier)	Why take a very long time (1-3 years) to implement BPR (involving major organisational changes, job profile changes and workflow path changes? Maybe 'too late' to take advantage of ERP
ERP before BPR	Why reinvent a wheel? Why not adopt proven world-class business practices, as a part of ERP solutions?	Why copy a best-in-class practice of the same industry, when each enterprise has a different market segment, customer expectations and overall unique business within the same industry? Why not copy best practices of other industries? (e.g., copy best distribution process of paint manufacturer for a detergent manufacturer)
BPR and ERP simultaneously module-by-module	Why aim for major changes (both BPR or ERP), which a company cannot sustain and risk failure (as clear from many not-so-successful implementation of both BPR and ERP)	Can we afford a long and time-consuming process of implementation (2-4 years) to face the 3C Crisis

Since all the arguments and counter-arguments are justified to some extent, there may not be one best answer to the problem of choosing the right route for the top management. However, if the main consideration for top management is **'Time'** and the **'Cost'** resource for implementation, it would be wiser to go in for a Big Bang approach (Route 2). This is because of the following reasons:

(a) The present state of ERP solutions, if honestly supported by the top management, with the help of business modelers and a good seasoned vendor, can be implemented in a 'short duration' of 5 to 7 months.
(b) The explicit and implicit benefits of ERP solutions start getting realised immediately.
(c) Most of the reputed ERP solutions are quite flexible, support ability to adopt to any re-engineered process quickly. Therefore, as processes are being re-engineered, those can be adopted as and when finalised, without having to change the ERP much. Even if it means changing of modules (incorporating a new one or obliterating the old one), it can be done in a live environment.
(d) Although adaptation of accepted best-in-class practices may not be the best for the unique requirement (and positioning) of the enterprise, if chosen judiciously, many of those accepted processes as a base may help to get the benefits faster. The process of BPR, however, can continue, process by process, depending upon the criticality and urgency of process.
(e) Having gained experience in business in Business modelling, best-in-class processes and ERP implementation, the organisation may not require external consultants for re-engineering the processes. It may be done by internal trained resources.

Thus, this approach (Route 2) can take advantage of both time and cost, and generally be more effective approach. It must be reiterated here that in order to meet 3C Crisis, the enterprises have to take benefits of both ERP and BPR. They have to view these as complementary to each other rather than conflicting with each other. But what is most critical in all this is that, in over-selling both BPR and ERP, top management should not overlook their limitations. Both BPR and ERP only help the tactical decision on **'how'** to do their business, rather than helping strategic decision on **what** business to do.

VENDORS OF ERP

Deciding on a vendor to supply ERP software is probably the most complex decision in the entire ERP Project. The reasons are many.

- ERP software being expensive, one cannot afford to make mistakes and correct it later.
- It is not clear as to who the decision-makers are and whose

benefits are addressed. If the decision-maker is a CEO who is genuinely looking for ERP for organisational benefits, he/she is likely to be pragmatic and look at the true benefits. Many CEOs today go for ERP because it is fashionable—a sort of "club membership".

- The genuine difficulty of non-availability of ERP consultants who are indeed knowledgeable about multiple products. Being complex products, ERP software takes years to master. Very few consultants have the luxury of knowing genuinely the strengths and weaknesses of many of the ERP software products. The knowledge of consultants is limited to vendors' brochure.

The prominent vendors of the ERP package are SAP, BaaN, Oracle, Ramco Marshal, Peoplesoft, IFS, BPCS, CBSI, IBM, TCS, etc., among others.

SAP

SAP R/3 is the hottest ERP software package in the computer industry. Its developer, SAP AG, ranks prominently among the worldwide software production industry. SAP R/3 was introduced in 1992 and has more than 300,000 users on 6000 installations. The 'R' in R/3 stands for 'Real Time', meaning the logical application links will concurrently update the related module. SAP AG produces two base products: SAP R/2 for the mainframe & SAP R/3 for the Client/Server environments. The SAP functionality is based on the 'business process concept'.

SAP is a market and technology leader in client/server enterprise application software, providing comprehensive solutions for companies of all sizes and all industry sectors. Cultivating innovative technologies on a solid foundation of business experience, SAP delivers scalable solutions that enable its customers to continually improve upon best business practices. SAP products empower people to respond quickly and decisively to dynamic market conditions, helping businesses achieve and maintain a competitive advantage.

Oracle

- Oracle is an American company
- It works on Windows NT platform
- It has 35 products for the following processes.

- Solution for Financial Management
- Supply Chain Management
- Manufacturing Project Systems
- Human Resource Management
- Sales Force Automation
- Clustering Technologies:
- Oracle Parallel server
- Oracle Fail-Safe
- Scalability for fast growing companies
- Global Functionality
- Great Flexibility
- Rapid Implementation.

BaaN

- It's a Dutch company
- Bundles workgroup servers with Informix
- Products: BaaN 4, Triton 2&3
- First-line support for DBMS: A single point of contact
- Competitive pricing and DBMS license
- Readily acceptable for middle markets: low cost and single point of contact
- One of their weakness is: not preferred for large users.

Peoplesoft

- It's a UK based company
- Product line
- Peoplesoft 6: SQL server, 32-bit architecture
- Peoplesoft 7: 2 & 3 tier architectures, internet-enabled with Web extensions, Java client, OLAP tools available
- Peoplesoft 7.5: Global payroll, manufacturing and distribution
- Delivery of People tools for: AS/400 & 11 new products
- Its strength: HR product (80% business)
- Its weakness: Financial services (20% business)
- Enhanced "Europeanisation":
- With more language translations
- More European functionality

IFS (Industrial and Financial Systems)

- Complete business system to handle: engineering, manufacturing, distribution, mantainence

- IFS applications: also financial, HRM
- Customers: large and medium size corporates.
- Long-term relations with customers:
- Responsible for implementation and mantainence
- Product development in cooperation with customers
- Revenues from consultancy/licensing (aiming at 50% of sales)
- Based on object-oriented technology and has 40 independent modules
- Due to open architecture, IFS applications may complement business systems from other vendors for specific process
- IFS product development: develop tools from Oracle, Microsoft, Centura, Sunsoft.

BPCS (Business Planning and Control System)

- It is a USA-based company
- 30 integrated products for manufacturing process industries, distribution and logistics, financial applications
- AS/SET: fully integrated
- EDI product line: customers/vendors through EDI network. Short cycle time
- Based on advance AS/400
- Over 5500 users worldwide
- Computer integrated manufacturing applications, integration with trading partners

RAMCO MARSHAL Systems

- It is an India-based company (Chennai)
- Largely in medium and small industries
- Invested Rs. 1000 million to develop MARSHAL 3.0
- More than 1400 man-years of effort have gone into building this world class product, completely using Microsoft tools and technology
- It is the only ERP package to be developed ground up on client server technology
- Launched by Madras cements, cumulative sales exceeding Rs. 1000 crores
- 44 sites in India, 100 companies from various industries worldwide (Steel, pharmaceuticals, healthcare, consumer electronics, engineering, power).

EVALUATION OF ERP

1. **Stages** should stem from a goal-directed approach with nearly sets out the areas of final usage and ROI. Thus **before giving in for ERP,** company should achieve the following objectives:
   - Overall alignment of the company's business plan to the project
   - Clear articulation of goals that are widely understood
   - A well-established communication plan that involves the whole organisation
   - Executive sponsor of the project process as well as process owners and managers
   - Understanding by users of the need to change to the new system and justification thereof.
2. **Platforms to be used**
   - SAP HP, NT and UNIX
   - ORACLE VMS and UNIX
   - BAAN UNIX
   - RAMCO AS/400
   - MARCAM AS/400

   Consideration here is the costs involved as compared to the probability, adaptability and compatibility required. This requires a sound decision from the IS manager after weighing all the pro's and con's
3. **Coding and interface environment**

   E.g., Marshal is written in visual C++ and is available on SARBAS and MS.SQ SAVER
4. **Database support offered**

   MS ODBC enables users to get to another database with ease
5. **Other needs specific to certain areas**

   Thus this requires a Distributed Object Computing architecture with which ensures flexibility in a business and the possibility for a re-configirability.

ERP thus does not make management as a skill absolutely obsolete, rather enables managers to manage better and more fully than before.

DILEMMA OF BUYING ERP

Key Issues

An of the-shelf ERP product like SAP appears more expensive than ERP applications. This is because many of the organisations do not have methods and practices to cost in house applications. Moreover, the price of the ERP refers to the price of the software product excluding the cost of the hardware. The second major concern relating to ERP is the non-availability of the competent professionals. Also, the key-users of the organisation need to be deputed full-time for the ERP implementation which puts the management in a dilemma.

Certain factors need to be considered while selecting an ERP product:

- Is the ERP product covering all aspects of the business?
- Is it truly integrated?
- Does it address the current trend in IT?
- How strong is the development team and what are the future plans?
- If the ERP vendor is located outside India, how committed is its plan for the Indian version?
- What is the installed base for the ERP product?
- How strong is the implementation team?
- Are the best global business practices covered and supported in the ERP?
- How close is the relationship between the ERP vendor and the hardware vendors?

Selecting and implementing an ERP product is an important and time-consuming exercise. Thus care should be taken in estimating the total ERP cost and visualising the post implementation scenario.

CASE STUDY A: ARVIND MILLS

Knitting Yarns of Success

With information available on-line, the senior management is now in a position to gauge and manage the performance better Company: Arvind Mills Ltd. Arvind Mills is the leading textile manufacturing company of the country and the third largest denim producer in the world. In keep with its growth objectives, the organisation is investing in new projects in high value Cotton Shirting fabric, Knit fabric and Bottom weight fabric.

Arvind Mills has always been a pioneer in using modern methods and techniques in management and technologies. It identified information technology as a major business driver and enabler for creating the right business infrastructure to cope with the pressures of globalisation and increased competition.

Industry: Textiles

Solution Area: Supply Chain Management

Problem: In keeping with the growth and diversification plans of Arvind Mills, Rs. 1200 crore was invested in state-of-the-art manufacturing facilities in Santej, about 20 km from Ahmedabad. These plants have the very latest and most advanced textile machinery available. The target customers from this Greenfield project are top international and domestic apparel Brands. Typically, these Brands come up with designs for the coming seasons and work in close co-ordination with the fabric supplier for developing and finalising the look of their end garments. It was felt that to service these discerning customers, Arvind Mills would require a world-class Supply Chain Management system.

Solution: A host of packages were evaluated, keeping in mind the above requirement. The global presence of their customers merited a world-class ERP package that supports all global practices and conventions in Supply Chain Management. SAP, the world's largest ERP package company, was felt to be the best choice. SAP is a very comprehensive package and has the maximum number of installations world-wide and has incorporated the best practices in enterprise processes in their package. SAP also agreed to assist Arvind Mills in addressing certain textile-specific requirements through innovative process-work around as well as in customisation.

Implementation: Coopers and Lybrand, one of the leading information technology management consultants in the country, was selected by SAP India to provide additional project management support and techniques. Arvind Mills' core team was taken through extensive SAP product training at the SAP facility at Bangalore. As required by SAP, Arvind Mills created a "war room" project facility, resulting in better focus and also ensured that Arvind Mills, SAP and Coopers and Lybrand worked as one team. The entire project was driven by the super-user and was implemented in just over 7 months. Since an ERP implementation involves change management in the organisation, the constant support of senior management is a key driver. Arvind Mills had excellent commitment right from the Chairman's level to all Presidents, Vice-Presidents and even practicing managers. A Project Steering Committee, consisting of senior members of Arvind Mills, SAP India and Coopers and Lybrand, was formed to closely monitor the progress of the implementation project.

Benefits: Major benefit was envisaged in strengthening the supply chain management and customer-order execution. It was felt that the planning for raw materials and for work-in-progress goods would show substantial improvement. SAP would help in standardising business processes in various functional areas within the organisation. Since it is an integrated package, it gives a lot of on-line information to the senior management regarding the performance of the organization. Earlier, the same information required some time in consolidation and compilation.

CASE STUDY B: CALCOM GROUP

Electronics Manufacturing

The company managed to reduce inventories and order fulfilment time by 20 per cent

Company: Calcom Group

Calcom was set-up in 1976 to manufacture calculators. Today it is a US $40 million (for the fiscal year 1997) company with over 1250 employees in seven factories spread in three states in India. Its activity today comprises manufacture of television, computer monitors, vacuum cleaners and audio products. The company's clients include brand names like Samsung, Philips, BPL, Sony, Videocon and ECTV. It exports to countries in the Middle East, South East Asia, Africa and South America. Its future plans include manufacturing picture tubes, automobile electronics, UPS, color and monochrome monitors.

Industry: Electronics

Solution Area: Inventory management, manufacturing planning and scheduling, distribution and the financial systems of the company.

Problem: Seven plants located in different states made planning extremely difficult.

Solution: Baan ERP. They felt BaanIV was stronger in manufacturing, planning and supply chain management. Initially the company implemented the electronics module with its multi-site capacity and then the manufacturing, warehousing and distribution management modules. In the next phase, the company will be implementing the finance module.

Implementation: The implementation was unique for two good reasons. To begin with, it had to grapple with the complexity of implementing the solution over seven plants in Wazirpur, Okhla 1, Okhla 2, Surajpur, Mohali, and Bhiwadi. The company kick-started the exercise by training its key-users group. For three months, module-wise training was given. It helped immensely, as it resulted in excellent cooperation from end users. Also, support from a strong top management helped considerably. Secondly, they implemented it in less than a year. They planned every activity well, impart training to as many users as possible. Basically, ERP is a management product and each functional head is the owner and the systems professional is only the facilitator.

They ensured that end-user time was monitored and also completed master data as quickly as possible. SIM-1, SIM-2 and the migration phase was over. The company successfully defined the process, mapped the functionality and workarounds were thought of. Actually, Calcom merely followed the Baan methodology. In SIM-1, it defined the process in detail and created the feedback loop and ensured no open-ended process. In SIM-2, however, master data was carefully structured and coded. The process was aided by the fact that DEM was easy to use and Reference Models reduced implementation time. Secondly, there was absolutely no resistance from key-users. The hardware was Windows NT server from Digital-Compaq.

Benefits: They wanted Baan's solution to streamline their planning process by integrating diverse information, ensuring smooth planning. Besides, they wanted on-line/on-time data visibility across the group to improve once they go live. In short, they should get better decision support. Most importantly, the Baan solution should help them cut burgeoning inventory costs yet service orders/order-fulfillment in time. The company managed to reduce order fulfilment time by a little over 20 per cent and the inventory costs by approximately 20 per cent. They have very good visibility of data for analysis, and there is synchronisation of data and planning across platforms.

SWOT ANALYSIS OF ERP

Strengths	Weaknesses
➢ Data is available on-line for managers on a relative basis for analysis ➢ Adaptable and flexible process oriented	➢ Costs are an area of concern. E.g., solution from SAP R/3 can cost upto $400,000 to implement and may take upto 3 years. Therefore, return on capitalised Investment and size of business is important. ➢ Software made by software programmers intended for people in radically different organisations. Therefore, require intermediary (functional area professional) as a link between coder and client ➢ Cannot work without complete top management.

(*Contd.*)

(*Contd.*)

	➢ Chance that employees may be demotivated after BPR, Downsizing, Headcount reduction, etc.
Opportunities	**Threats**
➢ Money Resource Planning ➢ Disintermediation: a process that re-configures activities, eliminating non-value tasks within an organization	➢ Total Quality Management which may create the illusion of a perfect business in some cases where ERP is really required. ➢ Each discontinuity or discrete technology gives birth to a new leader. The new software computing architecture along with the architecture of the ERP applications will need to support the way business transactions take place in the networked economy. ➢ In the long run, there will be a transition from transactional to relational jobs and systems which endanger the use of the ERP system.

CONCLUSION:THE FUTURE MRP III

Money Resource Planning

This brings together all the information into line. Each manager should be able to manage the whole company—this is especially necessary for companies that have long lead times as in Table 22.4.

Table 22.4 : MRP II and MRP III

MRPII	*MRPIII*
➢ Focusses on production capacity.	➢ Allows financial consequences for company to be viewed; this implies that manufacturing companies are more and more often considered as service companies.
➢ Facilitates the marketing and production/logistics part of organisations.	

MRP III thus says that it is no longer sensible to calculate 'Costs + Contribution'. Companies need to evaluate their position and develop a new version of integration and implement it successfully to be truly "ERPed".

To avoid the 'Costs + Contribution' which uses the historical data, as well as to understand that implementing BPR is not a ad-hoc process, rather it requires great courage to question the current business processes. Therefore, a benefit analysis is necessary where completely new as well as changed processes must be simulated at cash-flow level, both in the medium term and the long term.

A Newer Orientation

Most organisations do not carry out a cost simulation and their product profitability is expressed only in external parameters-in terms of the marketing mix (in the broadest sense of the word).

The need of the hour is thus to answer the question of internal product profitability, i.e., which product yields the most profit, allowing to the capacity of the organisation? The answer to this question in case of many companies may show that they fail to produce the most profitable product (looking at their BPR results) and probably this is why they gauge themselves on the basis of their external product profitability.

23

Supply Chain Management

COMMENTS

Supply chain management, of late, has been recognised as an issue of significant strategic importance. Although initially adopted by high tech, and consumer-packaged goods industries, it is now spreading to other industry segments. This report explains the concepts of supply chain management and the need for utilising these concepts—competitive advantage, shifting channel power, faster cycle-time-to-market, enhanced information technology applications and globalisation. An overview of supply chain management explains its functioning fully and points out the advantages of such an analysis. It is "typically viewed to lie between vertically-integrated firms and those where each channel member operates independently". Supply chain decisions can be strategic decisions or operational decisions, depending on the decision-area. The four major decision areas are:

1. *Location*
2. *Production*
3. *Inventory*
4. *Transportation (Distribution).*

The nature of these decisions is explained along with the factors that determine them.

Supply chain mapping is explained as the process of separating each supply chain into its separate elements. This will make it possible to look at separate functions within any particular element, hereby making the analysis much easier. It is explained that the aim is to look at all the 'supplier/customer' links throughout the chain. Certain terms used in the supply chain analysis are explained here, for example, stockholding, volume of the supply chain, critical pipelines, length of the supply chain,

---

Contributed by Anurag Kalra, Jatin Mahajan and Siddharth Gorewara.

etc. The aim of supply chain analysis is explained as "to discover means of shortening the chain and reducing its volume". A diagrammed flowchart illustrates a sample supply chain succinctly and clearly.

The report also discusses various strategies for maximising the effects of the supply chain. Five strategies have been discussed, although not all for the same function. These strategies are:

1. *Supply chain objectives should be globalised*
2. *The supply chain audit*
3. *Develop a vendor supply chain strategic matrix*
4. *Build a win-win vendor compliance programme*
5. *Recognise supply chain as an on-going process and not a project.*

Various ways of applying these strategies and their advantages are suggested and discussed. Also various performance measures in a supply chain are suggested, such as aggregate inventory volume, inventory to sales ratio, days of supply, inventory turnover etc. These measures, if applied and found to be increasing, will indicate that from a customer point of view in a 'make to order' scenario, the shorter the supply chain, the better.

Finally, three separate case studies have been presented. These case studies should help to clarify the points made in the main body of the report.

INTRODUCTION: THE NEED FOR SUPPLY CHAIN MANAGEMENT

Competitive Advantage

The discipline of logistics (transportation, warehousing and distribution) was not always considered to be of significant strategic importance. But in today's economy, the game has changed. In many industries, particularly in consumer goods, electronics, and automobiles, the logistics function is at the heart of a new strategic focus called supply chain management. Several market forces are dictating these changes:

Shifting Channel Power

Where powerful manufacturers once dictated the terms of business to retailers and other channel members, the tide has turned. Now, large retailers, spurred by value-conscious consumers, are able to dictate terms to manufacturers. This shift from a manufacturer push strategy to a consumer pull strategy dramatically alters the way in which manufacturers distribute and market their products.

Faster Cycle-time-to-market

Cycle-time-to-market is the time required to design, manufacture, and distribute a product. In today's economy, factors, such as, lowering technology costs and ever-demanding consumers have caused dramatic reduction in product cycle-time. Products quickly become outdated or are imitated by competitors. To remain competitive, manufacturers must embrace time-based competitive strategies.

Enhanced Information Technology Applications

The decreasing cost of computing capacity has fuelled a new wave of adopting information technology as a business-enabling tool. The application of information technology is a critical element in the quest to compete effectively in a global economy which places a premium on timely and efficient operation and communications.

Globalisation

With the end of the Cold War and the passage of trade treaties, such as, NAFTA and GATT, the global economy is more of a reality than ever. As companies look overseas for supply sources, manufacturing sites, and buyers of products and services, greater complexity is added to the logistics function.

SUPPLY CHAIN MANAGEMENT—AN OVERVIEW

The supply chain is a sequence of events intended to satisfy a customer. It can include procurement, manufacture, distribution and waste disposal, together with associated transport, storage and information technology. The language of supply chain management is based upon the following acronyms:

JIT—Just-in-Time manufacturing
EDI—Electronic Data Interchange
CRP—Continuous Replenishment Programme
POS—Point of Sale
EFT—Electronic Funds Transfer
ECR—Efficient Consumer Response
QR—Quick Response

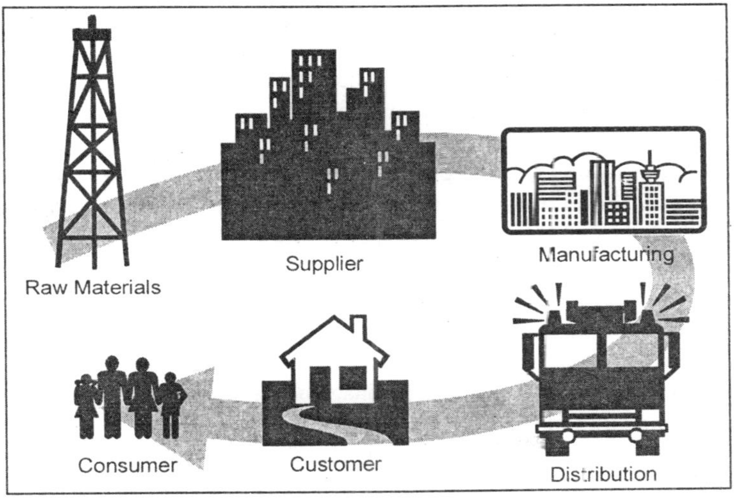

Fig. 23.1: Supply Chain Management

ADVANTAGES OF SUPPLY CHAIN MANAGEMENT

The advantages of supply chain management can be summarised as follows:

- Reduction of costs across the supply chain and more efficient management of working capital
- More efficient management of raw materials, work-in-process, and finished goods inventory
- Increased efficiency in the transactions between supply chain partners
- Enhanced customer value, often in the form of lower prices

A supply chain is a network of facilities and distribution options that performs the functions of procurement of materials, transformation of these materials into intermediate and finished products, and the distribution of these finished products to customers. Supply chains exist in both service and manufacturing organisations, although the complexity of the chain may vary greatly from industry to industry and firm to firm. Realistic supply chains have multiple end products with shared components, facilities and capacities. The flow of materials is not always along an arborescent network, various modes of transportation may be considered, and the bill of materials for the end items may be both deep and large.

Traditionally, marketing, distribution, planning, manufacturing, and the purchasing organisations along the supply chain operated independently. These organisations have their own objectives and these are often conflicting. Marketing's objective of high customer service and maximum sales conflict with manufacturing and distribution goals. Many manufacturing operations are designed to maximise throughput and lower costs with little consideration fór the impact on inventory levels and distribution capabilities. Purchasing contracts are often negotiated with very little information beyond historical buying patterns. The result of these factors is that there is not a single, integrated plan for the organisation—there were as many plans as businesses. Clearly, there is a need for a mechanism through which these different functions can be integrated together. Supply chain management is a strategy through which such an integration can be achieved.

Supply chain management is typically viewed to lie between fully vertically integrated firms, where the entire material flow is owned by a single firm, and those where each channel member operates independently. Therefore, coordination between the various players in the chain is key in its effective management.

DECISIONS FOR FORMULATING SIMPLY CHAIN

Supply chain decisions can be classified into two broad categories strategic and operational.

Strategic decisions are made typically over a longer time horizon. These are closely linked to the corporate strategy and guide supply chain policies from a design perspective.

Operational decisions are short term, and focus on activities over a day-to-day basis. The effort in these types of decisions is to effectively and efficiently manage the product flow in the "strategically" planned supply chain.

There are four major decision areas in *Supply Chain Management*:

1. Location
2. Production
3. Inventory
4. Transportation (Distribution).

There are both strategic and operational elements in each of these decision areas.

Location Decisions

The geographic placement of production facilities, stocking points and sourcing points is the natural first step in creating a supply chain. The location of facilities involves a commitment of resources to a long-term plan. Once the size, number, and location of these are determined, so are the possible paths by which the product flows through to the final customer. These decisions are of great significance to a firm since they represent the basic strategy for accessing customer markets, and will have a considerable impact on revenue, cost and level of service. These decisions should be determined by an optimisation routine that considers production costs, taxes, duties and duty drawback, tariffs, local content, distribution costs, production limitations, etc.

Production Decisions

The strategic decisions include what products to produce, and which plants to produce them in, allocation of suppliers to plants, plants to DC's, and DC's to customer markets. As before, these decisions have a big impact on the revenues, costs and customer service levels of the firm. These decisions assume the existence of the facilities, but determine the exact path(s) through which a product flows to and from these facilities. Another critical issue is the capacity of the manufacturing facilities—and this largely depends on the degree of vertical integration within the firm. Operational decisions focus on detailed production scheduling. These decisions include the construction of the master production schedules, scheduling production on machines, and equipment maintenance. Other considerations include workload balancing, and quality control measures at a production facility.

Inventory Decisions

These refer to means by which inventories are managed. Inventories exist at every stage of the supply chain as either raw materials, semi-finished or finished goods. They can also be in-process between locations. Their primary purpose is to buffer against any uncertainty that might exist in the supply chain. Since holding of inventories can cost anywhere between 20 to 40 per cent of their value, their efficient management is critical in supply chain operations. It is strategic in the sense that top management sets goals. The management of inventory from an operational perspective includes deployment strategies (push versus pull), control policies—the determination of the optimal levels of order quantities and reorder points, and setting safety stock levels, at each stocking location. These levels are critical, since they are primary determinants of customer service levels.

Transportation Decisions

The mode-choice aspects of these decisions are the more strategic ones. These are closely linked to the inventory decisions, since the best choice of mode is often found by trading off the cost of using the particular mode of transport with the indirect cost of inventory associated with that mode. While air shipments may be fast, reliable, and warrant lesser safety stocks, they are expensive. Meanwhile, shipping by sea or rail may be much cheaper, but they necessitate holding relatively large amounts of inventory to buffer against the inherent uncertainty associated with them. Therefore, customer service levels and geographic location play vital roles in such decisions. Since transportation is more than 30 per cent of the logistics costs, operating efficiently makes good economic sense. Shipment sizes (consolidated bulk shipments versus Lot-for-Lot), routing and scheduling of equipment are key in effective management of the firm's transport strategy.

SUPPLY CHAIN MAPPING

Each chain needs to be separated out into its separate elements. Typically these may be supplier, warehouse, factory, distribution warehouse, customer. (It will also then be important to look at the separate functions within any particular element; the aim is to look at all of the 'supplier/customer' links throughout the chain).

- Stock may be held at all of these points. The lead times, or time spent in processes between each point, need to be ascertained. The time will most probably be in terms of days, but faster chains may be measurable in hours.
- The sum of all these lead-times, from customer to supplier, is the length of the supply chain. It is a measure of the time to respond to an increase in demand beyond the limits of safety built into the stock buffers.
- This stockholding at each point also has to be taken into account. For each of these it should be expressed in days or hours, as appropriate.
- The volume of the supply chain is the sum of the lead times and

the stockholdings. It is a measure of the time needed to 'drain' the pipeline in response to a significant decrease in demand.

- The aim of supply chain analysis is to discover means of shortening the chain and reducing its volume. A mapping of each chain in this way identifies opportunities for change.
- Critical pipelines can be highlighted: the importance of replenishment frequency, and areas where no value is added can be pinpointed. By looking at the complete chain, particular obstacles to efficient trade will become clearer. This identification allows for further research to see whether improvements can be made.
- The mapping may result in a web-like diagram as different elements of the chain come into play. The viewing of the supply chain as a simple linear flow is only the beginning of an analysis.
- The flow of materials within an organisation and then between trading partners will depend on the material's characteristics. These will inevitably determine the improvements possible in the handling of these components. The main characteristics are:

    1. Seasonality of production/availability
    2. How well specified
    3. Degree of substitutability
    4. High or low volume
    5. Fast or slow moving
    6. Seasonality of demand
    7. Consistency of demand
    8. High or low value
    9. Durability.

INFORMATION AND DATA EXCHANGE

- The flow of materials takes place in response to the flow of information, which can be improved to assist in the planning and management of the constituent processes. Accurate trade data can be made available to reduce or eliminate the 'second guessing' that takes place by suppliers.
- Scanning can enable companies to track components (of known batches, lots, production dates, as appropriate) as well as finished products. The impact of European quality control standards will place increasing demands upon companies to provide full trace-ability. The implementation of systems to meet these standards, such as ISO 9000, is already taking place.

- EAN standards also provide for the identification of transport units containing a non-standard combination of products (i.e., a combination that may not be predefined or repeated). The EAN serial shipping container code enables merchandise that is picked and packed to meet individual orders to be identified and traced. This can support dispatch, distribution and receiving of non-standard packages.
- Demand and supply signals link each supplier and customer. If these are uncertain, extra buffer inventory will need to be held to protect against unexpected surges in demand. A better use and sharing of information can help reduce uncertainty about future demand, so allowing for stock holdings to be reduced while encouraging more responsive manufacturing.
- Attention will need to be given to other aspects of inventory control which may be unaffected by improved information. Certain raw materials (e.g., cocoa beans, wheat) have a seasonal production and little substitutability. Certain supply inflexibility cannot be reduced. Demand may also be such that it cannot be forecast and stocks have to be held indefinitely (e.g., antifreeze/ de-icing liquids, military equipment and ammunition).
- It is the holding of inventory that increases costs and reduces responsiveness. Companies should examine ways of improving demand and supply signals so that all value-added processes can be optimised.

MAXIMISING THE EFFECTS OF SUPPLY CHAIN THROUGH VARIOUS STRATEGIES

The following five strategies are offered in an effort to help minimise problems and maximise results in order to make the supply chain vision a reality.

Strategy 1: Supply Chain Objectives should be Globalised

The critical first step in designing, reengineering or optimising the supply chain is to assemble a team to define precisely what goals you want to achieve. Do you want to reduce inventory, improve service to your customers, or increase speed to market? What are your company's specific global supply chain goals? (Note that in this context, "global" refers to all of the comprehensive factors that relate to the organisation and supply chain, regardless of whether it is a domestic multinational or global supply chain). Make sure that the entire team understands and agrees on how to handle the basic elements of the Supply Chain, such as:

- Demand Planning
- Capacity Planning
- Strategic Scheduling
- Logistic Systems and Execution
- Service Strategies
- Regulatory Strategies
- Sales Analysis
- Sales Forecasting
- System Performance Measurement
- Feedback Adjustment Systems.

Managing a global supply chain is a little like playing five-dimensional chess. There are many components on the supply chain "chess board," and all the pieces must be moved strategically on different levels to achieve an optimal solution.

At some point, it is important to set numerical targets for the processes being implemented. It is not uncommon, for example, to take 10 to 15 per cent of the Supply Chain cost out of the system by process re-engineering. This is tangible, hard capital that can be saved and deployed towards other strategic projects while improving the organisation's sales and service positions.

The key is to examine all of the Supply Chain elements essential to the company and set achievable short-term and long-term goals that are in harmony with the organization's overall objectives.

Strategy 2: The Supply Chain Audit

With strategic goals in place, it is important to correctly understand the current Supply Chain system and the global factors that impact, impede and interact with the system. This audit is a critical first step in the process of moving from where one is today to where he/she wants to be tomorrow. The first phase of the Supply Chain audit is:

Collect the Data

- Where is the data located?
- How is it created?
- Who interacts with it?

The collected data consists not only of numbers but also information about people, procedures and processes.

Create a Supply Chain Model

Once the data is collected and analysed for consistency and accuracy, it can be compiled into a model representing the current Supply Chain. This model of the current Supply Chain represents a base-line that can be used to drill down to the next level of analysis, during which key areas can be studied to insure that:

- Current systems are consistent with the organisation's strategic goals and objectives.
- Customers' needs are being served by the current Supply Chain system.
- The Current Supply Chain system promotes efficiency.

Create a Matrix Analysis

After this review, the next step is to conduct a matrix analysis, using:

- Benchmark analysis
- Trend analysis
- Technology assessments
- Budget analysis
- Performance against target comparisons
- Chain dynamics tests.

These analyses will identify critical success factors and key performance indicators. It will also help assess problem areas and areas of vulnerability. The results of these assessments can then be inserted into the mix which can be called the "Global Supply Chain Optimiser" to help determine the proper direction for the design phase.

Strategy 3: Develop a Vendor Supply Chain Strategic Matrix

In today's world of ever-increasing velocities of change, few organisations want to commit to long-term relationships. The conventional wisdom suggests that if the organisation encumbers itself with long-term agreements, it will lose the flexibility to react to new opportunities. What if something better comes along tomorrow? The reality is that such an approach may be costing your organisation money.

Develop a Vendor Strategic Matrix

What is needed is the development of a Vendor Strategic Matrix (VSM) to help create the best mix of vendor relationships for optimal Supply Chain and organisational strategic success. A VSM is a tool that helps to determine the best type of relationship to have with each individual vendor. In the complex supply chain environment, a company must define the most favourable relationship to have with each vendor in order to optimize the supply chain. Any organisation that applies only one relationship structure to all vendors, either consciously or unconsciously, is most likely short changing itself.

The VSM methodology states that vendors fall into various categories. The four main groups are:

- Strategic-Collaboration
- Strategic-Cooperative
- Non-strategic-Limited
- Non-strategic-Commodity.

In a VSM, vendors will be evaluated according to how each contributes to the current or future success of the organisation. In this process, vendors will be categorised in terms of the uniqueness and scarcity of their products and services.

Collaborative Relationships

A vendor offering a strategic and unique, strategic and scarce, or strategic, unique and scarce product or service is a candidate for a collaborative relationship. This type of arrangement might entail long-term commitments by both parties to procure future production or investment. These vendors become strategic partners and the technological systems set up to share information with them are often critical components.

Cooperative Relationships

A vendor who offers a product or service that is strategic, but not necessarily unique or scarce, is a candidate for a cooperative relationship. In this type of relationship, one may want to provide an incentive for the vendor to invest in order to improve the Supply Chain. Such an incentive might take the form of reducing the number of vendors to increase the volume of business given to each vendor. Cooperative relationships may be short, medium or long term.

Non-strategic, Limited Relationships

A vendor who provides products or services that are not strategic to the organisation's success but are limited in supply is a candidate for a short-term non-strategic, limited relationship. In this situation, one should look at the quality and value provided by the vendor versus the availability of similar products and services from other vendors.

Non-strategic, Commodity Relationships

A vendor offering a product that is non-strategic and in plentiful supply will have a non-strategic, commodity relationship with the company. With these kinds of products or services, one can tolerate price variations without feeling critical impact on one's strategic products or services. These relationships are usually short-term, with minimal commitment.

When using the VSM methodology, it is important to review the vendor base on a periodic basis in order to determine if external or internal factors have changed enough to require that the vendor be moved to a different classification. By using a VSM, you can prevent some roadblocks to Supply Chain optimisation as one moves forward in improving your Supply Chain system.

Strategy 4: Build a Win-Win Vendor Compliance Programme

In today's global marketplace, vendor compliance programmes are complex and expensive, but the positive results of a good programme are worth the effort. What is necessary to create a "good" vendor compliance programme?

Five Steps to an Effective Vendor Compliance Programme

A winning environment for the organisation and the vendors must be created. Their needs and concerns must be determined.

1. ***Analyse the Vendor Base***

Some companies have the luxury of a small vendor base while others have tens of thousands of vendors. The vendor base must be analysed by number of vendors, volume, and strategic nature.

2. ***Apply Pareto Methodology to the Analyzed Vendor Base***

Matrix the strategic and volume factors, and select the vendors who are most strategic and have the largest volume. This selection process should be continued until eighty percent of the volume has been achieved. Is this group of manageable size for initial design, development and test scenarios? If not, start at the top of your list and move down until you have what you need for a test. Caution: If your vendor-base is large with small volumes contributed by each vendor, you probably need outside help with your test.

3. *Make Sure the Necessary Expertise is Available*

Management must understand that this is an undertaking that will cost time and money. It will be an on-going process rather than a project that comes to a finite end. The person assigned to the process must have the necessary expertise in areas such as UCC128, EAN, EDIFACT, EDI, label design, ID bar codes, 2D bar codes, laser readers, label printers, communication networks, testing techniques, international requirements, etc. Supply Chain vendor compliance is a complex area requiring lengthy study. If you do not have the necessary expertise in-house, it is better to seek outside help.

4. *Test a Pilot Project*

Once the optimal solution has been designed, it is time to develop, engineer and implement it. Select one or two members from each supply chain segment for a pilot. There is tremendous complexity to proper solutions. There will be numerous items that one did not think were critical or did not account for. It is important that the tests are deployed with a small group to provide for any other changes. Once tests have produced success, this very small group can be put into production. Once again, allow adequate time for verification and error correction.

5. *Re-group for the Hard Work*

If your pilot is a success, resist the temptation to roll out the new systems to all vendors. Be aware that you are now getting ready to move from the "A" students to the "F" students. Re-group and analyse your plan and the challenges you faced in the pilot. They will be much more severe in the upcoming rounds. You will face language, education, training and technology issues, to name a few. Make sure that you have adequate skills and resources available. Once the new plan is in place, implement it in successive waves of vendors from top to bottom, making sure to follow the above steps along the way as you fine-tune your supply chain system for success.

Strategy 5: Recognise Supply Chain as an On-going Process and not a Project

Your Supply Chain is a dynamic living organism. It consists of people, information, technology and systems (PITS). At today's maddening pace, it is critical that we keep in mind that our supply chain is a continuous process and not merely discreet elements. In order to be successful in today's global marketplace, we must adopt the process of Continuous Business Process Improvement (CBPI). Today we must learn the complex art of changing the tire on our automobile while we are still driving it. That is no small task, but it is one that will increasingly need to be mastered in the New World that beckons us.
In that New World, one will need to:

- Position your organisation effectively within the global supply chain in a manner that best optimises your organisation's short- and long-term goals.
- Integrate your organisation efficiently within the global supply chain to produce optimal results to your organization's bottom line.
- Create agility within your supply chain system through thoughtful design, so that your organisation can react quickly to rapidly changing market conditions.
- Monitor your Supply Chain system in ever increasing time slices to achieve maximum results today and adjust rapidly to the changing needs of tomorrow.

In the near future:

- Organisations will be increasingly challenged to manage a continuously changing environment.
- The definition of relevant marketplace will become more global.
- The business environment will become more demanding and complex.
- An organisation's supply chain web will need to be finely tuned to meet customers' needs, with the ability to react quickly to their changing requirements.
- Strong supply chain relationships will become increasingly more important. Organisational structures will be challenged by the shift from discreet to process management.
- Management will spend increasing amounts of effort on future supply chain considerations.

- Current compensation systems will be modified to represent new system structures. Information technology will continue to grow in strategic importance in the supply chain arena.

Amid the increasing complexity lies challenge and opportunity. Some organisations will ride the wave of change and prosper and grow in its wake.

PERFORMANCE MEASURES IN A SUPPLY CHAIN

There are a number of ways of measuring aggregate inventory performance within a business and these generally use financial figures or ratios as a common form of measurement.

There are four common approaches to monitor inventory within a business which are:

A. Aggregate Inventory Value

Aggregate Inventory Value which is the total inventory value at cost and an upper investment limit can be set for a range or class. Although such a measure is easy to use, it ignores the dynamic nature of inventory and its interactions with other business areas.

B. Inventory-to-Sales Ratio

This is the aggregate inventory value at cost divided by the annual sales. This approach recognises the dynamic relationship between these two figures but cost and/or selling price changes can cause substantial variations.

C. Days of Supply

Which is the total value of inventory at cost divided by the sales each day at cost? This is also a dynamic measure but can be subjected to some turbulence if the cost of sales is not maintained and controlled.

D. Inventory Turnover

This is the ratio of the annual sales at cost to the average inventory value and relates to the cycle of using and replacing material. It shows the number of turnovers achieved for a specified time period, usually a year, and high turnover reduces inventory investment and saves on stock holding costs.

These approaches enable easy comparison of performance across business areas, and between businesses in bench-marking exercises. It is obviously possible to compare the total cost of inventory within a supply-chain and to seek to use a chain that will minimise such costs. Similarly, provided sufficient data is available, total supply-chain costs can be minimised and many software routines seek to achieve this. Nevertheless, is it possible to compare the efficiency of different supply-chains by using some new form of aggregate measure, or ratio, such as the total supply-chain inventory costs to the total cycle time within the chain.

Such an approach would indicate that the higher the resultant figure the better since, from a customer point of view in a 'make to order' scenario, the shorter the overall chain the better.

CASE STUDY

The three companies described here run the gamut from big to small, consumer to industrial, but they all have one thing in common: Their Supply Chain initiatives are helping to increase profitability, boost customer service, and fuel skyrocketing sales.

A. DELL Computer

Problem: Excess inventory slowly sucks resources and money out of business. Realising this, Dell Computer is replacing inventory with information.

Objective: Right now, Dell carries about seven days of finished product. The goal is to count that already low figure in minutes.

To achieve this:

- The company is turning to the Internet to collaborate and conduct business with suppliers and customers at unprecedented levels.
- Dell recently created customised Web-pages for its top 30 suppliers, whose employees can log on to a secure, personalised site to view demand forecasts and other customer-sensitive information—such as, who Dell's customers are and how much equipment each is ordering—to help them better gauge demand. As a result suppliers can more easily match their production schedules to Dell's-making only what is needed, when it is needed.
- Dell is also passing on data about its defect rates, engineering changes, and product enhancements to these suppliers. Since both Dell and its suppliers are in constant communication, the

margin for error is reduced, allowing partners to collaborate in real time on product designs and enhancements.

- Suppliers share sensitive information with Dell, such as, their own quality problems. The suppliers also reap the benefits of faster cycle times, reduced inventory and improved forecasts. And ultimately, the customer gets a higher-quality product at a lower price.
- Dell is also using the Internet to create a community around its supply chain. The Web-sites all have links to bulletin boards where partners from around the world can exchange information about their experiences with Dell and its value chain.

Dell is also using the Internet to form tighter links with customers. For many of its business users, the company has created Premier Pages containing approved configurations, re-negotiated prices, and new work-flow capabilities, so when an employee requests a new computer, the order is automatically routed to the appropriate person within the buying organisation for approval

B. Herman Miller Incorporation

It is a company, which makes build to order line of office furniture and accessories. 80 per cent of the office systems manufacturers business is project-based and requires extensive coordination amongst its 5 plants and other operations.

Problem: Delayed delivery performance since some of the parts of a multipart order were always missing which led to poor customer satisfaction.

Objective: Being able to deliver what the customer wants and when does he want it without any timelag.

Solution: The company began a supply chain optimisation programme and incorporated software from i2 technologies and a Baan enterprise resource planning system to identify and manage the companies business constraints. This enabled the company to maintain customer identity throughout the manufacturing process, ensuring that the entire orders are completed simultaneously and shipped according to client specifications.

Results: The company has improved its on-time delivery from 76 to 99 per cent and decreased factory lead-time from 6 to 4 weeks. A 56 million-dollar in the cost of goods sold has allowed the company's sales force to pass along the savings in the form of deeper customer discounts.

C. Tum Yeto Incorporation

The company is the third largest manufacturer of skateboards and accessories.

Problem: The company's production was falling far short of consumer demand and what was required was a complete overhauling of its existing supply chain.

Objective: Inventory management to the maximum possible extent while keeping cost at their lowest level.

Solution: Just-in-time delivery and coordination amongst its various suppliers. Tum Yeto brought raw materials delivery in phases with production of its boards and other products, allowing the company to pay the suppliers from the sale of finished product. Along within this the also incorporated a separate department that made a 12 month demand forecast on an on-going basis. The company also started selling its products over the Internet where the customers could view the Tum Yeto catalogue, order merchandise and track delivery status via UPS online.

Results: In a market that showed 15 to 20 per cent growths, the sales volume of the company increased by almost 50 per cent. Along with this, the number of retailers of the company also doubled from 500 to 1000.

Index